The Primes Trilogy

The Primes Trilogy

The Complete Trilogy in One Volume

A Wheel Through the Primes · *A Spectrum Through the Primes* · *Scaling the Wall*

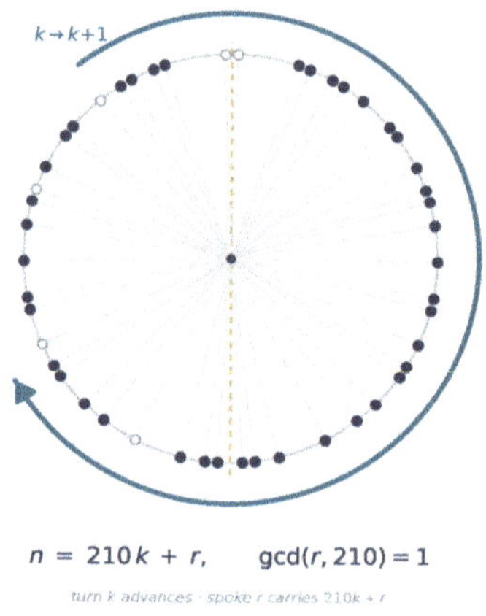

John Maggio

Efficient Intelligence, LLC

2026

Published by Efficient Intelligence, LLC

This omnibus edition collects the three books of *The Primes Trilogy*—*A Wheel Through the Primes*, *A Spectrum Through the Primes*, and *Scaling the Wall*—each previously published separately, in a single volume.

Omnibus edition (hardcover) ISBN: 979-8-9967325-9-3

First omnibus edition, 2026   Printed in the United States of America

Advance in Peace

About This Edition

This volume collects the three books of *The Primes Trilogy* in one binding: *A Wheel Through the Primes*, *A Spectrum Through the Primes*, and *Scaling the Wall.* They were written to be read in order, and they are gathered here in that order, but each remains a complete book with its own preface, and each can be entered on its own.

The three move outward from a single object. *A Wheel Through the Primes* builds the concrete machinery of analytic number theory on the mod-210 wheel—the sieve, the survivors, the characters, the first sight of the zeros. *A Spectrum Through the Primes* takes up the deepest dream in the subject, that the Riemann zeros are the spectrum of an undiscovered operator, and follows it honestly to where each attempt to build that operator is turned back. *Scaling the Wall* is the advanced survey at the summit: a dozen great roads to the Riemann Hypothesis, each climbed and each shown the exact place, and the exact reason, it stops.

One discipline holds across all three. Real mathematics is credited by the names of those who found it. Metaphor is kept strictly apart from mechanism. What is proven is kept apart from what is conjectured, and no road is described as crossing a wall it only reaches. Every longer computation is reproduced by deterministic code the reader can run rather than take on faith.

The trilogy does not solve the Riemann Hypothesis, prove the twin-prime conjecture, or make factoring easy. It ends, deliberately, at those three walls—because that is where the subject ends, and because seeing exactly where a wall stands is worth more than a story that pretends to pass it.

Book I is assumed by the two that follow; beyond that, the volume can be sampled. A reader new to the subject should begin at the wheel.

Contents

BOOK I

A Wheel Through the Primes

Preface

This is a book about the prime numbers, told through a single object: a wheel of forty-eight spokes, the residues coprime to $210 = 2 \cdot 3 \cdot 5 \cdot 7$. Everything in it — the way primes spread across the spokes, the faint bias in their race, the mirror that pairs each spoke with another, the diffraction grating that sifts them, the zeros of the zeta function reappearing as their spectrum, the gas of prime "modes," the climb up the primorials, the twin slots — is standard analytic number theory. The wheel is not a new theory. It is a place to stand where the standard theory can be seen at human scale.

Two things are worth saying at the outset. The book promises that the machinery is real and that you will meet it concretely, not in outline: Dirichlet characters as the harmonics of the wheel, the explicit formula as genuine wave interference, the Hardy–Littlewood singular series as fifteen surviving slots out of 210. And it promises the opposite of a triumphal ending. The tour does not solve the Riemann Hypothesis, prove the twin-prime conjecture, or make factoring easy. It ends, deliberately, at those three walls — because that is where the subject actually ends, and because seeing exactly where a wall stands, and why, is worth more than a story that pretends to pass it.

The wheel earns its place by being small. 210 is the smallest modulus that resolves the first four primes at once, so it is the smallest wheel on which the whole early sieve is visible in a single turn. That legibility is the entire source of its specialness. The structures it reveals belong to the integers, not to 210, and the book closes by saying so plainly.

How to read this

The chapters build, but loosely. Chapter 1 sets up the wheel and is assumed everywhere after it; beyond that, each part can be sampled on its own. A reader who wants the spectral heart of the subject can go from Chapter 1 straight to Part III; a reader who wants the elementary combinatorics can stay in Parts I and IV. The figures are meant to be read, not skipped — much of the argument lives in them. Each figure is accompanied in the text by the short Python program that produced it, printed in full so it can be rerun rather than trusted; this book ships no separate code files. A glossary at the back collects the recurring vocabulary.

Acknowledgements

The author gratefully acknowledges the use of Anthropic's Claude as a computational and drafting assistant during the preparation of this work. Any mathematical results or interpretations presented here remain entirely the responsibility of the author. — J.M.

Part I

The object

Chapter 1

The Mod-210 Wheel

This is the opening chapter of a book that studies the prime numbers through a single concrete object: the wheel modulo $210 = 2 \cdot 3 \cdot 5 \cdot 7$*. The book claims no new result; its aim is unity and clarity — to gather a scattered body of classical number theory onto one stage, so that each idea can be seen as a feature of the same picture. Here we build the stage. The integers coprime to* 210 *form a group of* $\varphi(210) = 48$ *residue classes — the "spokes" of the wheel — and every prime larger than* 7 *lands on one of them. We record the structure of this group,* $(\mathbb{Z}/210\mathbb{Z})^\times \cong \mathbb{Z}/2 \times \mathbb{Z}/4 \times \mathbb{Z}/6$*; the wheel as a sieve, with the surviving density* $\varphi(M)/M = \prod_{p|M}(1 - 1/p)$*; and the reflection* $r \mapsto 210 - r$ *that returns in every later chapter.*

The later chapters then act on this stage. Dirichlet's theorem spreads the primes evenly across the 48 *spokes, while a closer look exposes the persistent Chebyshev tilt — a prime race among the classes. The wheel's seven real characters arrange into a conductor lattice split by parity, and the reflection* $r \mapsto 210 - r$ *is unmasked as the elementary shadow of the functional equation* $s \mapsto 1 - s$*. The primes are then read off the zeros of the zeta function as a diffraction pattern; the zeros themselves repel like the eigenvalues of a random matrix; and the squarefree integers assemble into a supersymmetric "primon gas" whose bookkeeping is the Möbius function. Climbing the primorials yields a conservation law and Mertens' theorem; the repeating decimals of the wheel — a structure already in Gauss — split the spokes into two cyclic families; and the book closes on the twin primes, where the same wheel that organizes everything so cleanly runs headlong into the parity barrier that no elementary method has crossed.*

Everything is elementary and classical. The wheel is a lens, not a law: it renders a great deal of known mathematics visible at once, and it is candid about where the difficulty actually lives — not in the structure of the wheel, which is settled, but in how the primes distribute across it, where the oldest questions remain open. This first note fixes the stage cleanly, before the others act on it.

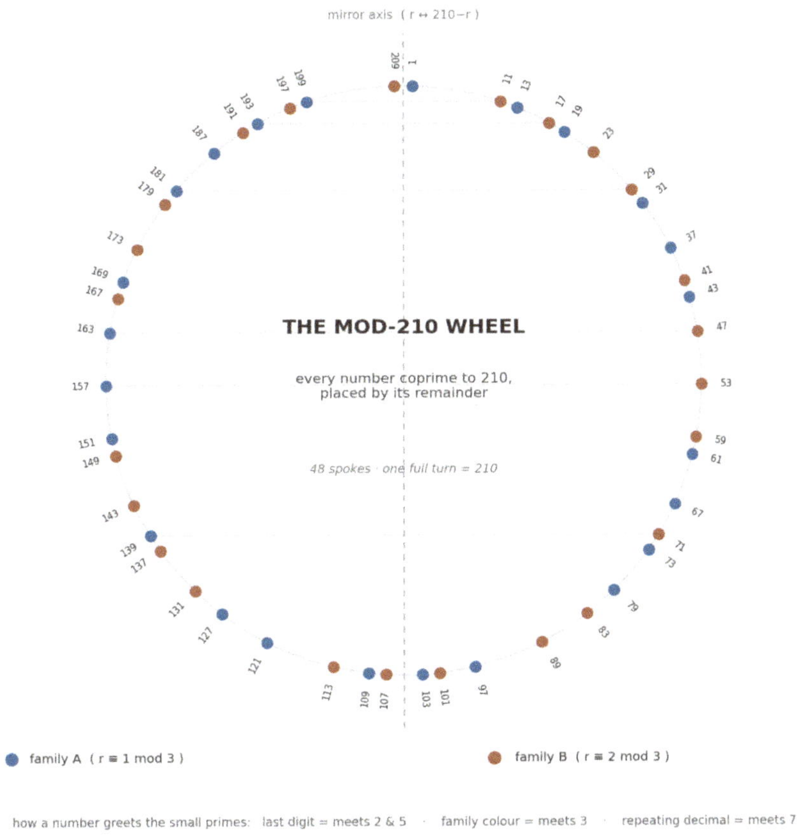

Figure 1.1: The whole subject on a single face. Every number coprime to 210 is placed by its remainder; color marks the family ($r \equiv 1$ versus $r \equiv 2 \bmod 3$), and the vertical line is the mirror $r \mapsto 210 - r$, across which the two families swap. The line at the foot decodes the rest: a number's last digit records how it meets 2 and 5, its color how it meets 3, and its repeating decimal how it meets 7. Everything in this chapter and the ones that follow is a closer look at one feature of this one picture.

1.1 Scope

This chapter is foundational and entirely classical. It introduces no result; it fixes notation and the basic structure of the mod-210 wheel so that the chapters that follow — on the prime race, the real characters and their conductor lattice, the zeros as a diffraction pattern, and the supersymmetric primon gas — can each act on a stage that has already been described.

1.2 The modulus and its units

The modulus is

$$210 = 2 \cdot 3 \cdot 5 \cdot 7, \tag{1.1}$$

the product of the first four primes — the fourth *primorial.* The residues modulo 210 that are coprime to it form the unit group $(\mathbb{Z}/210\mathbb{Z})^\times$, whose size is Euler's totient

$$\varphi(210) = \varphi(2)\,\varphi(3)\,\varphi(5)\,\varphi(7) = 1 \cdot 2 \cdot 4 \cdot 6 = 48. \tag{1.2}$$

These 48 classes are the spokes of the wheel (Fig. 1.2). By the Chinese Remainder Theorem the group factors as

$$(\mathbb{Z}/210\mathbb{Z})^\times \;\cong\; (\mathbb{Z}/2)^\times \times (\mathbb{Z}/3)^\times \times (\mathbb{Z}/5)^\times \times (\mathbb{Z}/7)^\times \;\cong\; \mathbb{Z}/2 \times \mathbb{Z}/4 \times \mathbb{Z}/6, \tag{1.3}$$

of order $2 \cdot 4 \cdot 6 = 48$. It is not cyclic — 210 has no primitive root — which is why the wheel carries several independent quadratic characters rather than one; that multiplicity is the subject of the conductor-lattice note.

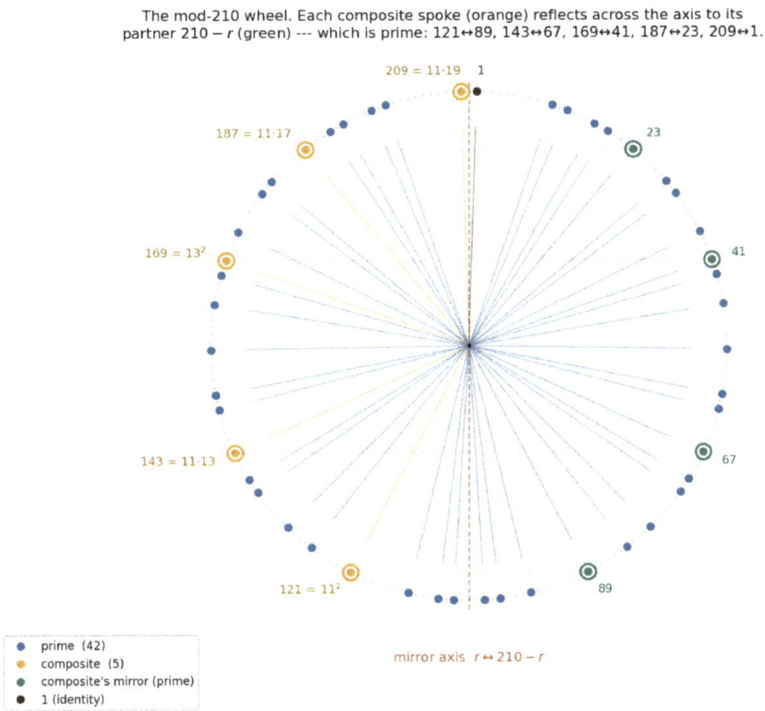

Figure 1.2: The mod-210 wheel. The 48 residues coprime to 210 are the spokes; the remaining positions (multiples of 2, 3, 5, or 7) are the faint dots on the rim. The spokes are colored by type: of the 48, one is the identity, 42 are primes, and five are composite (orange)—the products of two primes ≥ 11 that remain below 210, namely $121 = 11^2$, $143 = 11 \cdot 13$, $169 = 13^2$, $187 = 11 \cdot 17$, and $209 = 11 \cdot 19$. Every prime greater than 7 is a spoke, but so is each of these composites. The dashed axis is the reflection $r \mapsto 210 - r$ of Sec. 1.4. Reflecting each composite across it gives its partner $210 - r$ (green, with faint chords): $121 \leftrightarrow 89$, $143 \leftrightarrow 67$, $169 \leftrightarrow 41$, $187 \leftrightarrow 23$, and $209 \leftrightarrow 1$. Every one of those partners is prime (or the identity) — the reflection keeps a residue coprime to 210 but does not keep it composite.

1.3 The wheel as a sieve

The reason the primes live here is immediate: a prime $p > 7$ is divisible by none of $2, 3, 5, 7$, so $\gcd(p, 210) = 1$ and p occupies one of the 48 unit classes. The first four primes $2, 3, 5, 7$ are exactly the teeth of the sieve — they are the primes *removed* to build the wheel — so apart from those four, the entire prime sequence is distributed among the 48 spokes. This is the elementary content of *wheel factorization*: to hunt for primes one need only examine the residues coprime to 210, a little under a quarter of all integers.

The spokes are not all prime, and it would misrepresent the wheel to draw them as if they were. Of the 48, one is the identity 1, forty-two are the primes between 11 and 199, and five are composite: the products $121 = 11^2$, $143 = 11 \cdot 13$, $169 = 13^2$,

$187 = 11 \cdot 17$, $209 = 11 \cdot 19$ of two primes at least 11 that stay below 210 (orange in Fig. 1.2). The wheel selects every residue coprime to 210 whatever its factorization; in this range the primes are simply the large majority.

The exact fraction is the density of coprime residues,

$$\frac{\varphi(210)}{210} = \frac{48}{210} = \frac{8}{35} \approx 0.229 = \prod_{p|210}\left(1-\frac{1}{p}\right) = \left(1-\tfrac{1}{2}\right)\left(1-\tfrac{1}{3}\right)\left(1-\tfrac{1}{5}\right)\left(1-\tfrac{1}{7}\right). \quad (1.4)$$

Adding more primes to the primorial thins the wheel further (Fig. 1.3); by Mertens' theorem the density $\prod_{p\le y}(1-1/p)$ decays like $e^{-\gamma}/\log y$, slowly but without bound. The choice of 210 is a pragmatic sweet spot: large enough that the four smallest primes are swept away and the structure is rich, small enough that all 48 classes fit on a single page.

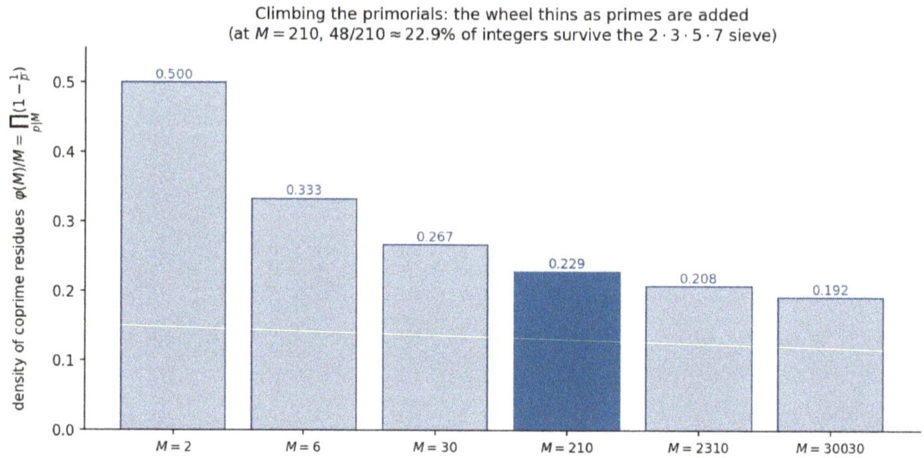

Figure 1.3: The density $\varphi(M)/M = \prod_{p|M}(1-1/p)$ of residues coprime to successive primorials. At $M = 210$ (highlighted) just under 23% of integers survive the $2 \cdot 3 \cdot 5 \cdot 7$ sieve. The decay continues without bound, at the slow Mertens rate.

1.4 The mirror

One symmetry of the wheel deserves to be named at the outset, because it recurs in every later chapter. The reflection

$$r \longmapsto 210 - r \pmod{210} \quad (1.5)$$

is an involution that preserves coprimality, since $\gcd(210 - r, 210) = \gcd(r, 210)$. Its only fixed points are $r = 0$ and $r = 105$, and neither is a unit; so the 48 spokes split cleanly into 24 mirror pairs $\{r, 210 - r\}$. In Fig. 1.2 this is the reflection across the dashed vertical axis.

The reflection keeps a residue coprime to 210, but it need not keep it prime: it sends each of the five composite spokes to a prime, $121 \leftrightarrow 89$, $143 \leftrightarrow 67$, $169 \leftrightarrow 41$,

$187 \leftrightarrow 23$, and $209 \leftrightarrow 1$. The mirror is a fact about residues, not about factorization — a distinction worth keeping in view, since the later notes lean on it.

This single reflection wears several costumes downstream. In the prime-race note it is the axis along which Chebyshev's bias is measured, separating the quadratic residues from the non-residues. In the character note it is the parity $\chi(-1) = \pm 1$, because $\chi(210 - r) = \chi(-r) = \chi(-1)\chi(r)$, sorting the characters into even and odd. And in the deepest reading it is the shadow of the functional equation's reflection $s \mapsto 1 - s$. Seeing it first as a plain geometric fold on the wheel makes its later appearances easier to recognize.

1.5 The wheel turning

The wheel of Fig. 1.2 is a single turn. The integers do not stop at 210: the spoke at residue r carries the number $210k + r$ on turn k, so the same 48 spokes are reused turn after turn, with only the leading multiple of 210 advancing. The later turns are worth a look, because the pattern on the spokes both persists and quietly changes.

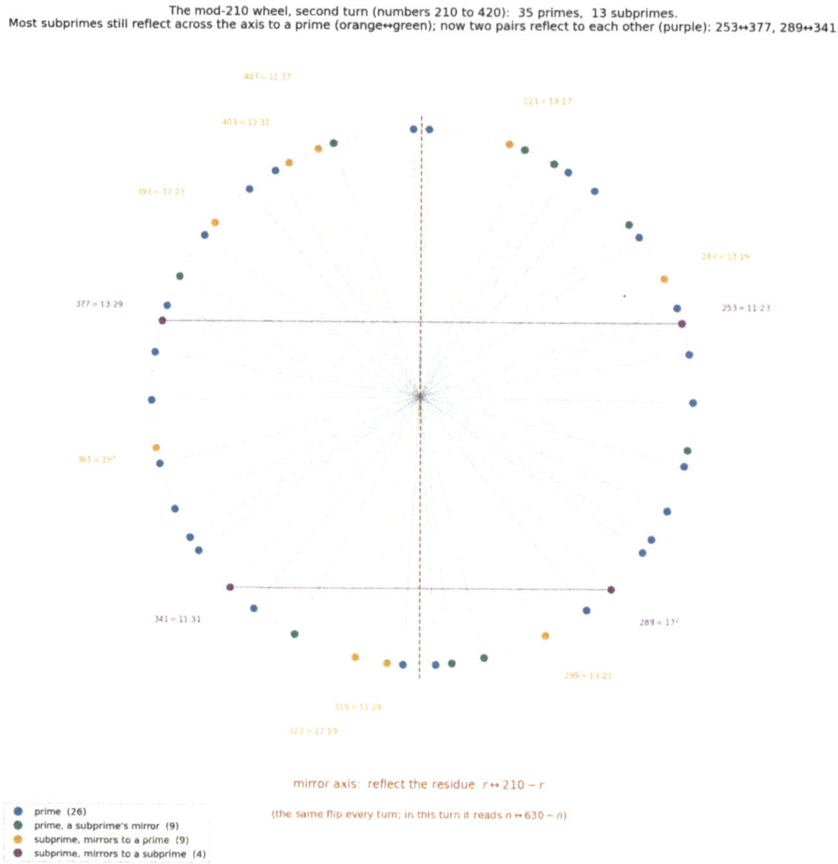

Figure 1.4: The wheel's second turn, the numbers 210 to 420 (the spoke at residue r now carries $210 + r$). The primes have thinned to 35 and the subprimes risen to 13. As on the first turn most subprimes still reflect across the axis onto a prime (orange onto green); the two purple pairs, $253 \leftrightarrow 377$ and $289 \leftrightarrow 341$, are the first composite spokes whose mirror is *also* composite.

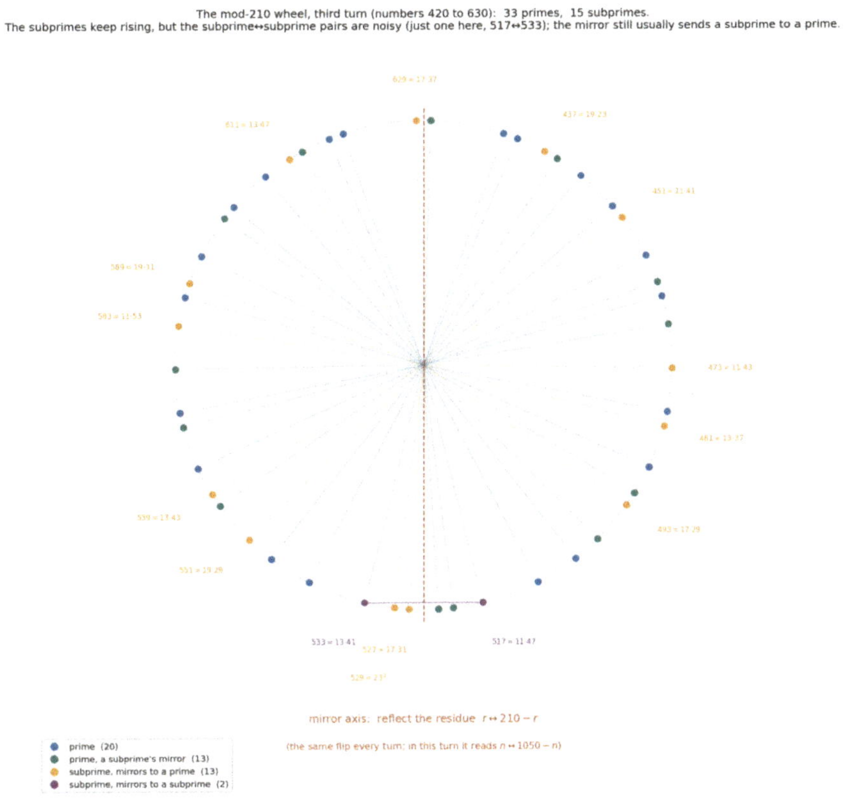

Figure 1.5: The third turn, 420 to 630: 33 primes and 15 subprimes. The thinning continues, while the subprime-onto-subprime pairs are noisy from turn to turn (just one here, 517 ↔ 533). The mirror still usually carries a subprime to a prime.

Two things happen as the wheel turns. First, the primes thin: 42 primes on the first turn, 35 on the second, 33 on the third (Figs. 1.4 and 1.5), with the subprimes filling the vacated spokes. This is the prime number theorem seen one turn at a time, the prime density falling like $1/\log$. Second, the mirror of Sec. 1.4 keeps acting, but its character clarifies. On the first turn every composite spoke reflected onto a prime, which can make the reflection look as though it inverts compositeness. From the second turn onward, pairs of composite spokes begin reflecting onto *each other* (Fig. 1.4). The mirror is not inverting primality and never was; it is simply indifferent to it, sometimes sending a composite to a prime and sometimes to another composite. Primality is not among its invariants.

Beneath the thinning lies a strictly periodic mechanism. Fix a spoke r and ask when a prime $p \geq 11$ divides the number sitting on it. Since $p \nmid 210$, we have $210 \equiv c_p \pmod{p}$ for some nonzero c_p, so the residue $210k + r \bmod p$ advances by c_p each turn and reaches 0 once every p turns. Each prime is therefore a metronome with its own period: 11 strikes a given spoke every 11 turns, 13 every 13, 17 every 17. Placing the spokes at their true values straightens these beats into exact diagonals of slope $-c_p$, namely -1, -2, and -6 for 11, 13, and 17 (Fig. 1.6). A spoke is composite on a given turn exactly when at least one metronome strikes it, and prime only when it dodges every one up to $\sqrt{210k + r}$.

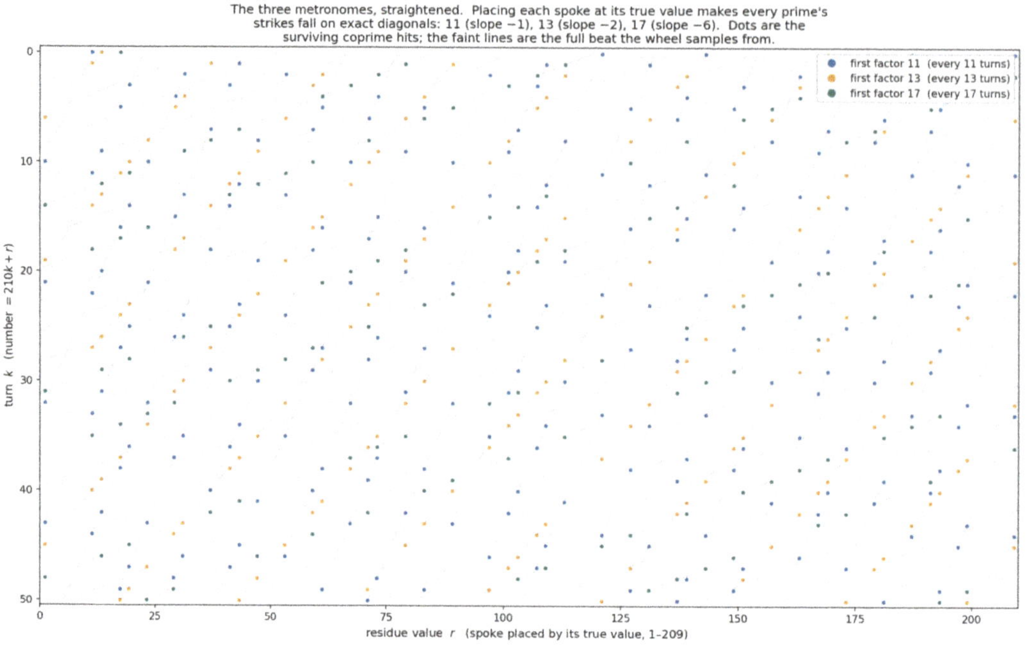

Figure 1.6: The three smallest metronomes across fifty turns. With each spoke placed at its true residue value, the multiples of 11, 13, and 17 fall on exact diagonals of slope $-(210 \bmod p)$. The dots are the coprime hits the wheel keeps; the faint lines are the full periodic beat they are sampled from. The blank space between the stripes is where the primes live.

Pulled apart one metronome to a panel (Fig. 1.7), the same beats become unmistakable. Each prime rides a single family of parallel diagonals, and the larger its step c_p, the faster it sweeps across all 48 spokes: 17, stepping by 6, lays down flatter and more tightly spaced stripes than 11, which steps by 1. Each panel is one prime's entire contribution to the sieve, drawn in isolation.

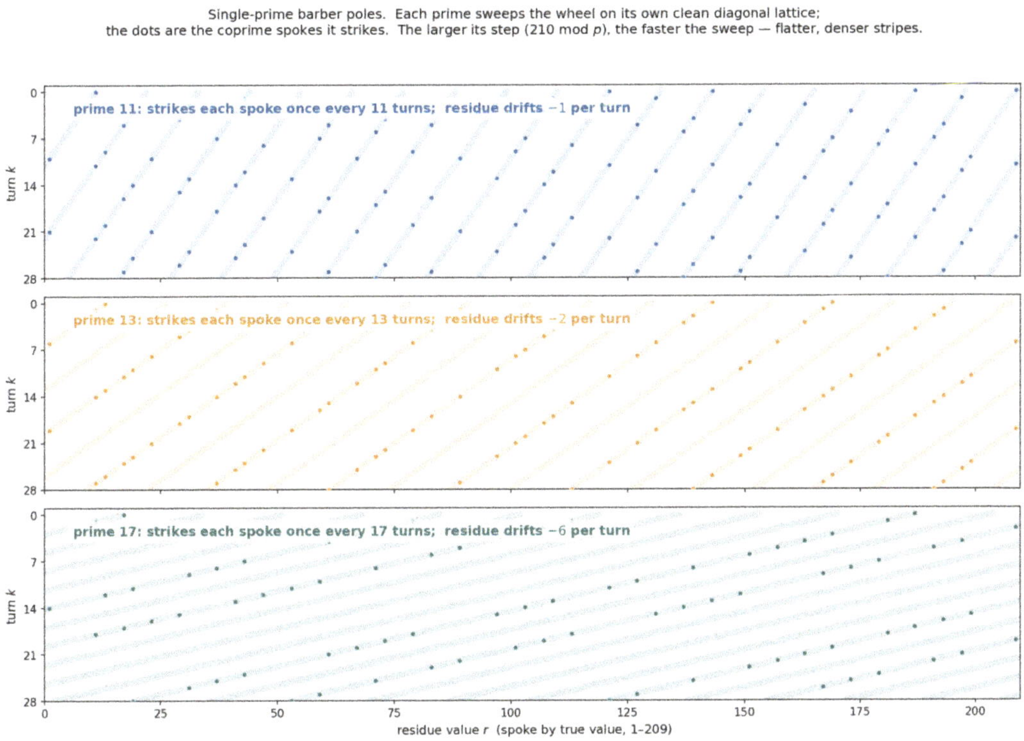

Figure 1.7: The three smallest metronomes separated one to a panel. Each prime p strikes a coprime spoke (the dots) once every p turns, drifting by $-(210 \bmod p)$ spokes per turn, so its hits ride a single family of parallel diagonals — the faint bands are the full periodic beat they are sampled from. The step grows with p, namely -1, -2, and -6 for 11, 13, and 17, so the larger primes sweep the wheel faster: flatter, more crowded stripes.

This is the boundary the wheel keeps drawing, now laid out in time. Any *fixed* set of metronomes is perfectly periodic: 11 and 13 together repeat every $11 \cdot 13 = 143$ turns, and admitting one more prime is precisely the step from one primorial modulus to the next, $210 \to 2310 \to 30030$, which is the subject of the chapter on climbing the primorials. But genuine primality requires every metronome up to $\sqrt{n}$, and that set never stops growing; a new, incommensurate beat keeps joining, so the full pattern never closes into a finite period. The composites are pure clockwork — one could mark every one of them in advance from the metronomes alone. The primes are the spokes no metronome ever claims, and that leftover is the one thing all this periodic machinery cannot itself put a period to. The wheel turns systematically; what it leaves behind does not.

1.6 Two sides of one coin

Every spoke on the wheel, in every turn, carries exactly one number, and that number is either prime or a product of primes that are themselves at least 11; call the latter the *subprimes* of the wheel. The two sets are exact complements: a spoke is prime precisely when it is not a subprime, so the primes are nothing more than the gaps the subprimes leave, and naming either set names the other (Fig. 1.8). This is what a sieve is — Eratosthenes finds the primes by marking every composite and keeping what survives. The only number on the wheel that is neither is the unit 1, sitting alone on the first spoke.

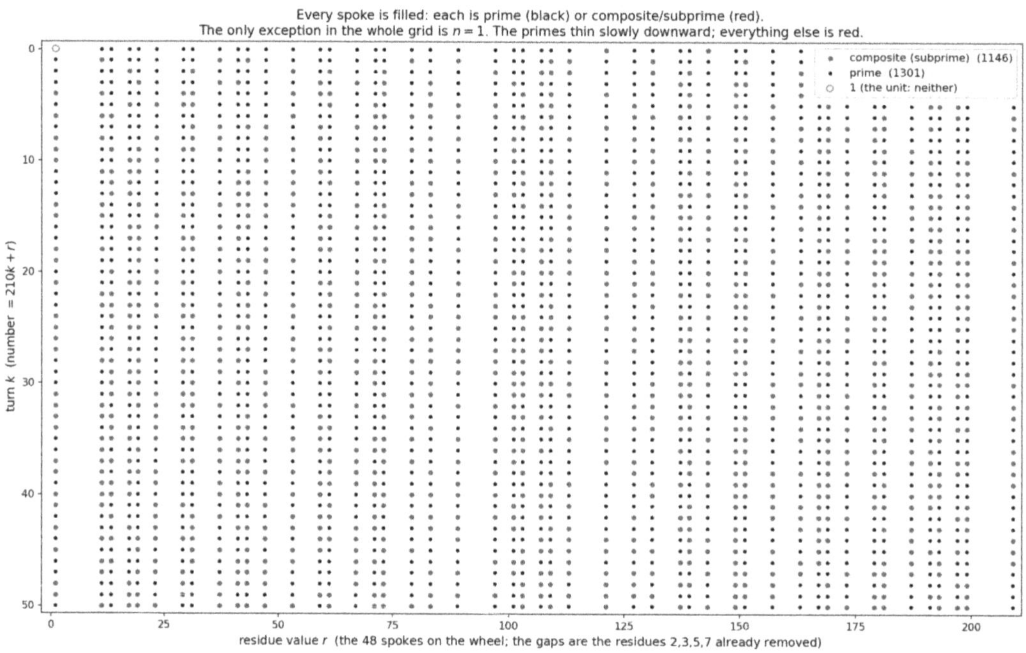

Figure 1.8: Fifty turns of the wheel, every spoke filled: prime (black) or subprime (red). The vertical bands of empty space are not missing spokes; they are the residues divisible by 2, 3, 5, or 7, removed before the wheel begins. Along the 48 true spokes nothing is ever blank, and $1301+1146+1 = 2448 = 48\times 51$ accounts for every number. Red gains slowly on black going down: subprimes grow more common as the supply of small prime factors widens, the same $1/\ln n$ thinning of the primes seen from the other side.

The two faces of this coin are not peers, however. The subprimes are *generated*: each is a product of smaller wheel-primes, so the whole red set can be built forwards by multiplication (11^2, $11 \cdot 13$, $13^2, \ldots$), which is exactly the periodic clockwork of the metronomes above. The primes carry no such rule; they are defined only negatively, as the numbers that are *not* such products. The red side is therefore built out of the black side — the primes are the source and the subprimes their shadow. That lopsidedness is the familiar asymmetry of the subject: forming a subprime is a moment's multiplication, while recovering the primes a given subprime was built from is integer factorization, for which no efficient method is known. The duality is perfect; the direction is not.

Stripped of its cheap structure, the leftover looks startlingly like noise, and this too is a precise phenomenon. The primes are *pseudorandom*: fully deterministic, yet passing most statistical tests one can put to them. Cramér made the resemblance quantitative by modeling each integer n as prime independently with probability $1/\ln n$; on the wheel, where the residues coprime to 210 are already $210/48 = 4.375$ times denser in primes, the matched rate is $(210/48)/\ln n$. Figure 1.9 sets the real wheel beside a single draw of this coin-flip model at that rate. The two are hard to tell apart, which is the point: the model knows nothing about which numbers are actually prime, only their density, yet reproduces their texture. What it cannot reproduce is their rigidity. The real primes are not random — they are forced — and the gap between *behaves like a coin flip* and *is a coin flip* is precisely where the Riemann Hypothesis and the twin-prime conjecture live, and why they remain open.

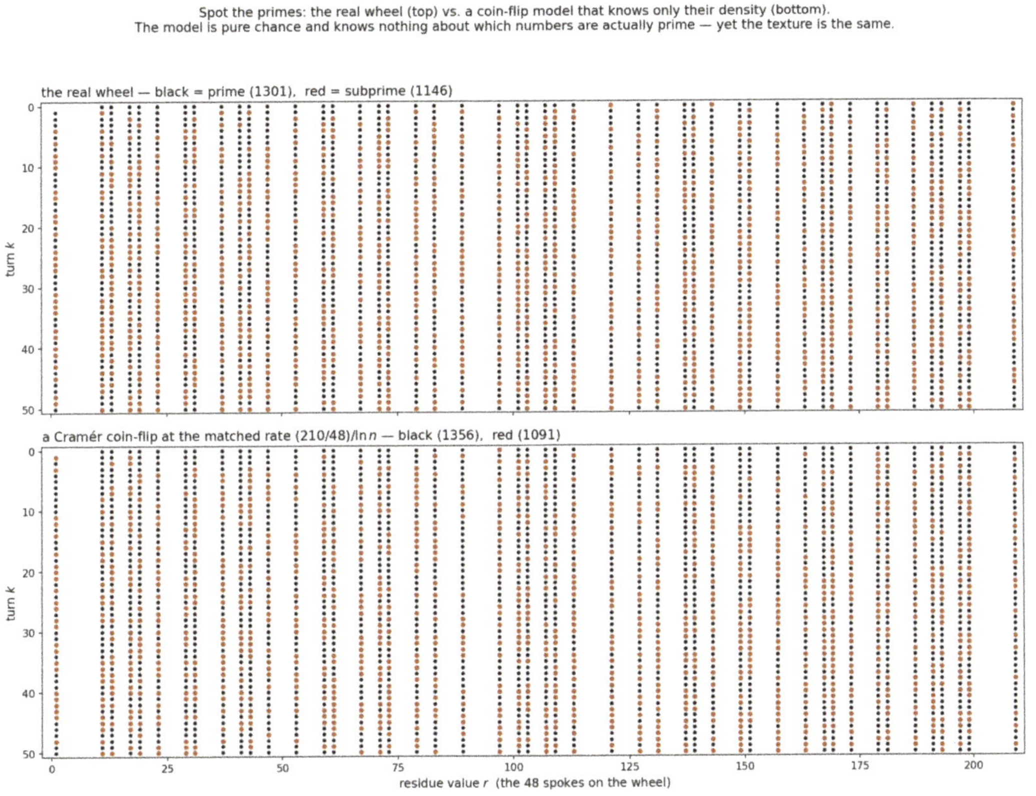

Figure 1.9: The real wheel (top) beside one draw of a Cramér coin-flip (bottom), each number coloured prime with probability $(210/48)/\ln n$ and capped at 1. The model produced 1356 primes against the wheel's 1301 — a match in density, and to the eye a match in texture, though the model is pure chance and the wheel is wholly determined. The visible difference is statistical, not structural; the model's one genuine failure is that it cannot know the primes are not, in fact, being flipped.

This same density reaches further than the two counts. Applied to the wheel's mirror pairs (Sec. 1.4), it fixes how the two reflected residues fall by primality, turn after turn—a small effective model, with the prime number theorem as its only input, set out in the chapter on the mirror-pair populations.

1.7 What this sets up

With the stage fixed, each later chapter puts something in motion on it. The prime race watches the primes distribute themselves across the 48 spokes and measures the slight, persistent bias in that distribution. The conductor-lattice chapter studies the quadratic characters the non-cyclic group carries, and organizes them by their conductors. The diffraction chapter passes to the spectral side, drawing the Riemann and Dirichlet zeros and showing their Fourier transform lands on the primes. The supersymmetric-gas chapter recovers the Möbius function as the partner of those same L-functions. The mirror-pair chapter stays closest to this stage, carrying the Cramér density just above into a count of how the reflection's prime–prime, prime–subprime, and subprime–subprime pairs populate as the primes thin. Every one of them returns, sooner or later, to the 48 spokes and the one mirror described here.

Reproducing the figures

```
from math import gcd
M = 210
units = [r for r in range(M) if gcd(r, M) == 1]
print(len(units)) # 48
print(units[:6], units[-3:]) # [1, 11, 13, 17, 19, 23]... [197, 199, 209]

# density of coprime residues for the primorials
prod = 1.0
for p in [2, 3, 5, 7, 11, 13]:
 prod *= (1 - 1/p)
 print(p, round(prod, 4)) # 0.5, 0.3333, 0.2667, 0.2286, 0.2078, 0.1918

# the mirror pairs
pairs = sorted({tuple(sorted((r, (M - r) % M))) for r in units})
print(len(pairs)) # 24 mirror pairs

# metronome diagonals: the coprime spokes a prime p strikes on turn k
def strikes(p, k):
 step = M % p # 11 -> 1, 13 -> 2, 17 -> 6
 return [r for r in units if (r + step * k) % p == 0]
for p in (11, 13, 17):
 print(p, strikes(p, 0)) # 11 [11,121,143,187,209]; 13 [13,143,169]

# Cramer coin-flip model, matched to the wheel's prime density
import math, random
random.seed(7)
density = 210 / 48 # = 4.375; primes are this much denser
def model_prime(n):
 return random.random() < min(1.0, density / math.log(n))
nums = [210*k + r for k in range(51) for r in units if 210*k + r > 1]
print(sum(model_prime(n) for n in nums)) # 1356 (seed 7)
```

Chapter 2

The Mirror-Pair Populations

On the mod-210 wheel the reflection $r \mapsto 210 - r$ sorts the 48 coprime spokes into 24 mirror pairs, and on any given turn each pair falls into one of three classes by the primality of its two members: prime–prime, prime–subprime, or subprime–subprime. This chapter records a single, deliberately modest fact about how those classes are populated. Turn after turn, their counts are reproduced—to within per-turn noise, across the first six hundred turns—by the prime density alone: if p is the fraction of coprime spokes that are prime on a turn, the three classes hold $24p^2$, $48p(1-p)$, and $24(1-p)^2$ pairs. No correlation between the two partners, and no arithmetic beyond the density, is needed or detectable. We present this as an effective description of the texture, in the spirit of Cramér's model: the agreement is itself the content, because it shows the pair census carries nothing the prime number theorem does not already supply. It is not a new signal, a mechanism, or a spectrum, and we are explicit about that throughout.

2.1 Scope

This chapter is entirely classical and introduces no result. It quantifies one descriptive feature of the mod-210 wheel set up in the opening chapter: how the wheel's mirror pairs distribute among three primality classes as the wheel turns, and why that distribution is fixed by the prime density and nothing else. The purpose is partly bookkeeping—to capture the pattern correctly, so that it is not mistaken for structure it does not have.

2.2 The three pair classes

The mirror $r \mapsto 210 - r$ splits the 48 coprime spokes into 24 pairs $\{r, 210 - r\}$. On turn k the pair occupies the integers $210k + r$ and $210k + (210 - r)$, and each is either prime or a product of primes ≥ 11 (a *subprime*); the lone exception, the unit 1, sits on turn 0 and is set aside. Each of the 24 pairs therefore belongs, on each turn, to exactly one of three classes:

$$\text{prime–prime (PP)}, \qquad \text{prime–subprime (PS)}, \qquad \text{subprime–subprime (SS)}. \quad (2.1)$$

On the first turn the primes are dense and PP dominates; as the wheel turns and the primes thin, PP gives way to SS, with PS in between. The question is whether anything governs that handover beyond the thinning itself.

2.3 The density model

Let $p = p(k)$ be the fraction of the 48 coprime spokes carrying a prime on turn k. Near $n = 210k$ the wheel's prime density is $1/\ln n$ lifted by the factor $210/\varphi(210) = 210/48$, so

$$p(k) \approx \frac{210/48}{\ln(210k)}. \tag{2.2}$$

If the two members of a mirror pair are each prime with probability p, *independently*—the Cramér-style heuristic, the partners sharing only their coprimality to 210—then the 24 pairs split binomially:

$$\langle \mathrm{PP} \rangle = 24\,p^2, \qquad \langle \mathrm{PS} \rangle = 48\,p(1-p), \qquad \langle \mathrm{SS} \rangle = 24\,(1-p)^2. \tag{2.3}$$

Figure 2.1 sets these forms against the counts read directly off the wheel. The left panel follows the three classes turn by turn: the binned counts track Eq. (2.3) as p falls from about 0.73 to 0.39 over six hundred turns, PP and SS crossing near $p = \frac{1}{2}$. The right panel removes the turn axis entirely and plots the same counts against p; they land on the three curves of Eq. (2.3) with only the scatter expected from 24 pairs per turn. The model has no free parameters.

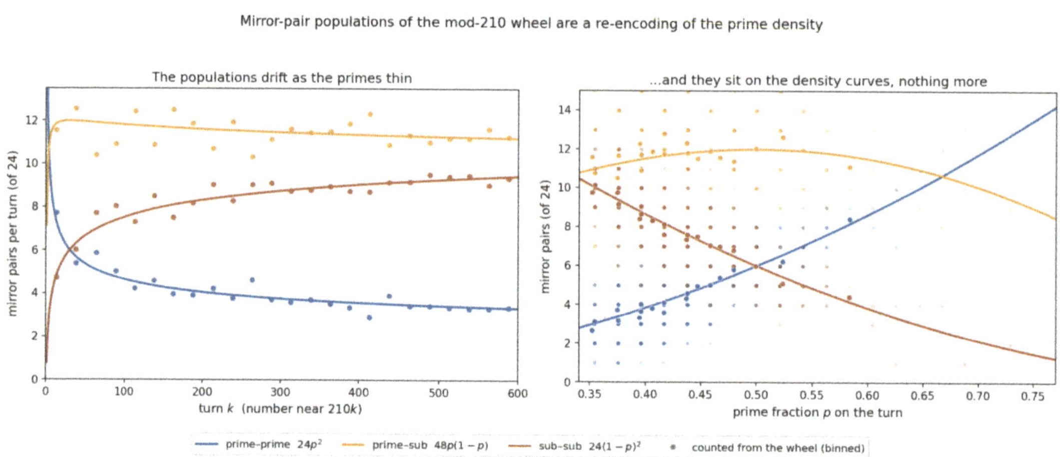

Figure 2.1: Mirror-pair populations of the wheel against the density model, Eq. (2.3). **Left:** the three classes versus turn k; lines are $24p^2$, $48p(1-p)$, $24(1-p)^2$ with $p = (210/48)/\ln n$, dots are counted from the wheel and binned. As the primes thin, prime–prime pairs give way to subprime–subprime, with prime–subprime holding near its maximum. **Right:** the same counts plotted against the prime fraction p on each turn (faint points per turn, solid points binned) fall onto the three curves. The populations are a re-encoding of the prime density; there is no further structure.

2.4 What the agreement means

The fit is the point, and it cuts against reading anything more into the classes. That the counts match Eq. (2.3) to within the per-turn scatter says the two members of a mirror pair behave as independent draws at the local density: there is no detectable correlation between a residue's primality and its mirror's, beyond the coprimality both already carry. Whatever "damping" distinguishes the three classes—the steady decline of PP, the rise of SS—is therefore not a property of the pairing at all. It is the prime number theorem, the $1/\ln n$ thinning, seen through the mirror. Naming a separate damping for each class would only be renaming $p(k)$ three times.

This places the pair census firmly on the descriptive side of the ledger. It is an effective model of texture, of exactly the kind Cramér's coin-flip is for the primes themselves: faithful to the statistics, silent on the mechanism. It reproduces no zeros, defines no operator, and isolates no signal; it cannot, because it contains only the density, and the density is the smooth backbone the primes and subprimes already share. In the language of the diffraction chapter, the subprimes are the periodic clockwork and the primes the aperiodic remainder, but this particular statistic sees neither the clockwork's comb nor the remainder's zeros—only the slowly falling envelope common to both. That is why it is so well behaved, and why it is not, on its own, telling us anything the prime number theorem has not already told us.

A final word on status. The independence in Eq. (2.3) is a heuristic, not a theorem; it is of a piece with the Hardy–Littlewood conjectures on correlations among primes in residue classes, which remain open. The agreement here is a finite-window observation over the first six hundred turns, offered as a clean description and nothing stronger.

Counting the classes

The three populations need no heuristic to *count*—only the wheel and a primality test:

```
from sympy import isprime
from math import gcd
units = [r for r in range(1, 210) if gcd(r, 210) == 1]
pairs = sorted({tuple(sorted((r, 210 - r))) for r in units}) # 24 mirror pairs
def classes(k): # (PP, PS, SS) on turn k
 PP = PS = SS = 0
 for a, b in pairs:
 pa, pb = isprime(210*k + a), isprime(210*k + b)
 if pa and pb: PP += 1
 elif pa or pb: PS += 1
 else: SS += 1
 return PP, PS, SS
print(classes(1), classes(100), classes(300)) # (13, 9, 2) (7, 10, 7) (3, 9, 12)
```

Part II

Distribution on the spokes

Chapter 3

Equidistribution

This expository chapter records the zeroth-order law of the primes on the wheel modulo 210*: the law that the later chapters measure deviations from. Two classical theorems do the work. Dirichlet's theorem (1837) guarantees that each of the* 48 *residue classes coprime to* 210 *contains infinitely many primes; its engine is the non-vanishing* $L(1,\chi) \neq 0$ *of the Dirichlet L-functions. The prime number theorem for arithmetic progressions then makes the stronger statement that the primes are* equidistributed *among those* 48 *classes — each receives, asymptotically, the same share* $1/48$*. We illustrate this directly: the primes up to* 10^7 *fall across the* 48 *classes with a spread of only* 0.26%*, and that spread shrinks as the cutoff grows. The small residual wobble around exact equality is precisely where Chebyshev's prime race lives. Everything here is classical; this chapter simply fixes the leading-order picture before the race chapter studies the correction.*

3.1 Scope

This chapter is expository and entirely classical. It states the two results—Dirichlet's theorem and the prime number theorem for arithmetic progressions—that govern the coarse distribution of primes across the wheel, and it shows them numerically. It is the foundation the prime-race chapter builds on: that chapter's entire subject, Chebyshev's bias, is the second-order fluctuation around the equidistribution established here. The opening chapter describes the wheel itself.

3.2 The forty-eight rays

Before the formulas, a picture. Fix a coprime residue r and follow the spoke it sits on through every turn of the wheel: the numbers r, $r + 210$, $r + 420, \ldots$ all share the same angle and march outward, so over infinitely many turns the spoke fills in to a solid ray running from the centre to the rim. There are exactly 48 such rays, one per coprime residue, and a ray is nothing other than an arithmetic progression $\{\,210k + r : k \geq 0\,\}$ drawn radially (Fig. 3.1). The rim, reached only in the limit, plays the role of infinity.

This single image carries the two theorems of this chapter in advance. Dirichlet's theorem says no ray is ever starved: each of the 48 holds infinitely many primes. The prime number theorem for progressions says more — the rays are *balanced*, each collecting the same limiting share $1/48$ of all primes — so the boundary measure the rays induce at infinity is simply uniform on 48 points. The geometry also makes the wheel's mirror literal. The map $r \mapsto 210 - r$ is reflection across the vertical diameter; it carries each ray to its partner and sorts the 48 into 24 mirror pairs (the horizontal rungs in the figure). Whether a Dirichlet character is symmetric or antisymmetric under that reflection is exactly its parity $\chi(-1) = \pm 1$, the organizing distinction taken up in the conductor-lattice chapter.

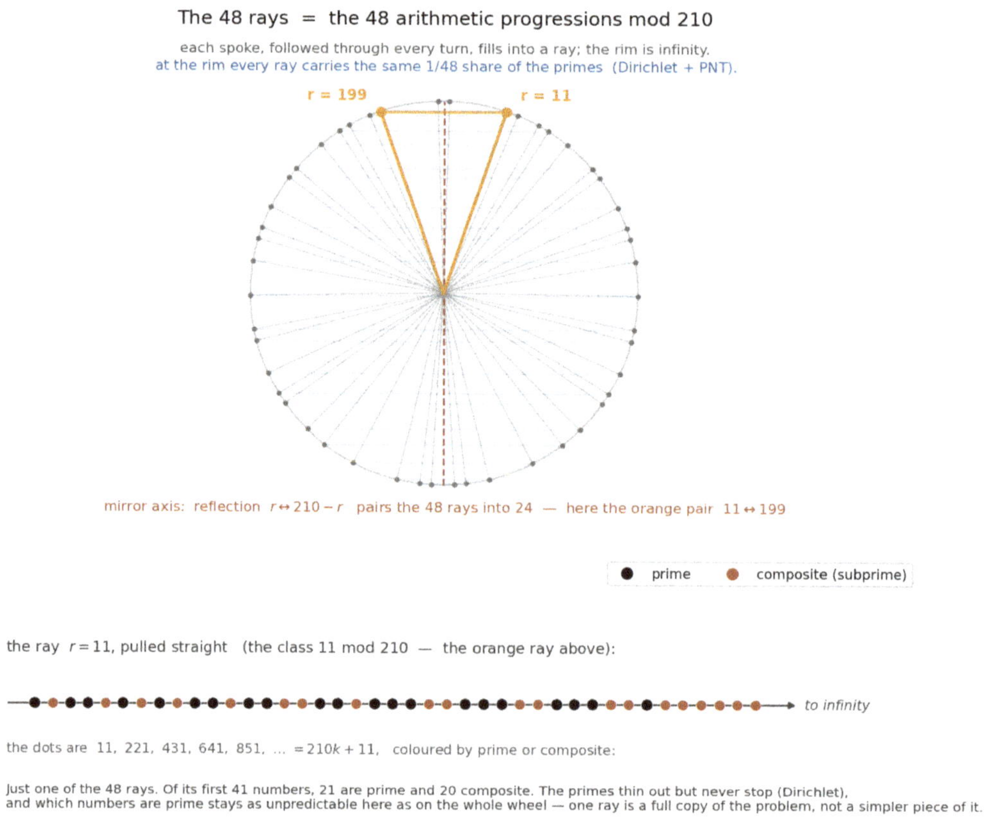

Figure 3.1: Top: the 48 rays, one per coprime residue, with the 24 horizontal rungs marking the mirror pairs $r \leftrightarrow 210 - r$ and the vertical dashed line the axis of that reflection. At the rim — the limit of infinitely many turns — each ray carries the same $1/48$ share of the primes. Bottom: a single ray unrolled, the class 11 mod 210 (the numbers $11, 221, 431, \cdots = 210k + 11$), with its primes (black) and subprimes (red) over the first 41 numbers. Zooming into one ray does not simplify anything: the primes along it are as pseudorandom as on the whole wheel, only sparser.

Two cautions keep the picture honest. First, the 48 rays are an organizing device, not a simplification: each ray is a full-size copy of the same hard problem, carrying a thinned-out but equally pseudorandom stream of primes, so no single ray is the easy one. Second, the natural way to work with the rays is not one at a time but through the Dirichlet characters mod 210 — specific weighted blends of all the rays that diagonalize the wheel's symmetry. A single character touches every ray at once, and recovering one ray means summing over all the characters by orthogonality; this is precisely the mechanism Dirichlet's proof uses, and the subject of the next sections. (The "boundary at infinity" here is a faithful metaphor for the limit, not a hidden geometry: the deep boundary-at-infinity of number theory, where modular forms live, is a separate and far more structured construction.)

3.3 Dirichlet's theorem

The wheel sorts every prime $p > 7$ into one of the 48 residue classes coprime to 210. The first question is whether any class is ever starved of primes. Dirichlet answered it in 1837: for any a with $\gcd(a, m) = 1$,

$$\#\{\, p \text{ prime} : p \equiv a \pmod{m} \,\} = \infty. \tag{3.1}$$

Applied to $m = 210$, every one of the 48 coprime classes holds infinitely many primes; none runs dry.

The proof introduced the objects that organize the entire subject. To each character χ of $(\mathbb{Z}/m\mathbb{Z})^\times$ one attaches the L-function $L(s, \chi) = \sum_{n \geq 1} \chi(n) n^{-s}$, and the primes in a class are isolated by averaging these over χ. The argument succeeds because, for every non-principal χ,

$$L(1, \chi) \neq 0. \tag{3.2}$$

This single non-vanishing is the load-bearing fact: were any $L(1, \chi)$ to vanish, the primes could pile up away from some class and leave it finite. The characters of the wheel, and the seven real ones in particular, are the subject of the conductor-lattice chapter; here we need only that their L-functions stay clear of zero at $s = 1$.

3.4 Equidistribution

Dirichlet's theorem says every class is infinite; it does not yet say the classes are *balanced*. The prime number theorem for arithmetic progressions, established by de la Vallée Poussin in 1896 — the prime number theorem itself having been proved that year independently by Hadamard and de la Vallée Poussin — supplies the balance. For $\gcd(a, m) = 1$,

$$\pi(x; m, a) \sim \frac{1}{\varphi(m)} \frac{x}{\log x} \sim \frac{\pi(x)}{\varphi(m)}, \tag{3.3}$$

so each coprime class captures the *same* asymptotic fraction $1/\varphi(m)$ of all primes. On the wheel, $\varphi(210) = 48$, and every one of the 48 classes should hold about $\pi(x)/48$ of the primes up to x.

Figure 3.2 shows this for $x = 10^7$. There are $\pi(10^7) = 664{,}579$ primes; discarding $2, 3, 5, 7$ leaves 664,575 on the wheel, an average of 13,845 per class. The actual counts run from a minimum of 13,757 to a maximum of 13,922 — a coefficient of variation of just 0.26% about the mean. To the eye the 48 bars are simply level: equidistribution is not a subtle limiting statement here but a visible fact already at 10^7.

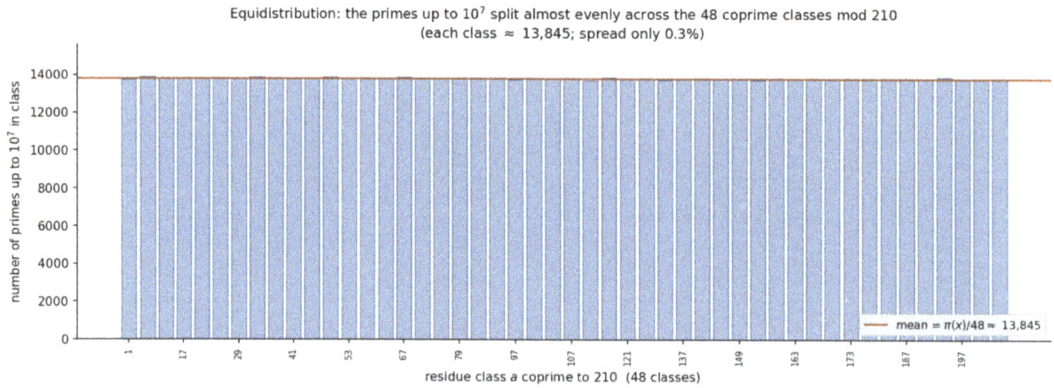

Figure 3.2: The primes up to 10^7 distributed across the 48 residue classes coprime to 210. Each bar is a class; the red line is the mean $\pi(x)/48 \approx 13{,}845$. The counts vary by only 0.26% from class to class — the equidistribution promised by Eq. (3.3), plainly visible.

3.5 The balance is asymptotic — and that is where the race lives

Equation (3.3) is a statement about the limit, and the approach to it is gradual. Measuring the spread across the 48 classes by their coefficient of variation (standard deviation over mean), one finds it falling steadily as the cutoff grows (Fig. 3.3):

$$6.5\% \ (x = 10^4) \longrightarrow 2.5\% \ (10^5) \longrightarrow 0.8\% \ (10^6) \longrightarrow 0.26\% \ (10^7). \tag{3.4}$$

The classes are pulling level, but they are never exactly level at finite x. That residual imbalance is not noise to be discarded; it is the entire subject of the next chapter. When one resolves the small deviations $\pi(x; m, a) - \pi(x)/\varphi(m)$ they are found to be *systematic*: classes of one arithmetic type tend to lead, classes of another to lag. That persistent tilt is Chebyshev's bias, and its fine structure is controlled by the zeros of the same L-functions whose non-vanishing at $s = 1$ gave us Dirichlet's theorem in the first place.

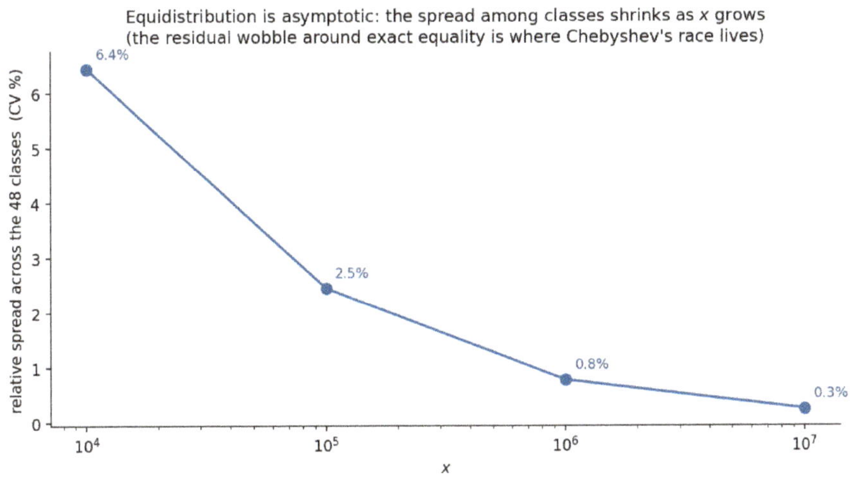

Figure 3.3: The relative spread among the 48 classes (coefficient of variation) as a function of the cutoff x. It shrinks toward zero, confirming that equidistribution is an asymptotic law. The small wobble that remains at any finite x is the deviation the prime-race chapter measures.

3.6 What this sets up

The picture to carry forward is a two-term one. To leading order the primes are blind to which spoke they land on: all 48 classes are equal, by Eq. (3.3). The interesting arithmetic is in the correction. The race chapter measures the systematic part of that correction; the conductor chapter studies the characters that sort the classes into the competing teams; and the diffraction chapter traces the correction back to the L-function zeros. All three presuppose the flat zeroth-order law established here.

Reproducing the figures

```
import numpy as np
from math import gcd

N = 10_000_000
sieve = np.ones(N+1, bool); sieve[:2] = False
for i in range(2, int(N**0.5)+1):
 if sieve[i]: sieve[i*i::i] = False
primes = np.nonzero(sieve)[0]

units = [r for r in range(210) if gcd(r, 210) == 1] # the 48 classes
pp = primes[primes > 7]

# Figure 1: counts per class at x = 1e7
res = pp % 210
counts = np.array([np.count_nonzero(res == r) for r in units])
print(counts.sum(), counts.mean(), counts.min(), counts.max(),
```

```
 100*counts.std()/counts.mean()) # 664575 13845.3 13757 13922 0.26

# Figure 2: spread (CV%) vs cutoff
for X in [1e4, 1e5, 1e6, 1e7]:
 c = np.array([np.count_nonzero((pp[pp <= X]) % 210 == r) for r in units], float)
 print(X, round(100*c.std()/c.mean(), 2)) # 6.45, 2.46, 0.79, 0.26

# Figure 3: one ray unrolled -- the class 11 mod 210 over its first 41 numbers
from sympy import isprime
ray = [(k, isprime(210*k + 11)) for k in range(41)]
nprime = sum(p for _, p in ray)
print(nprime, 41 - nprime) # 21 primes, 20 subprimes
```

Chapter 4

The Prime Race

This is an expository *chapter. We make a classical phenomenon visible by computing it directly on a single small modulus: we sort primes into two competing families of residue classes modulo* 210*, count them block by block, and watch the race. The lead changes hands endlessly, the cumulative advantage wanders at the characteristic* $\sqrt{N}$ *rate rather than accumulating, the prime "population" of each block thins as the local form of the prime number theorem, and—underneath all of it—the fluctuations are governed by the zeros of Dirichlet L-functions. None of this is new: the bias was first noticed by Chebyshev (1853), the modern framework is due to Rubinstein and Sarnak (1994), and a celebrated accessible account is Granville and Martin's* Prime Number Races *(2006). For this particular (mirror-symmetric) split we also pin down the leading direction of the bias—the lower team is favored—and isolate the one genuinely remaining computation, the exact logarithmic density from the L-function zeros, honestly as exactly that. The aim here is only clarity—to let a reader* watch *the zeros act through an elementary count—and to mark honestly where the small experiment connects to the deepest open question in the subject.*

4.1 Scope

This chapter contains no new theorem and makes no claim outside number theory. The computations are elementary sieving; the explanations are textbook. Its value, if any, is pedagogical.

4.2 The wheel

Fix the modulus $M = 210 = 2 \cdot 3 \cdot 5 \cdot 7$, with the $\varphi(210) = 48$ live spokes and the mirror involution $r \mapsto 210 - r$ of the opening chapter (Fig. 4.1). Every integer n lands on spoke $n \bmod 210$; advancing by one *period* (one block $\mathcal{B}_k = [210k,\, 210k + 210)$) returns to the same spoke. The mirror pairs each spoke with another ($11 \leftrightarrow 199$, etc.) and exchanges the lower semicircle $(0, 105)$ with the upper $(105, 210)$, each half holding 24 live spokes.

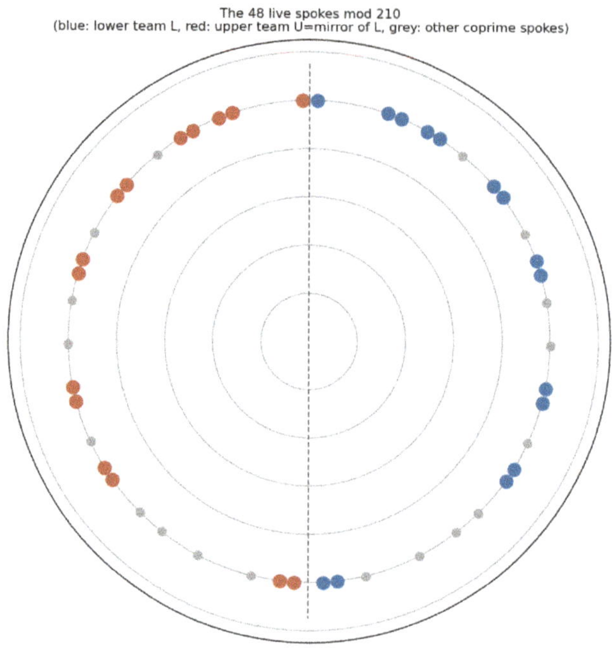

Figure 4.1: The 48 live spokes modulo 210. Blue: the lower team $\mathcal{L}$; red: the upper team $\mathcal{U}$, its mirror image; grey: the remaining coprime spokes. The two teams are reflections of each other across the 0–105 axis (dashed).

4.3 Two complementary teams

Among the live spokes, single out the *twin-admissible* ones: residues r for which r and $r+2$ are *both* coprime to 210 (the only positions where a twin-prime pair can sit). Split these by the mirror into two equal teams:

$$\mathcal{L} = \{1, 11, 13, 17, 19, 29, 31, 41, 43, 59, 61, 71, 73, 101, 103\}, \tag{4.1}$$

$$\mathcal{U} = \{107, 109, 137, 139, 149, 151, 167, 169, 179, 181, 191, 193, 197, 199, 209\}, \tag{4.2}$$

fifteen spokes each, with $\mathcal{U} = \{210 - r : r \in \mathcal{L}\}$. Most appear in visible twin pairs $(11, 13), (17, 19), (29, 31), \ldots$, and one pair wraps the seam: spoke 1 is twin-admissible because 209 and $211 \equiv 1$ are both coprime, so the twin $(419, 421)$ straddles a block boundary. Nothing is special about "twin" spokes here; they merely give one clean, symmetric way to form two equal teams, and any balanced split behaves identically.

4.4 The race

For each period k, count the primes landing on each team,

$$A_{\mathcal{L}}(k) = \#\{p \text{ prime} : p \in \mathcal{B}_k,\ p \bmod 210 \in \mathcal{L}\}, \qquad A_{\mathcal{U}}(k) \text{ likewise}, \tag{4.3}$$

and form the per-period asymmetry $\Delta(k) = A_{\mathcal{L}}(k) - A_{\mathcal{U}}(k)$. Does either team build a durable lead?

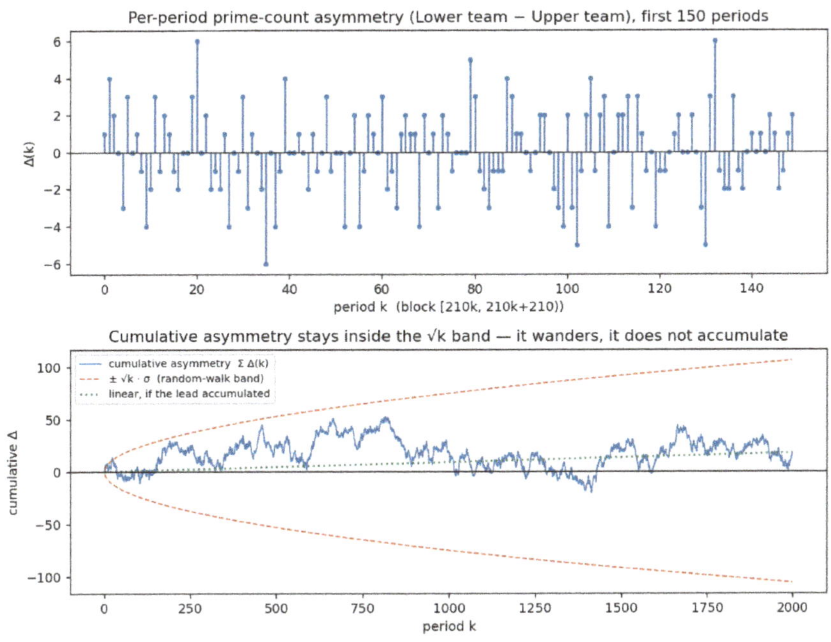

Figure 4.2: Top: the per-period lead $\Delta(k)$ oscillating around zero (first 150 periods). Bottom: the cumulative lead stays inside the $\pm\sqrt{k}\,\sigma$ random-walk band (red dashed) and nowhere near the straight line a genuinely accumulating bias would follow (green dotted).

4.5 What the race does

Computed over the first 2000 periods (integers up to $\sim 4.2 \times 10^5$), the answer is vivid (Figs. 4.2 and 4.3):

- **The lead flips constantly.** $\Delta(k)$ changes sign every few periods; over 500 periods there are about 208 sign changes, with $\mathcal{L}$ ahead in 213, $\mathcal{U}$ ahead in 195, and 92 ties. The per-period spread is small, $\sigma \approx 2.4$.

- **The cumulative lead does not run away.** $\sum_k \Delta(k)$ is +26 after 500 periods and +18 after 2000—it *shrinks* back toward zero with more data. Against the random-walk envelope $\pm\sqrt{N}\,\sigma$ (± 53 at $N{=}500$, ± 106 at $N{=}2000$), the cumulative stays comfortably inside the band throughout.

- **No net bias is detectable here.** The cumulative is consistent with an unbiased $\sqrt{N}$ wander. Whether a genuine, tiny bias exists for *this* team split is exactly the question the Rubinstein–Sarnak analysis settles (it depends on how many spokes in each team are quadratic residues mod $3, 5, 7$); raw counting at this scale cannot resolve it directly, and Sec. 4.9 extracts the direction analytically instead.

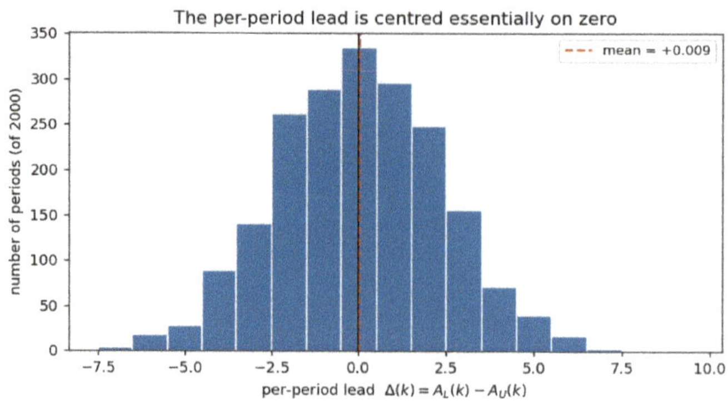

Figure 4.3: Distribution of the per-period lead $\Delta(k)$ over 2000 periods: centred essentially on zero (mean +0.009).

4.6 The thinning: the prime number theorem on one wheel

A second, slower motion runs alongside the race. In each block the 48 live spokes split between *primes* and *subprimes* (composites coprime to 210, i.e. with smallest prime factor ≥ 11). In the first block only five subprimes fit—$121, 143, 169, 187, 209$—because the smallest possible one is $11^2 = 121$; that block is 42 primes, 5 subprimes, and the unit. As k grows, primes thin and subprimes occupy the freed spokes (Fig. 4.4). Measured directly: the prime count per block tracks $210/\log(210k)$ with correlation $\approx$0.55—simply the prime number theorem read off block by block, the residual being the same fluctuation as in the race; the two reach parity (24 each) near $k = 19$ ($N \approx 4000$); and the mean changeover rate, fit over the first $\sim$110 periods, is about +0.077 subprimes per period. This "interchange" is a smooth classical thinning buried in arithmetic fluctuation, carrying no structure beyond $\pi(x) \sim x/\log x$. The fill-in is moreover *uniform across the wheel*: at $X = 2 \times 10^6$ the subprime-to-prime ratio is essentially constant over all 48 classes—mean 2.07, coefficient of variation under 1% (range 2.04–2.11)—and the two teams receive equal shares (2.06 versus 2.07). The ratio itself grows slowly with X as composites accumulate, but it grows the *same* in every class, so the subprimes obey the same equidistribution the primes do (Sec. 4.7), filling the freed spokes evenly rather than favoring either team.

4.7 Why the race cancels: equidistribution

Both motions trace to one classical fact, established in the equidistribution chapter: the primes are *equidistributed* across the 48 classes coprime to 210, each receiving its fair share $1/\varphi(210)$ in the limit. Two equal teams therefore receive equal shares, so the team totals balance and $\sum_k \Delta(k)$ has no linear drift. The lead cannot run away because the field is, in the limit, perfectly level. What remains is the *rate of approach* to that level field—the fluctuation—and that is where the depth lives.

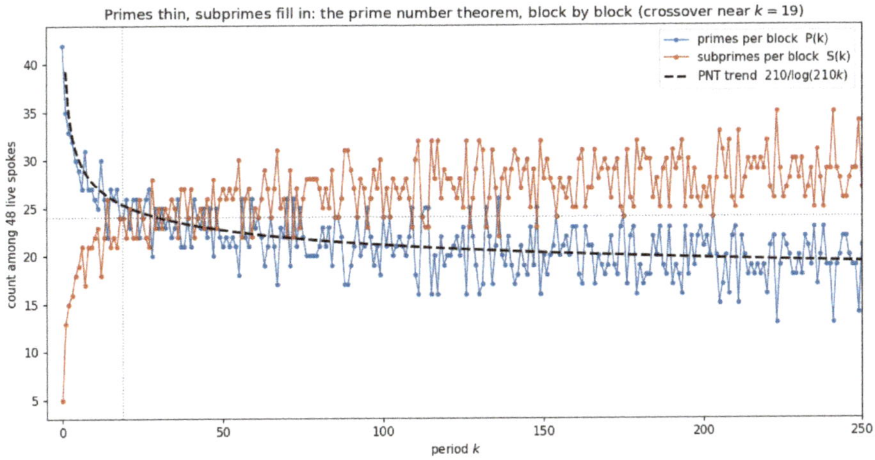

Figure 4.4: Primes thin, subprimes fill in, block by block. The dashed curve is the prime-number-theorem trend $210/\log(210k)$; the two populations cross near $k = 19$.

4.8 Why it can only wander: the zeros

The fluctuation is not arbitrary. By the explicit formula—the bridge Riemann opened in 1859, in its form for arithmetic progressions—the error in counting primes in a residue class is a sum over the nontrivial zeros $\rho = \frac{1}{2} + i\gamma$ of the relevant Dirichlet L-functions,

$$\text{(prime count in class)} - \text{(expected)} \approx -\sum_{\rho} \frac{x^{\rho}}{\rho}. \tag{4.4}$$

Each zero contributes an oscillation; their sum is the wobble in $\Delta(k)$. Two consequences seen directly above: under the Riemann Hypothesis every $\operatorname{Re}\rho = \frac{1}{2}$, so the error is of order $\sqrt{x}$—precisely the $\sqrt{N}$ band the cumulative race lived inside; and because infinitely many zeros keep contributing, the lead changes hands forever, which is the content of the Chebyshev-bias / Rubinstein–Sarnak analysis and its limiting logarithmic density. The small experiment on the 210 wheel is, underneath, a direct sighting of the zeros of L-functions—the same objects whose spacing statistics famously resemble random-matrix eigenvalues and whose location is the Riemann Hypothesis. That connection is the genuine prize, and it is a question *in mathematics.*

4.9 The leading bias: which way the race tilts

The direct count could not name a winner—at this scale the tilt is buried under the $\sqrt{N}$ fluctuations—but the Rubinstein–Sarnak framework extracts it anyway, and the mirror symmetry of our teams makes the leading term clean.

First, a simplification. Because $\mathcal{U} = \{210 - r : r \in \mathcal{L}\}$, every *even* Dirichlet character ($\chi(-1) = +1$) contributes equally to both teams and cancels in the difference.

The race between two mirror-image teams is therefore governed entirely by the *odd* characters modulo 210.

The bias itself is the Chebyshev effect: residue classes that are perfect squares mod q are prime-deficient (the squares of primes accumulate there), so a team carrying more quadratic residues wins less often. Modulo 210 there are exactly six coprime quadratic residues, $\{1, 79, 109, 121, 151, 169\}$, each with eight square roots. Of these the lower team contains only one (the residue 1), while the upper team contains three $(109, 151, 169)$:

$$\sum_{a\in\mathcal{L}} c(a) = 8, \qquad \sum_{a\in\mathcal{U}} c(a) = 24, \qquad c(a) = \#\{t \bmod 210 : t^2 \equiv a\}. \tag{4.5}$$

The upper team is the more prime-deficient, so the race is *not* fair: the lower team is favored, and its Rubinstein–Sarnak logarithmic density of leading exceeds $\frac{1}{2}$. This agrees in direction with the finite-X observation that the lower twin-corridor half is the "Chebyshev-favored" half, and with the block-0 lean of 14 primes to 13 seen above.

Two honest caveats close the loop. First, this persistent leading bias coexists with the count ratio tending to 1 (equidistribution): the cumulative totals equalize while one side still leads more than half the time—that coexistence *is* the Chebyshev paradox. Second, the *direction* ($\mathcal{L}$ favored) is now settled, but the *exact* density requires the variance contributed by the nontrivial zeros of the odd Dirichlet L-functions modulo 210. That is the one genuinely remaining computation—a known method applied to a non-canonical example, conditional on GRH and the linear-independence hypothesis—and we leave it open here as exactly that.

4.10 What this is and is not

This chapter re-derives by hand, on one wheel, the prime-race / Chebyshev-bias picture. The findings—oscillating lead, $\sqrt{N}$ wander, prime-number-theorem thinning, control by the zeros—are textbook consequences of theory that is 30 to 170 years old. It is not a new method or result, and makes no claim outside number theory. Its only ambition is to let a reader watch the zeros act through a simple count. The rigorous next step—computing the exact Rubinstein–Sarnak bias density for a chosen team split—is a real, specialized calculation within the existing framework. The broader study of unexpected biases in the distribution of primes remains active; for a striking modern instance in a different guise—biases between *consecutive* primes—see Lemke Oliver and Soundararajan.

Reproducing the computation

```
import numpy as np, math
M = 210
Z = [r for r in range(1, M) if math.gcd(r, M) == 1] # 48 live spokes
Zs = set(Z); T = set()
for r in Z: # twin-admissible spokes
 if (r + 2) % M in Zs:
```

```
 T.add(r); T.add((r + 2) % M)
L = sorted(r for r in T if r < M/2) # lower team (15)
U = sorted((M - r) % M for r in L) # upper team (15)

Nper = 2000; top = M*(Nper + 1)
spf = np.zeros(top + 1, dtype=np.int64) # smallest-prime-factor sieve
for i in range(2, top + 1):
 if spf[i] == 0:
 spf[i::i] = np.where(spf[i::i] == 0, i, spf[i::i])
isprime = lambda n: n >= 2 and spf[n] == n

d = np.array([sum(isprime(M*k+r) for r in L) - sum(isprime(M*k+r) for r in U)
 for k in range(Nper)])
C = np.cumsum(d)
print("cumulative:", int(C[-1]), " sqrt-band:", round(np.sqrt(Nper)*d.std(), 1))
```

The leading bias term (Sec. 4.9) and the subprime equidistribution (Sec. 4.7) reproduce as follows.

```
# leading bias term: quadratic residues mod 210 (each has 8 square roots)
QR = {(x * x) % M for x in Z}
c = lambda a: 8 if a in QR else 0
print("sum c over L:", sum(c(a) for a in L), # 8 (one QR: residue 1)
 " over U:", sum(c(a) for a in U)) # 24 (three QRs: 109, 151, 169)

# subprime equidistribution: per-class subprime/prime ratio at X = 2e6
X = 2_000_000
s = np.zeros(X + 1, dtype=np.int64)
for i in range(2, X + 1):
 if s[i] == 0:
 s[i::i] = np.where(s[i::i] == 0, i, s[i::i])
ratio = []
for r in Z:
 start = r if r > 1 else r + M # skip the unit at n = 1
 prime = sum(1 for n in range(start, X + 1, M) if s[n] == n)
 sub = sum(1 for n in range(start, X + 1, M) if n > 1 and s[n] != n)
 ratio.append(sub / prime)
ratio = np.array(ratio)
print("ratio mean %.3f CV %.4f" % (ratio.mean(), ratio.std() / ratio.mean()))
```

Chapter 5

The Conductor Lattice

This chapter is an expository *extension of the diffraction chapter, "Seeing the Primes in the Zeros," which drew the Riemann zeros as a quasicrystal whose diffraction is the primes, and sketched the same picture for two characters of the mod-210 wheel. Here we carry it out for all seven* real *Dirichlet characters modulo 210 and make the organizing structure explicit. The seven characters are the nontrivial divisors of* $105 = 3 \cdot 5 \cdot 7$*, which form a Boolean lattice; each is a Kronecker symbol attached to a fundamental discriminant; each L-function's zeros form a quasicrystal on the same prime lattice; and the conductor of a character switches off exactly its own peaks among* $\{\log 3, \log 5, \log 7\}$*, leaving* $\log 2$ *and every prime from 11 upward always present. The even/odd grading of the characters—the wheel's mirror* $r \mapsto 210 - r$*—turns out to be precisely the parity of* $|c \cap \{3, 7\}|$*. Everything here is classical and fully explicit: nothing beyond Kronecker symbols and the explicit formula is required, and no new result is claimed.*

5.1 Scope

This chapter proves nothing new. Its content is standard Dirichlet–Kronecker theory, drawn concretely on a single modulus and paired with the diffraction picture from the chapter on the zeros. The value, if any, is that every statement is exact and every figure is computed from the actual L-function zeros.

5.2 The seven real characters

The Dirichlet characters modulo 210 form a group dual to $(\mathbb{Z}/210\mathbb{Z})^\times$. Among them, the *real* characters are those of order dividing two—the quadratic characters—and the nontrivial real *primitive* ones are exactly the Kronecker symbols

$$\chi_D(\cdot) = \left(\frac{D}{\cdot}\right), \tag{5.1}$$

where D runs over the fundamental discriminants whose conductor divides 210. Because 2 does not ramify here, these conductors are the nontrivial divisors of $105 = 3{\cdot}5{\cdot}7$,

Table 5.1: The seven real primitive characters modulo 210. "Parity" is $\chi(-1)$; "dark primes" are the primes $p \mid c$, for which $\chi(p) = 0$ and the corresponding diffraction peak is absent.

χ	conductor c	D	symbol	$\chi(-1)$	$\lvert c \cap \{3,7\}\rvert$	dark primes
$\chi_{3.2}$	3	-3	$(-3/\cdot)$	-1 (odd)	1	$\log 3$
$\chi_{5.4}$	5	$+5$	$(5/\cdot)$	$+1$ (even)	0	$\log 5$
$\chi_{7.6}$	7	-7	$(-7/\cdot)$	-1 (odd)	1	$\log 7$
$\chi_{15.14}$	15	-15	$(-15/\cdot)$	-1 (odd)	1	$\log 3,\ \log 5$
$\chi_{21.20}$	21	$+21$	$(21/\cdot)$	$+1$ (even)	2	$\log 3,\ \log 7$
$\chi_{35.34}$	35	-35	$(-35/\cdot)$	-1 (odd)	1	$\log 5,\ \log 7$
$\chi_{105.104}$	105	$+105$	$(105/\cdot)$	$+1$ (even)	2	$\log 3,\ \log 5,\ \log 7$

namely

$$c \in \{3,\ 5,\ 7,\ 15,\ 21,\ 35,\ 105\}. \tag{5.2}$$

Each prime contributes a *prime discriminant*

$$p^* := (-1)^{(p-1)/2}\, p, \qquad \text{so} \qquad 3^* = -3, \quad 5^* = +5, \quad 7^* = -7, \tag{5.3}$$

the sign being $-$ for $p \equiv 3 \pmod 4$ and $+$ for $p \equiv 1 \pmod 4$. The discriminant of a composite conductor is the product of the prime discriminants of its prime factors, e.g. $15 \to (-3)(5) = -15$ and $21 \to (-3)(-7) = +21$. Table 5.1 lists the seven characters, their discriminants, parities, and the primes each one extinguishes.

5.3 The conductor lattice

The seven conductors are the nontrivial divisors of 105, and under divisibility they form the Boolean lattice on the set $\{3, 5, 7\}$—the subset lattice, drawn as a cube in Fig. 5.1. The bottom vertex is the trivial conductor 1 (the principal character, not one of the seven); the three atoms are the prime conductors; the three middle vertices are the products of two primes; the top is 105 itself.

The parity of each character reads straight off the lattice. Since $\chi_D(-1) = \operatorname{sgn}(D)$, and D is the product of the prime discriminants $\{3^*, 5^*, 7^*\} = \{-3, +5, -7\}$, the sign of D flips once for each prime $\equiv 3 \pmod 4$ dividing the conductor—that is, once for each of $\{3, 7\}$ it contains. Hence

$$\chi \text{ is even} \iff D > 0 \iff |c \cap \{3,7\}| \text{ is even}. \tag{5.4}$$

This gives the even set $\{5, 21, 105\}$ and the odd set $\{3, 7, 15, 35\}$ (three even, four odd), shown by colour in Fig. 5.1. The same even/odd split is the wheel's mirror symmetry $r \mapsto 210 - r$: even characters are fixed by it, odd characters are negated by it. The grading that in the prime-race chapter decides which characters carry the bias is, here, the parity of a two-element intersection.

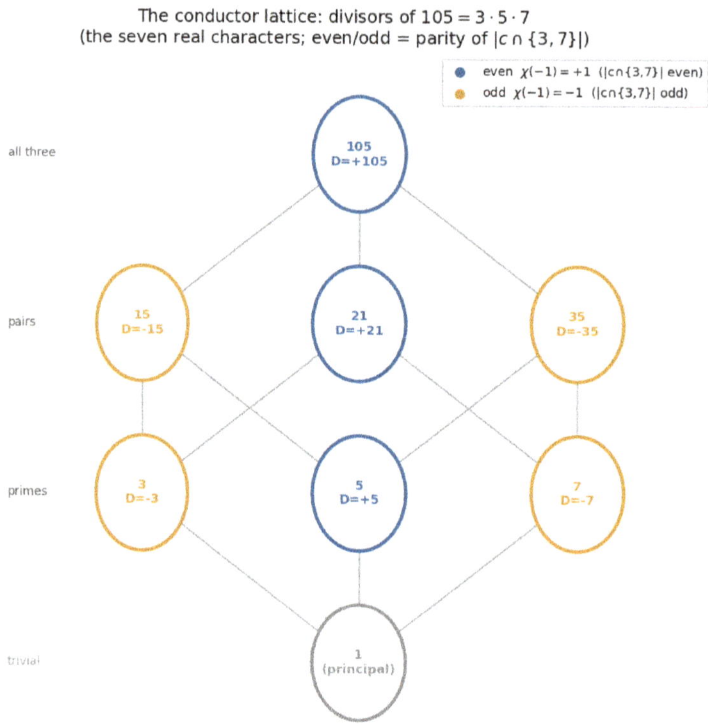

Figure 5.1: The conductor lattice: the divisors of $105 = 3 \cdot 5 \cdot 7$ as a Boolean cube. The seven nontrivial divisors are the seven real characters; colour denotes parity, which equals the parity of $|c \cap \{3, 7\}|$. The trivial conductor 1 (principal character) sits at the bottom.

5.4 Each character as a quasicrystal

Each real character χ of conductor c has an L-function $L(s, \chi)$, and the Guinand–Weil explicit formula for $L(s, \chi)$ is again a Fourier duality between its nontrivial zeros $\rho = \frac{1}{2} + i\gamma_\chi$ and the primes: for a suitable test function,

$$\sum_{\gamma_\chi} h(\gamma_\chi) = (\text{smooth terms}) - \sum_p \sum_{m \geq 1} \frac{\chi(p^m) \log p}{p^{m/2}} \widehat{h}(m \log p) . \tag{5.5}$$

The right-hand side differs from the ζ case only by the weight $\chi(p^m)$. So the diffraction of the zeros of $L(s, \chi)$,

$$D_\chi(u) = \sum_{\gamma_\chi} w(\gamma_\chi) \cos(\gamma_\chi u), \tag{5.6}$$

peaks at the same positions $u = \log(p^m)$ as before, but the sign of each peak is the Kronecker symbol $\chi(p) = \pm 1$, and a prime $p \mid c$ contributes nothing at all, because $\chi(p) = 0$.

Two facts make the picture clean (Fig. 5.2). First, 2 divides none of the conductors, so the $\log 2$ peak is always present. Second, no prime ≥ 11 divides 105, so every peak from $\log 11$ upward is always present. The only peaks the conductor can switch off are therefore those of $\{3, 5, 7\}$—and it switches off exactly the ones dividing c. Climbing the lattice extinguishes more of $\{\log 3, \log 5, \log 7\}$, until at $\chi_{105.104}$ all three are gone.

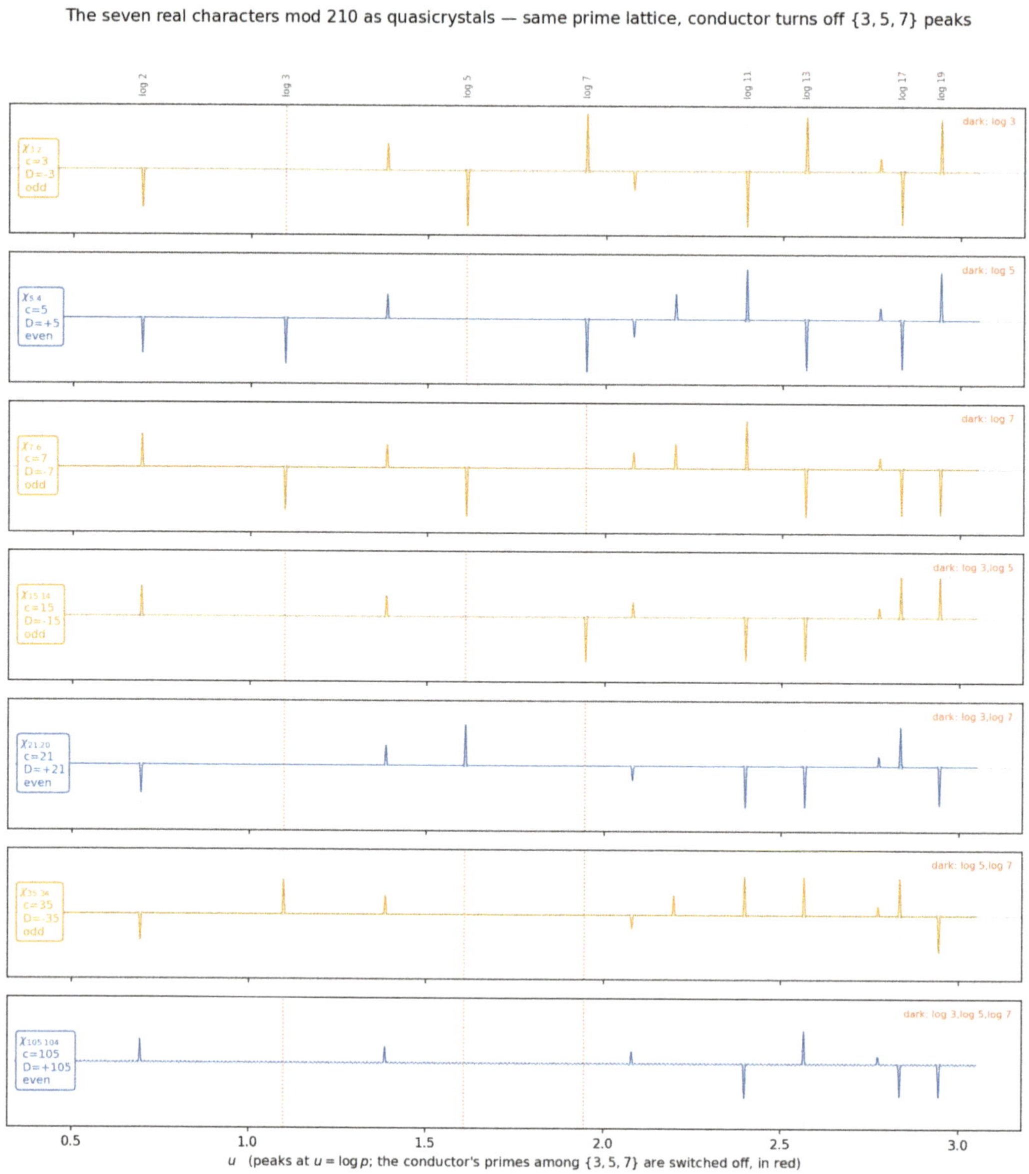

Figure 5.2: The seven real characters modulo 210, each drawn as the diffraction of its L-function zeros (computed from the actual zeros up to height $T = 1500$). All share one prime lattice; the sign of each live peak is $\chi(p)$. Red dotted lines mark the peaks switched off by the conductor (the primes $p \mid c$). Reading top to bottom, the conductor climbs the lattice and extinguishes more of $\{\log 3, \log 5, \log 7\}$; $\chi_{105.104}$ (bottom) is the maximally filtered member, with $\log 3$, $\log 5$, and $\log 7$ all dark and only $\log 2$ and the primes from 11 up surviving.

5.5 The family

Put together, Figs. 5.1 and 5.2 say the same thing two ways. The seven characters are one family of quasicrystals built on a single prime lattice; what distinguishes them is only (i) which of $\{3,5,7\}$ is extinguished—a position on the Boolean cube—and (ii) the sign pattern $\chi(p)$ on the surviving primes. The vertical axis of the cube is a filtration: the higher the conductor climbs, the more of $\{3,5,7\}$ goes dark, with $\chi_{105.104}$ at the top stripped down to $\log 2$ and the primes ≥ 11. The colouring across the cube is the parity, equivalently the wheel's mirror. Nothing in the family escapes the prime lattice underneath; the conductor only chooses which spokes of it to dim.

Two of these seven are the repeating-decimal families. The chapter on the wheel's repeating decimals sorts the 48 residues into two *families* according to the periodic block of $r/210$. That split is not an independent structure: it is one of the seven characters tabulated above, and *which* one is fixed entirely by the base in which the wheel is written. Writing the wheel in base 10, the family of a residue r is the value of the Legendre symbol $\left(\frac{r}{3}\right)$—that is, the conductor-3 character $\chi_{3.2}$. Writing it in base 60 (a clock's base), the family is $\left(\frac{r}{7}\right)$—the conductor-7 character $\chi_{7.6}$. The mechanism is the coset arithmetic of the repeating period: base 10's powers generate all of $(\mathbb{Z}/7)^\times$ but fix $(\mathbb{Z}/3)^\times$ (since $10 \equiv 1 \pmod 3$), so $\langle 10\rangle$ is the kernel of $\left(\frac{\cdot}{3}\right)$ and the family is $\left(\frac{\cdot}{3}\right)$; base 60 instead has $3 \mid 60$ (genuinely absorbed) and $\langle 60\rangle = \{1,2,4\}$, the squares modulo 7 — the kernel of $\left(\frac{\cdot}{7}\right)$ — so the family is $\left(\frac{\cdot}{7}\right)$. Both correspondences are exact—every base-10 period block consists of residues sharing a single value of $\left(\frac{\cdot}{3}\right)$, and every base-60 block a single value of $\left(\frac{\cdot}{7}\right)$. So the two "families" of the decimals chapter are precisely the two lowest-conductor members of this lattice, $\chi_{3.2}$ and $\chi_{7.6}$, surfaced one at a time by the choice of base: a reminder that the digits are a lens and these characters are the object underneath.

5.6 What this is and is not

Everything above is classical and explicit. The real characters as Kronecker symbols, the fundamental discriminants, the conductor as a Boolean lattice of divisors, and the explicit formula (5.5) are textbook results; the diffraction reading is the one from the diffraction chapter, itself a rendering of the explicit formula in the quasicrystal language of Dyson. This chapter simply organizes the seven and computes their diffractions.

It is worth being plain about what the conductor structure does and does not do. It *organizes* the family beautifully—it is the exact dial that says which of $\{3,5,7\}$ each character dims, and its parity is the mirror. But it is not a hidden key. A direct per-character study of these same seven characters finds no clean prime-versus-composite or conductor-magnitude signal; the only load-bearing distinction across the family is the parity, which is the mirror we already have. Computing the zeros of any one of these L-functions is a few seconds of standard software; the open problem is the analytic one, and it is the Generalized Riemann Hypothesis, one L-function at a time. The conductor lattice arranges the seven quasicrystals on a shelf; it does not crack any of them. And the bright peaks under all seven are, still, the primes.

Reproducing the figures

The diffraction of any character is the same cosine sum used in the diffraction chapter, applied to that character's zeros:

```
import numpy as np
def diffraction(gamma, u): # gamma: imaginary parts of L(s,chi) zeros
 g = gamma[gamma > 0]
 w = 0.5*(1 + np.cos(np.pi*g/g.max())) # Hann taper
 return -(w[:,None]*np.cos(np.outer(g, u))).sum(0)/len(g)

u = np.linspace(0.45, 3.05, 3500) # peaks appear at u = log(p^m)
```

The character table regenerates from Kronecker symbols. With `D` the fundamental discriminant and `p` an odd prime, $\chi_D(p)$ is the Kronecker symbol (D/p), equal to $+1, -1$, or 0; a character is even iff $D > 0$, and the dark primes are those dividing the conductor. The seven discriminants are $\{-3, +5, -7, -15, +21, -35, +105\}$, each the product of the prime discriminants $\{-3, +5, -7\}$ of the primes in its conductor.

Part III

Mirrors and the spectrum

Chapter 6

The Two Mirrors

The reflection $r \mapsto 210 - r$ has run through every chapter so far as a plain geometric fold of the wheel. This chapter connects it to a second, deeper reflection: the $s \mapsto 1 - s$ symmetry of the functional equation. The hinge between them is the parity of a character, $\chi(-1) = \pm 1$. The wheel's mirror sorts the characters into even (their values symmetric across the axis) and odd (antisymmetric); that very parity chooses the gamma factor, $\Gamma(s/2)$ or $\Gamma(\frac{s+1}{2})$, that completes $L(s, \chi)$ and fixes the shape of its functional equation, including where its trivial zeros fall. The two mirrors are the same fact wearing two costumes. Everything here is classical—Riemann's functional equation for ζ and its extension to Dirichlet L-functions; this chapter only places the wheel's reflection beside the analytic one.

6.1 Scope

This chapter is expository and classical. It states the functional equation for the wheel's L-functions and shows how the geometric mirror of the opening chapter and the even/odd grading of the conductor-lattice chapter are tied to it through one quantity, the parity $\chi(-1)$. It introduces no result; the functional equation is a theorem more than a century and a half old. The Riemann Hypothesis, which concerns the one feature of the analytic mirror this chapter does *not* settle, is flagged in Sec. 6.5 and otherwise left to the diffraction chapter.

6.2 The wheel's mirror and the parity of a character

The wheel's reflection is $r \mapsto 210 - r \equiv -r \pmod{210}$. A Dirichlet character interacts with it in the simplest possible way:

$$\chi(210 - r) = \chi(-r) = \chi(-1)\,\chi(r), \tag{6.1}$$

so the reflection multiplies every value by the single sign $\chi(-1)$. A character is called *even* if $\chi(-1) = +1$ and *odd* if $\chi(-1) = -1$, and on the wheel the names are literal: an even character takes equal values at r and its mirror $210 - r$, an odd one takes opposite values. Figure 6.1 shows both, the quadratic character modulo 5 (even) and modulo 3 (odd): the even wheel is symmetric in color across the axis, the odd wheel is color-reversed.

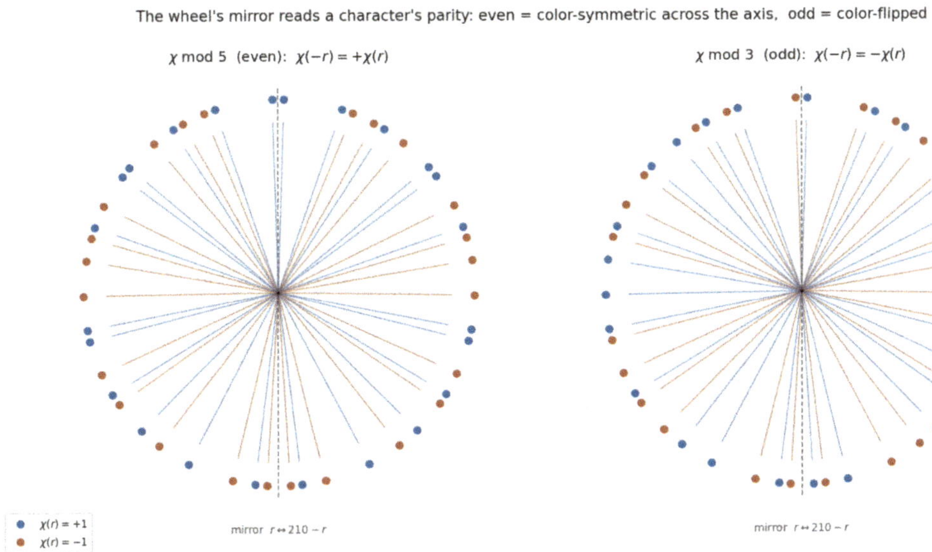

Figure 6.1: The wheel's mirror reading a character's parity. Left: the even character χ mod 5, whose values satisfy $\chi(-r) = +\chi(r)$, so the coloring is symmetric across the dashed mirror axis. Right: the odd character χ mod 3, with $\chi(-r) = -\chi(r)$, so the coloring flips across the axis. Blue marks $\chi(r) = +1$ and red $\chi(r) = -1$.

For the seven real characters of the wheel this sorts them three-and-four. The even ones are the characters of conductor $5, 21, 105$ (fundamental discriminant $D > 0$); the odd ones have conductor $3, 7, 15, 35$ ($D < 0$). The parity is exactly the even/odd grading the conductor-lattice chapter found from the discriminant sign: the wheel's mirror and the sign of D are the same partition.

6.3 The functional equation and its mirror

The deeper reflection lives not on the wheel but in the complex plane of the variable s. For a primitive character χ modulo q, set $a = 0$ if χ is even and $a = 1$ if χ is odd—that is, $\chi(-1) = (-1)^a$—and form the completed L-function

$$\Lambda(s,\chi) = \left(\frac{q}{\pi}\right)^{(s+a)/2} \Gamma\!\left(\frac{s+a}{2}\right) L(s,\chi). \tag{6.2}$$

It satisfies the functional equation

$$\Lambda(s,\chi) = \varepsilon(\chi)\,\Lambda(1-s,\overline{\chi}), \qquad |\varepsilon(\chi)| = 1, \tag{6.3}$$

relating the value at s to the value at its mirror image $1-s$. The reflection $s \mapsto 1-s$ is a point reflection through $s = \frac{1}{2}$, which leaves the critical line $\mathrm{Re}(s) = \frac{1}{2}$ invariant — the axis of the symmetry, and the mirror line of the s-plane (Fig. 6.2).

The parity enters Eq. (6.2) through the single integer a, and it is decisive. An even character ($a = 0$) is completed by $\Gamma(s/2)$; an odd one ($a = 1$) by $\Gamma(\frac{s+1}{2})$. Because the completed function is entire and non-vanishing away from the nontrivial zeros, $L(s, \chi)$ must vanish wherever its gamma factor has a pole, and those poles sit at different places for the two parities:

$$\text{even: trivial zeros at } s = 0, -2, -4, \ldots; \qquad \text{odd: trivial zeros at } s = -1, -3, -5, \ldots \tag{6.4}$$

The Riemann zeta function is the even prototype: $\xi(s) = \pi^{-s/2}\Gamma(s/2)\zeta(s)$ (with a factor $\frac{1}{2}s(s-1)$ to absorb the pole) obeys $\xi(s) = \xi(1-s)$, and ζ's trivial zeros are the even ones, $-2, -4, \ldots$ So the choice of gamma factor, made by the parity, sets both the precise form of the $s \mapsto 1-s$ symmetry and the location of the trivial zeros on the negative real axis.

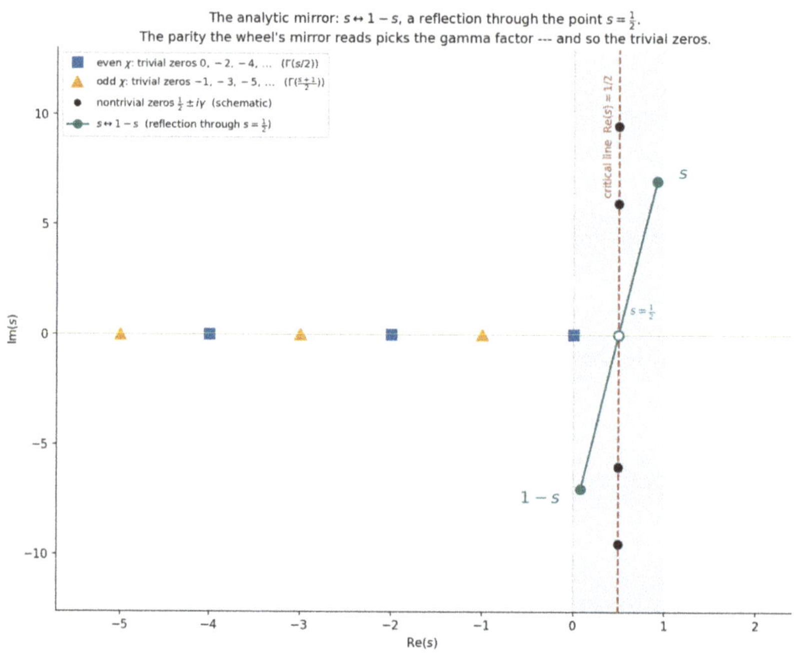

Figure 6.2: The analytic mirror. The functional equation (6.3) sends a point s to its image $1 - s$ (green), a point reflection through $s = \frac{1}{2}$ on the critical line $\text{Re}(s) = \frac{1}{2}$ (red). Even characters are completed by $\Gamma(s/2)$ and carry trivial zeros at $0, -2, -4, \ldots$ (blue squares); odd characters by $\Gamma(\frac{s+1}{2})$, with trivial zeros at $-1, -3, -5, \ldots$ (orange triangles). The nontrivial zeros (black, schematic) lie in the critical strip; the Riemann Hypothesis is the conjecture that they lie exactly on the mirror axis.

6.4 The two mirrors are one

The chain is short and it closes. The wheel's reflection $r \mapsto 210 - r$ acts on a character as multiplication by $\chi(-1)$; that sign is the parity; the parity fixes the integer a; a

selects the gamma factor $\Gamma(\frac{s+a}{2})$ in Eq. (6.2); and that gamma factor determines the exact functional equation (6.3) under $s \mapsto 1-s$. The geometric fold on the wheel and the analytic fold in the s-plane are therefore not two coincidental symmetries but two readings of one quantity. When the opening chapter named the mirror "a fact about residues, not factorization," this is where the fact reappears: the same reflection that recolors the wheel decides the shape of the L-function's reflection.

6.5 What this is and is not

The functional equation is a theorem, proved by Riemann for ζ in 1859 and extended to the Dirichlet L-functions by the classical methods recorded in any text. The parity dependence of the gamma factor is equally standard. The contribution of this chapter is only the framing: to set the wheel's reflection beside the analytic one and identify the hinge. The "two mirrors" is a pedagogical packaging of a real correspondence, not a new mathematical statement.

One feature of the analytic mirror is emphatically not settled here. The functional equation places the nontrivial zeros symmetrically about the critical line; it does not place them *on* it. That the zeros lie exactly on the mirror axis $\mathrm{Re}(s) = \frac{1}{2}$ is the Riemann Hypothesis, open for ζ and for every L-function on the wheel. This chapter establishes the symmetry of the mirror, not the location of what it reflects.

Parity of the seven real characters

```
# A real character mod 210 has conductor a divisor of 105 = 3*5*7.
# Even iff |conductor cap {3,7}| is even (i.e. fundamental disc D > 0).
for cond in [{3},{5},{7},{3,5},{3,7},{5,7},{3,5,7}]:
 parity = "even" if len(cond & {3,7}) % 2 == 0 else "odd"
 print(sorted(cond), parity)
# {3} odd {5} even {7} odd {3,5} odd
# {3,7} even {5,7} odd {3,5,7} even
# -> even: conductors 5, 21, 105; odd: conductors 3, 7, 15, 35
```

Chapter 7

Seeing the Primes in the Zeros

This is an expository *chapter paired with the prime-race chapter, "A Prime Race on the Mod-210 Wheel." Where that chapter watched the primes compete in the* counting *domain—who is ahead, by how much—this one looks at the same objects in the* spectral *domain. Riemann's explicit formula is a Fourier duality between the nontrivial zeros of $\zeta(s)$ and the primes. We make it visible the most direct way possible: we Fourier-transform a bare list of the first thousand zeros and watch sharp peaks appear at exactly the logarithms of the primes and prime powers. Viewed this way the zeros are, conditionally on the Riemann Hypothesis, a one-dimensional* quasicrystal *whose diffraction pattern is the primes. We then draw the same picture for the Dirichlet characters of the mod-*210 *wheel: each character's L-function gives its own quasicrystal on the same prime lattice, with the sign of each peak fixed by $\chi(p)$, and the wheel's mirror $r \mapsto 210 - r$ appearing as the even/odd grading of the characters. None of this is new. The duality is the explicit formula (Riemann 1859; Guinand; Weil); the metaphor is the "music of the primes" (Berry; Mazur–Stein); the quasicrystal framing is Dyson's (2009). The only ambition is to* see *it, concretely, on one wheel.*

7.1 Scope

This chapter contains no new theorem and makes no claim outside number theory. The picture it draws is one of the most reproduced in the subject. Its value, if any, is pedagogical: to let a reader watch a classical duality happen on real data. The single piece here that is a fresh *rendering* rather than a textbook classic is Sec. 7.6, where the characters of one specific modulus are drawn side by side; even there the *structure* is entirely standard.

7.2 The duality

Riemann's explicit formula, in the form due to Guinand and Weil, equates a sum over the nontrivial zeros $\rho = \frac{1}{2} + i\gamma$ of $\zeta(s)$ with a sum over the primes: schematically, for a suitable test function h with Fourier transform $\widehat{h}$,

$$\sum_{\gamma} h(\gamma) = \text{(smooth archimedean terms)} - \sum_{p} \sum_{m \geq 1} \frac{\log p}{p^{m/2}} \widehat{h}(m \log p) . \tag{7.1}$$

The right-hand side is supported on the numbers $\log(p^m)$. So the zeros and the primes are a Fourier pair: each is, in the language of harmonic analysis, the spectrum of the other. This is the exact sense of the popular metaphor that the zeros carry the *music of the primes*; each zero $\rho = \frac{1}{2} + i\gamma$ contributes to the prime distribution an oscillation $x^{\rho} = \sqrt{x}\, e^{i\gamma \log x}$ of frequency γ.

7.3 The same primes, in real space

The duality just stated—zeros and primes as a Fourier pair—has two concrete pictures, one in each domain. This section draws the spatial one; the rest of the chapter draws the spectral one.

Lay the integers on the mod-210 wheel and let it turn, so that the number $210k + r$ sits at residue r on turn k. Each prime $p \geq 11$ then marks the multiples it owns on a single family of evenly spaced parallel lines—a *grating* of slope $-(210 \bmod p)$, striking a given spoke once every p turns (Fig. 7.1). Three gratings already resemble an optical element; laid over one another (Fig. 7.2) they make moiré interference, and the clear channels no line crosses are the spokes where the primes survive. It is the sieve of Eratosthenes, drawn as overlapping diffraction gratings.

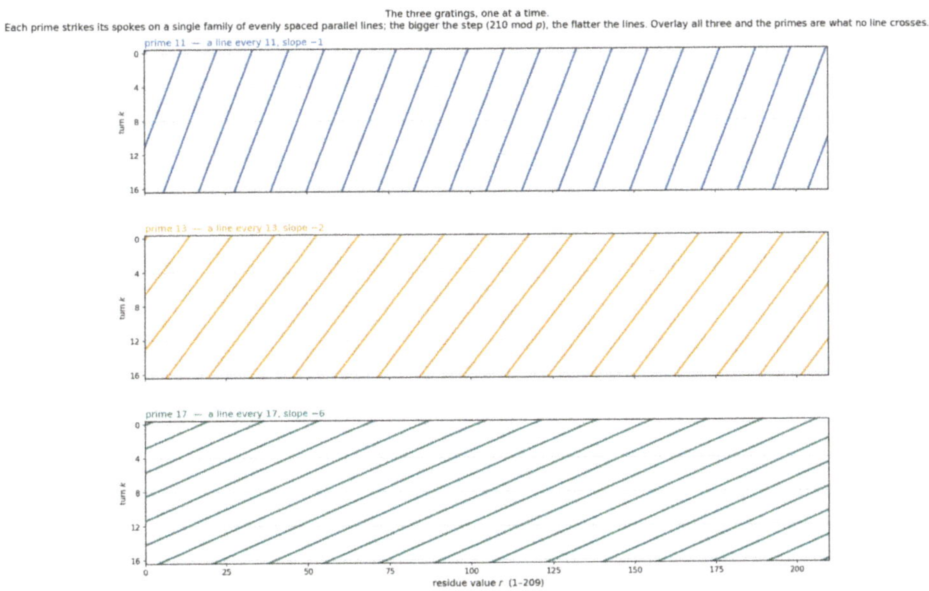

Figure 7.1: The three smallest gratings of the wheel, one per panel. Each prime p marks its multiples on a single family of parallel lines, spaced p apart and sloped by $-(210 \bmod p)$; the larger that step, the flatter the lines. These are the periodic beats of the sieve, shown separately before being overlaid.

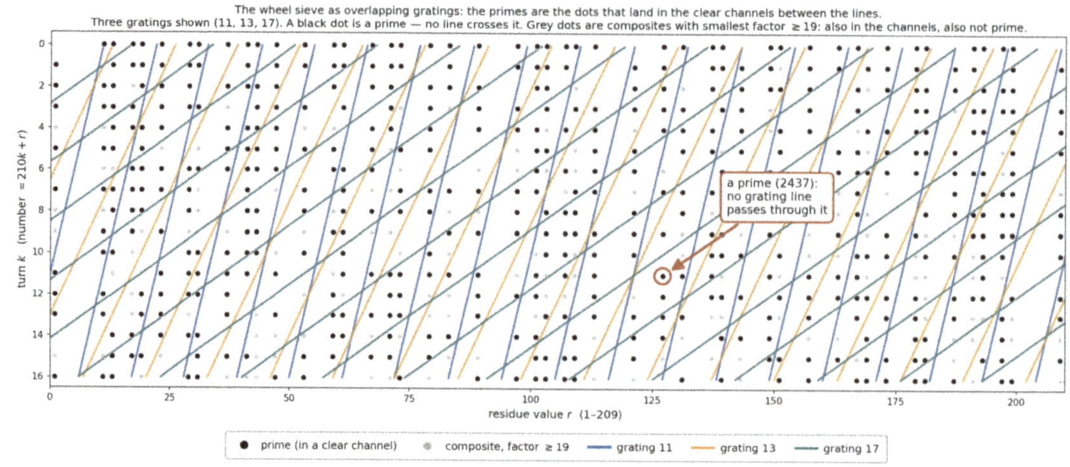

Figure 7.2: The same three gratings overlaid: the wheel's sieve as moiré interference. A black dot is a prime—a spoke no line crosses. Grey dots are composites whose smallest factor is ≥ 19: also in the open channels, also not prime, caught only by gratings not drawn here. A true prime survives the gratings of *every* prime up to $\sqrt{n}$.

This real-space picture carries a single honest seam—the one the whole chapter circles. Every grating is perfectly periodic, and any finite stack of them repeats with the period of a primorial: the composites are clockwork an optical bench could reproduce. The primes are what remains once all that periodicity is removed—the channels surviving every grating up to $\sqrt{n}$—and that remainder has no period at all. In real space, physics grips the gratings completely and what they leave behind not at all.

By Eq. (7.1), the spectrum of that same prime distribution is the set of zeros, so the diffraction picture of the next section (Fig. 7.3) shows these very primes from the spectral side. The two are companion renderings, not literal transforms of one another—the overlay is an analogy for the sieve, the explicit formula is the exact duality—but together they make one point: the periodic part of the prime story is the part that behaves like physics, and the primes themselves are the aperiodic remainder.

7.4 The picture

Equation (7.1) can be run as an experiment. Take only the imaginary parts of the first $N = 1000$ zeros—a list of bare numbers $14.13, 21.02, 25.01, \ldots$ with no reference to primes anywhere in it—and form the cosine sum

$$D(u) = \sum_{n=1}^{N} w_n \cos(\gamma_n u), \tag{7.2}$$

where w_n is a smooth taper that suppresses truncation ringing. Plotted against u (Fig. 7.3, top), $D(u)$ is flat and small almost everywhere and erupts into sharp peaks at exactly $u = \log 2,\ \log 3,\ \log 5,\ \log 7,\ \log 11, \ldots,$ with smaller peaks at the prime powers $\log 4,\ \log 8,\ \log 9$. The peak heights follow $\log p \,/\, p^{m/2}$ from Eq. (7.1), largest around $p \sim 5$–13 and decaying thereafter. The primes were not put in; they fall out of the zeros.

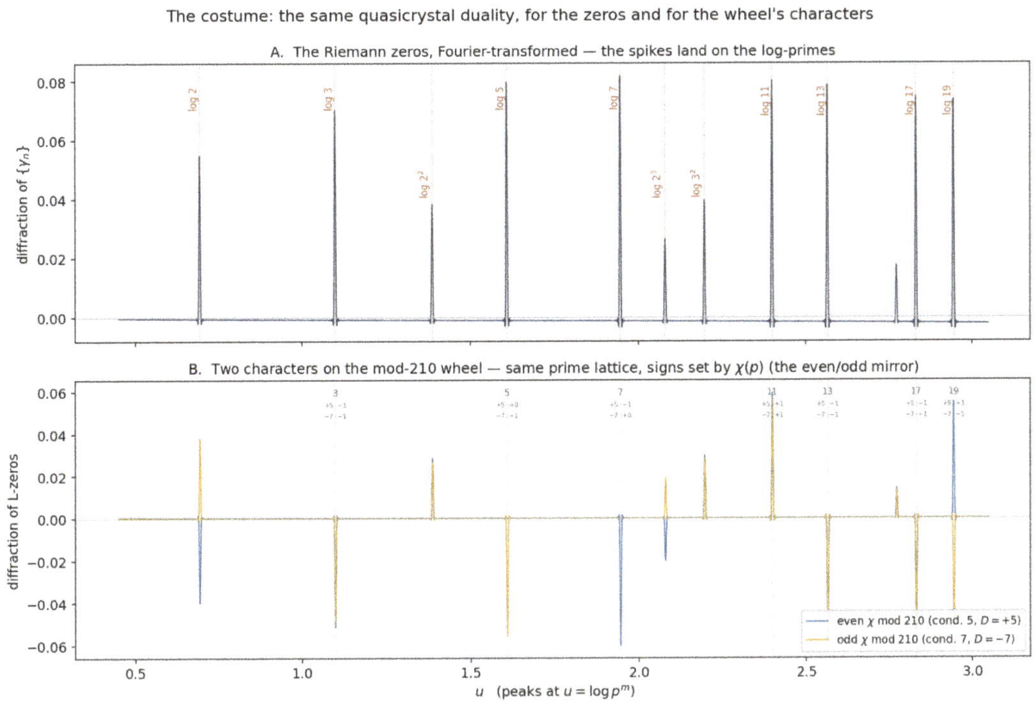

Figure 7.3: **Top:** the Fourier transform (7.2) of the first 1000 zeros of $\zeta(s)$. The peaks land on the logarithms of the primes and prime powers (dotted red). The zeros, transformed, are the primes. **Bottom:** the same construction for two Dirichlet characters of the mod-210 wheel, computed from their L-function zeros—one even (χ of conductor 5, $D = +5$) and one odd (χ of conductor 7, $D = -7$). The peaks sit on the same prime lattice, but each peak's *sign* is $\chi(p)$; a prime dividing the conductor goes dark. The even/odd grading is the wheel's mirror $r \mapsto 210 - r$.

7.5 The quasicrystal

A one-dimensional quasicrystal, in Dyson's sense, is a discrete set of points on a line whose Fourier transform is *also* a discrete set of points. Figure 7.3 is exactly that statement: a discrete set of points (the zeros, lying on the critical line) whose diffraction is a discrete set of points (the log-primes). Quasicrystals are real materials—Shechtman's diffraction patterns with crystallographically forbidden symmetry won a Nobel Prize—so this is the most literal sense in which the Riemann spectrum could be said to appear "in nature": a hypothetical one-dimensional atomic arrangement whose X-ray diffraction would show bright spots at the zeros.

Two honest qualifications. The peaks-at-log-primes of Fig. 7.3 are *unconditional*—they are Eq. (7.1), true whether or not the Riemann Hypothesis holds. What requires the Hypothesis is the word *quasicrystal* itself: to call the zeros a pure point set lying *on a single line* is to assert $\operatorname{Re}\rho = \frac{1}{2}$ for all of them. So the diffraction is solid ground; the crystallographic name is the conditional dressing. And the picture, beautiful as it is, settles nothing: it is a faithful redescription of the primes, not an independent object that pins them. Dyson's proposal—classify all one-dimensional quasicrystals and find the zeros among them—remains a program, because the quasicrystal whose diffraction is the log-primes simply *is* the primes.

7.6 The wheel's characters

The mod-210 wheel enters through its Dirichlet characters. Each character χ modulo 210 has an L-function $L(s,\chi)$, and each L-function carries its own explicit formula—a version of Eq. (7.1) in which the prime p is weighted by $\chi(p)$. So the zeros of $L(s,\chi)$ form their own quasicrystal, on the *same* prime lattice as before, but with each peak's sign set by $\chi(p)$. Figure 7.3 (bottom) draws two of them, computed from their L-function zeros: an even character (conductor 5) and an odd one (conductor 7). Where $\chi(p) = +1$ the peak points with ζ; where $\chi(p) = -1$ it flips; where p divides the conductor the peak vanishes entirely. The grading of the characters into even ($\chi(-1) = +1$) and odd ($\chi(-1) = -1$) is exactly the wheel's mirror $r \mapsto 210 - r$—the same mirror that, in the prime-race chapter, sends the contest onto the odd characters. The wheel does not light up its own spokes; it indexes a *family* of quasicrystals, one per character, all sharing the prime lattice and differing only by a character's sign pattern.

7.7 What this is and is not

The phenomenon in Fig. 7.3 is classical and has been drawn many times: as the "music of the primes", as Fourier-duality plots of the zeros against the primes, and as a standalone computation of the same transform. The quasicrystal reading is Dyson's. The spacing statistics of the zeros that make them look like a physical spectrum—the random-matrix universality shared with quantum-chaotic systems—are Montgomery's and Rubinstein–Sarnak's territory, not pursued here. This chapter adds none of that; it only renders the duality on one modulus and notes, in passing, that the wheel's characters give a family of these crystals. It bears on the Riemann Hypothesis not at all: the diffraction is a consequence of the primes, and the bright spots are the primes. It is the primes, drawn in crystal.

Reproducing the picture

```
import numpy as np
from mpmath import zetazero # generate the zeros (no data file)
gamma = np.array([float(zetazero(k).imag) for k in range(1, 1001)])
g = gamma[gamma > 0]
w = 0.5*(1 + np.cos(np.pi*g/g.max())) # Hann taper -> clean peaks
u = np.linspace(0.45, 3.05, 4000)
D = -(w[:,None]*np.cos(np.outer(g, u))).sum(0)/len(g) # the diffraction, Eq. (2)
```

```
import math # peaks should sit at log(p^m):
for p in [2,3,5,7,11,13,17,19]:
 print(p, round(math.log(p),3)) # 0.693, 1.099, 1.609, 1.946,...
```

For a character χ (Fig. 7.3, bottom), replace `gamma` with the imaginary parts of the zeros of $L(s, \chi)$; the peaks appear at the same $\log p$ with sign $\chi(p)$, no extra bookkeeping required.

The real-space companion (Figs. 7.1, 7.2) needs no zeros at all—each prime's grating is one line of modular arithmetic:

```
from math import gcd
units = [r for r in range(210) if gcd(r, 210) == 1] # the 48 spokes
def strikes(p, r, k): # does prime p's grating hit spoke r on turn k?
 return (210*k + r) % p == 0 # i.e. p divides 210k + r
def open_channels(k, ps=(11, 13, 17)): # spokes no drawn grating crosses
 return [r for r in units if all(not strikes(p, r, k) for p in ps)]
print(len(open_channels(0))) # 40: unit 1 plus 39 primes, 19..199
```

Chapter 8

The Zeros as a Spectrum

The diffraction chapter read the zeros as the Fourier-dual of the primes. This chapter looks instead at the zeros among themselves. *Once their slowly growing density is removed, the nontrivial zeros of ζ—and of the seven L-functions of the wheel—carry the statistics of the eigenvalues of a large random Hermitian matrix from the Gaussian Unitary Ensemble: they repel at short range, almost never falling close together, and their pair correlation follows Montgomery's law* $1 - (\sin \pi u/\pi u)^2$. *For the first thousand zeros, no normalized gap falls below* 0.1, *and only* 7% *fall below* 0.5 *where an uncorrelated sequence would put* 39%. *These are the statistics of a quantized chaotic system, and they are the principal evidence for the Hilbert–Pólya idea that the zeros are the spectrum of some self-adjoint operator. The statistical agreement is established; the operator is not known, and the Riemann Hypothesis is not thereby proved. This chapter is expository.*

From here the tour borrows a vocabulary it has not used before. This chapter and the next read the wheel's zeros and primes through ideas from physics—random-matrix statistics here, a gas of particles in the chapter that follows—and the words can look like a change of subject. They are not. Every object stays one already built: the zeros of the wheel's L-functions, the explicit formula, the Möbius function. Only the lens changes. Where a term from physics is needed it is given a plain meaning before it is named, and the names are kept so they can be looked up; nothing here requires physics you do not meet on the page.

8.1 Scope

This chapter is expository and reports established mathematics. It describes the local statistics of the zeros and their match to random-matrix theory, illustrates the match with the first thousand zeros of ζ, and states the quantum-chaos and Hilbert–Pólya context that the match suggests. It proves nothing. The one thing this circle of ideas would most like to deliver—an operator whose spectrum is the zeros, hence a proof of the Riemann Hypothesis—is exactly what remains open, and Sec. 8.6 is careful about that line. The diffraction chapter treats the complementary picture, the zeros against the primes.

8.2 Unfolding

The nontrivial zeros $\frac{1}{2} + i\gamma_n$ are not uniformly dense: near height T they crowd together with mean density $\frac{1}{2\pi}\log(T/2\pi)$, so the average gap shrinks as one climbs the critical line. To compare the zeros with a universal statistical law one first removes this trend—*unfolds* them—by applying the Riemann–von Mangoldt counting function

$$\widetilde{\gamma}_n = N(\gamma_n), \qquad N(T) = \frac{T}{2\pi}\log\frac{T}{2\pi} - \frac{T}{2\pi} + \frac{7}{8}, \tag{8.1}$$

which stretches the sequence to unit mean spacing. The unfolded gaps $s_n = \widetilde{\gamma}_{n+1} - \widetilde{\gamma}_n$ then have mean 1 (here, mean 1.000, standard deviation 0.38 over the first thousand zeros), and it is their *distribution* that carries the arithmetic.

8.3 Level repulsion

The first and most visible feature is that the zeros avoid one another. Figure 8.1A is the histogram of the unfolded gaps. It rises from zero, peaks near $s \approx 0.9$, and falls—there is almost no probability of a very small gap. This is *level repulsion*, and it is quantitative: among the first thousand zeros not a single normalized gap is below 0.1, and only 7.3% lie below 0.5. An uncorrelated (Poisson) sequence with the same density would crowd 39% of its gaps below 0.5 and pile up at $s = 0$ (the dashed curve e^{-s}). The data instead tracks the Gaussian Unitary Ensemble surmise

$$p_{\mathrm{GUE}}(s) = \frac{32}{\pi^2}\, s^2\, e^{-4s^2/\pi}, \tag{8.2}$$

the spacing law of eigenvalues of large random Hermitian matrices, whose s^2 prefactor is the repulsion.

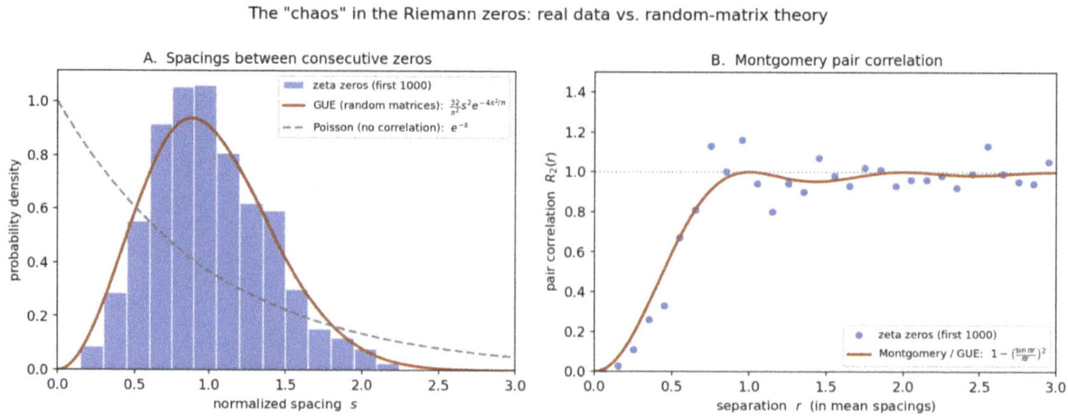

Figure 8.1: The zeros of ζ (first thousand, unfolded) against random-matrix theory. **A:** the nearest-neighbor spacing distribution. The data (bars) follow the GUE surmise (red), with its characteristic repulsion at small s, and look nothing like the uncorrelated Poisson law e^{-s} (dashed). **B:** the pair correlation. The data (points) follow Montgomery's law $1 - (\sin \pi u/\pi u)^2$ (red), including the deep suppression at small separation—the same correlation hole seen in the GUE.

8.4 Pair correlation

The second feature is the two-point statistic. Montgomery, studying the spacing of pairs of zeros in 1973, found that their pair correlation function is

$$R_2(u) = 1 - \left(\frac{\sin \pi u}{\pi u}\right)^2, \tag{8.3}$$

the very function that governs eigenvalue pairs in the GUE—a coincidence Freeman Dyson recognized on the spot. Figure 8.1B shows the first thousand zeros tracking Eq. (8.3): the correlation is strongly suppressed for separations below one mean spacing (zeros do not bunch), rises to 1, and oscillates gently around it. Montgomery proved the agreement for a restricted class of test functions; Odlyzko's high-precision computations, reaching enormous heights on the critical line, made the match overwhelming numerically; and Rudnick and Sarnak extended the agreement to higher correlations and to general L-functions, including the Dirichlet L-functions of the wheel.

8.5 Quantum chaos and the Hilbert–Pólya idea

It is worth stating the question this section answers in plain words before the names arrive. The zeros, we have just seen, carry the statistics of a spectrum—the energy levels of *something*. The natural question is: the spectrum of what? Three answers have been offered below, each attached to a name, and none has been delivered. What follows is the catalogue of those answers and the honest report that the object they reach for is still missing.

These two laws—level repulsion and the sine-kernel pair correlation—are the signature statistics of a particular kind of physical system. The Bohigas–Giannoni–Schmit conjecture holds that the quantized energy levels of a classically chaotic system with broken time-reversal symmetry obey GUE statistics. The zeros, statistically, look exactly like such a spectrum. That resemblance revives a very old guess, attributed to Hilbert and Pólya: that the numbers γ_n are the eigenvalues of some self-adjoint operator H, so that $\frac{1}{2} + i\gamma_n$ are forced onto the critical line by the reality of a quantum spectrum. Berry and Keating proposed that the underlying classical dynamics might be the simple Hamiltonian $H \sim xp$; Connes gave a spectral realization on the *adele class space*—a single space that records every way of measuring the size of a rational number at once, the ordinary absolute value together with one scale for each prime—in which the zeros appear as an absorption spectrum. Universality ties this back to the wheel: the seven L-functions are not statistical outliers but members of the same ensemble, so each carries the same "spectrum" of zeros.

8.6 What this is and is not

What is established is the *statistics*. The local correlations of the zeros match those of the GUE; this is supported by theorems for restricted test functions and by numerics of extraordinary precision. It is genuine mathematics and it is not in doubt.

What is not established is the *operator*. There is no known quantum system—no Hamiltonian, no chaotic billiard, no dynamics—whose energy levels are the Riemann zeros. The Hilbert–Pólya statement is a conjecture supported by the statistical resemblance, not a construction; the proposals of Berry–Keating and Connes are programs, not realizations. "The zeros as a spectrum" is therefore the most physical-looking face of the subject and, at the same time, an analogy: the zeros behave as though they were a spectrum, and no one has exhibited the system for which they are one. The Riemann Hypothesis, which a genuine self-adjoint operator would settle at a stroke, remains open—for ζ and for every L-function on the wheel. This chapter shows that the zeros keep the statistics of a chaotic spectrum; it does not produce the chaos.

Reproducing the figure

```
import numpy as np
from mpmath import zetazero # generate the zeros (no data file)
gamma = np.sort(np.array([float(zetazero(k).imag) for k in range(1, 1001)]))

# unfold by the Riemann-von Mangoldt counting function -> unit mean spacing
def N(T): return T/(2*np.pi)*np.log(T/(2*np.pi)) - T/(2*np.pi) + 7/8
u = N(gamma)
s = np.diff(u)
print(s.mean(), s.std()) # 1.000, 0.38
print(np.mean(s < 0.1), np.mean(s < 0.5)) # 0.000, 0.073 (Poisson: 0.39)

# Panel A: histogram of s vs GUE surmise and Poisson
# p_GUE(s) = (32/pi**2) s**2 exp(-4 s**2/pi); p_Poisson(s) = exp(-s)
```

```
# Panel B: pair correlation of u vs 1 - (sin(pi r)/(pi r))**2
```

Chapter 9

The Supersymmetric Primon Gas

This is an expository *capstone to the earlier chapters on the mod-*210 *wheel. It assembles a single object from pieces that are all classical. A bosonic gas with one mode per prime, twisted by a Dirichlet character χ, has partition function the L-function $L(s,\chi)$ — the same object whose zeros the earlier chapters drew as a quasicrystal. Its fermionic superpartner, in which Pauli exclusion restricts the states to the squarefree integers and the fermion number is graded by $(-1)^F$, has partition function $1/L(s,\chi) = \sum_n \chi(n)\mu(n)\, n^{-s}$. The two are joined by supersymmetry, and the boson–fermion cancellation is exactly the (twisted) Möbius inversion identity; the grading $(-1)^F$* is *the twisted Möbius function $\chi\mu$. On the wheel this unifies the bosonic gas, the seven real L-functions, and the Möbius function as one supersymmetric system, and the conductor of each character silences its primes in both the diffraction picture and the Möbius picture — the same lattice indexing both. None of this is new: the construction is the supersymmetric primon gas of Spector and Julia, with the arithmetic-gas twisting of Bakas and Bowick and its recent L-function generalization. The output is the Möbius function; the spectrum is the primes; the Riemann zeros do not appear.*

Before any of the physics, here is the whole chapter in one plain sentence. The wheel's L-function is a sum over the integers; its reciprocal is the same sum with Möbius signs attached; the two multiply to 1, and that statement *is* Möbius inversion. Everything that follows—the gas, the bosons and fermions, the supersymmetry, the Witten index—is a physical costume worn by that one identity. The costume is real physics and it is the reason the construction is beautiful, but it is a costume: the content is the inversion, an object the wheel already owns from the Möbius function. A reader who wants only the mathematics can take $Z_B = L$, the signed $Z_F = 1/L$, and their product 1, and has lost nothing. The rest of the chapter dresses those three facts and ties the dressing back to the conductor lattice.

9.1 Scope

This chapter states no theorem and claims nothing new. It is a synthesis: it puts together the bosonic primon gas (Julia), its supersymmetric completion and the resulting Möbius function (Spector), the arithmetic-gas twisting machinery (Bakas and

Bowick), and the by-now-standard fact that a Dirichlet L-function is the partition function of a character-twisted primon gas. The only thing local to this chapter is that the assembly is carried out on the mod-210 wheel and tied to the earlier chapters on the prime race, the diffraction picture, and the conductor lattice. A reader who wants the physics in full should see the review of Schumayer and Hutchinson.

9.2 The bosonic primon gas

Take a bosonic Fock space with one mode per prime p. A basis state is a finite multiset of primes, equivalently a positive integer $n = \prod_p p^{a_p}$, and we assign it energy $E_n = \log n$, so that $e^{-\beta E_n} = n^{-s}$ with $s = \beta$. The partition function is then

$$Z_B(s) = \sum_{n\geq 1} n^{-s} = \prod_p (1 - p^{-s})^{-1} = \zeta(s), \tag{9.1}$$

the Euler product being unique factorization read as "each mode independently occupied any number of times." This is Julia's free Riemann gas: the bosonic Fock space over the primes, with the integers as its occupation-number states.

Twisting each mode by a Dirichlet character—giving the primon of prime p a charge $\chi(p)$—replaces ζ by the L-function,

$$Z_B(s,\chi) = \prod_p (1 - \chi(p)p^{-s})^{-1} = \sum_{n\geq 1} \chi(n)\, n^{-s} = L(s,\chi), \tag{9.2}$$

valid because χ is completely multiplicative. The L-functions of the diffraction and conductor-lattice chapters are therefore nothing other than the partition functions of character-twisted primon gases.

9.3 The fermionic superpartner

Now give each prime mode a fermionic partner subject to the Pauli exclusion principle: each prime may occupy its fermionic mode at most once. The accessible states are then the *squarefree* integers, and the partition function is

$$Z_F(s,\chi) = \prod_p (1 + \chi(p)p^{-s}) = \prod_p \frac{1 - \chi(p)^2 p^{-2s}}{1 - \chi(p)p^{-s}} = \frac{L(s,\chi)}{L(2s,\chi^2)}. \tag{9.3}$$

The construction that matters for the marriage is the *signed* fermionic sum, in which each state is weighted by the fermion parity $(-1)^F$. Since a squarefree $n = p_1 \cdots p_k$ carries k fermions, its weight is $(-1)^k\chi(n) = \chi(n)\mu(n)$, and

$$\prod_p (1 - \chi(p)p^{-s}) = \sum_{n\geq 1} \chi(n)\mu(n)\, n^{-s} = \frac{1}{L(s,\chi)}. \tag{9.4}$$

So the signed fermionic gas is the reciprocal L-function, and its Dirichlet coefficients are the *twisted Möbius function* $\chi(n)\mu(n)$, supported on the squarefree integers and signed by the parity of the number of prime factors. This is Spector's observation, twisted by a character.

9.4 The marriage: supersymmetry

The grading $(-1)^F$ sounds like physics and means something elementary: count each state with a plus sign if it is built from an even number of distinct primes and a minus sign if from an odd number—exactly the sign the Möbius function already assigns. With that bookkeeping the boson and fermion partners cancel in pairs, and only the empty state survives.

Join the two gases as superpartners and grade the combined trace by $(-1)^F$. Because the bosonic states multiply the signed fermionic states, the graded partition function factorizes,

$$\mathrm{Tr}\big[(-1)^F e^{-\beta H}\big] = Z_B(s,\chi)\cdot\sum_n \chi(n)\mu(n)\,n^{-s} = L(s,\chi)\cdot\frac{1}{L(s,\chi)} = 1. \tag{9.5}$$

Every state above the vacuum cancels against a partner of opposite statistics; only the vacuum ($n = 1$) survives, giving Witten index 1. Read coefficient by coefficient, Eq. (9.5) is the (twisted) Möbius inversion identity

$$\sum_{d|n} \chi(d)\,\chi(n/d)\mu(n/d) = \chi(n)\sum_{d|n} \mu(d) = [\,n = 1\,]. \tag{9.6}$$

The boson–fermion cancellation *is* Möbius inversion, and the grading $(-1)^F$ *is* the twisted Möbius function. Table 9.1 records the three identities, evaluated numerically at $s = 2$ for the trivial character and for two of the wheel's real characters; the partition functions and their reciprocals multiply to 1 to machine precision, and the per-integer cancellation is exact.

Table 9.1: Numerical check at $s = 2$. The bosonic partition function (Euler product) equals $L(s,\chi)$; the signed fermionic sum $\sum_n \chi(n)\mu(n)n^{-s}$ equals $1/L(s,\chi)$; their product is 1. The trivial character recovers $\zeta(2) = \pi^2/6$ and $1/\zeta(2) = 6/\pi^2$.

character	$L(2,\chi)$	$\sum_n \chi(n)\mu(n)/n^2$	$1/L(2,\chi)$	product
trivial (ζ)	1.644934	0.607927	0.607927	1.000000
$\chi_{5.4}$ ($D{=}{+}5$)	0.706211	1.416007	1.416007	1.000000
$\chi_{7.6}$ ($D{=}{-}7$)	1.151925	0.868112	0.868112	1.000000

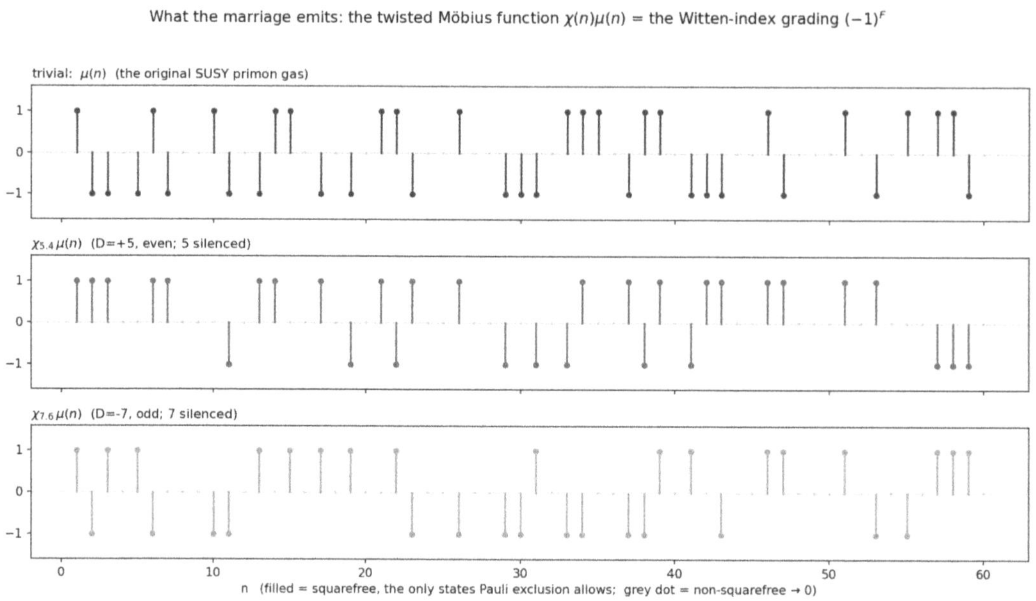

Figure 9.1: What the marriage emits: the twisted Möbius function $\chi(n)\mu(n)$, which is the fermion-parity grading $(-1)^F$ and the Dirichlet coefficient of $1/L(s,\chi)$. It is nonzero only on the squarefree integers — the only states Pauli exclusion allows — and signed by χ and the parity of the number of prime factors. Top: the untwisted Möbius function $\mu(n)$ (Spector's original gas). Middle and bottom: the twists by the conductor-5 and conductor-7 characters; each silences the multiples of its own conductor prime.

9.5 On the wheel

Each of the seven real characters modulo 210 thus carries its own supersymmetric primon gas: a bosonic gas whose partition function is its L-function, and a fermionic superpartner whose signed partition function is the reciprocal, with coefficients the twisted Möbius function. The seven L-functions of the conductor-lattice chapter are the bosonic halves; the seven twisted Möbius functions are the fermionic halves.

The conductor reappears in the same role it played in the diffraction picture. There, a prime dividing the conductor switched off its diffraction peak because $\chi(p) = 0$; here, the same vanishing removes that prime from the twisted Möbius function (Fig. 9.1: $\chi_{5.4}\mu$ has no support on multiples of 5, $\chi_{7.6}\mu$ none on multiples of 7). The conductor lattice indexes both pictures at once: it says which primes are silenced, whether one is looking at the spectral side (the zeros' diffraction) or the arithmetic side (the Möbius function). The quasicrystal and the supersymmetric gas are the same wheel seen from two directions.

It is worth saying which composites each side actually sees, since the two are easy to confuse. The diffraction peaks sit at prime powers—the support of the von Mangoldt function Λ—so that picture registers the prime *squares* $121 = 11^2$ and $169 = 13^2$ as faint echoes near $2\log 11$ and $2\log 13$, and is blind to the squarefree composites. The Möbius function μ has exactly the complementary support, the *squarefree* integers, so the gas is blind to those two squares and instead registers the three squarefree composites $143 = 11 \cdot 13$, $187 = 11 \cdot 17$, $209 = 11 \cdot 19$ (each with $\mu = +1$). The wheel's five composite spokes thus divide cleanly between the pictures—$121, 169$ seen only in the diffraction, $143, 187, 209$ only in the gas—while the primes are untouched by either subtlety, being the full-strength peaks of the one and the $\mu = -1$ generators of the other.

9.6 What this is and is not

The construction is classical throughout. The bosonic gas is Julia's; the supersymmetric completion and the Möbius function as Witten index are Spector's; the twisting and boson–fermion equivalence are Bakas and Bowick's; the identification of a Dirichlet L-function with a character-twisted primon partition function is theirs as well, recently generalized. This chapter only assembles them on one modulus and ties them to the earlier chapters. It is a reformulation, not a result: we supplied the primon gas (known) and the L-functions (known), and out came the twisted Möbius function (the known corollary). Nothing emerged that was not placed.

In particular it does not approach the Riemann zeros. The spectrum of the gas, bosonic or supersymmetric, is the set of energies $\{\log n\}$ — the primes and their products. The partition functions are $L(s,\chi)$ and $1/L(s,\chi)$; the nontrivial zeros live in the analytic continuation of those functions, not among the energy levels of the Hamiltonian. So the marriage manufactures the Möbius function, which is itself built from prime factorization, and unifies the bosonic and fermionic descriptions of the wheel beautifully — and it remains entirely on the elementary side of the wall. Producing the zeros as an actual spectrum is the separate, unsolved problem of Hilbert and Pólya, untouched here.

Reproducing the check

```
import numpy as np
N = 2_000_000; s = 2.0
mu = np.ones(N+1, dtype=np.int64); mu[0]=0 # linear/Eratosthenes Mobius sieve
sieve = np.ones(N+1, bool); sieve[:2]=False
for i in range(2, int(N**0.5)+1):
 if sieve[i]: sieve[i*i::i]=False
for p in np.nonzero(sieve)[0]:
 mu[p::p] *= -1; mu[p*p::p*p] = 0

def chi(conductor, plus): # real character by residue mod conductor
 idx = np.arange(N+1) % conductor
 tab = np.array([0 if np.gcd(r,conductor)!=1 else (1 if r in plus else -1)
 for r in range(conductor)])
 return tab[idx]

inv = np.zeros(N+1); inv[1:] = np.arange(1,N+1)**(-s)
```

```
for name, c, plus in [("zeta",1,set()), ("chi_5.4",5,{1,4}), ("chi_7.6",7,{1,2,4})]:
 x = np.ones(N+1) if c==1 else chi(c, plus)
 L = (x[1:]*inv[1:]).sum() # bosonic: L(s,chi)
 invL = (x[1:]*mu[1:]*inv[1:]).sum() # signed fermionic: sum chi*mu/n^s
 print(name, L, invL, L*invL) # product -> 1
```

Part IV

Climbing and counting

Chapter 10

Climbing the Primorials

The opening chapter fixed one wheel, $M = 210$. This chapter asks what happens to its 48 spokes as one walks along the number line, and as one climbs to larger primorials. Two classical facts answer it. An exact conservation law holds in every period: the unit, the primes, and the coprime composites always sum to $\varphi(M)$, so that as the primes thin out (by the prime number theorem in progressions) the composites fill the spokes they vacate. And the wheel's density itself thins at the rate fixed by Mertens' third theorem, $\prod_{p\leq X}(1-1/p) \sim e^{-\gamma}/\log X$, which a short identity ties to the Hardy–Littlewood twin constant of the twins chapter. Everything is classical—Mertens (1874), the prime number theorem in progressions—and none of it touches the Riemann Hypothesis. This chapter is expository.

10.1 Scope

This chapter is expository and classical. It extends the opening chapter, which described the single wheel $M = 210$, in two directions—along the integers (the per-period census) and up the primorials (the density)—and it connects forward to the twin constant of the twins chapter. It states no new result. As Sec. 10.5 records, the content here is exactly the kind that depends only on the prime number theorem and Mertens' theorems, with no bearing on the Riemann Hypothesis.

10.2 The census: a conservation law

Fix a primorial M and walk through the integers one period at a time, the kth period being the block $(kM, (k+1)M]$. Each integer in the block that is coprime to M is one of exactly three things: the unit 1 (only in the first block), a prime, or a composite coprime to M. Writing $U(k)$, $P(k)$, $S(k)$ for their counts, this trichotomy is a conservation law,

$$U(k) + P(k) + S(k) = \varphi(M) \qquad \text{for every } k, \tag{10.1}$$

exact in every period because each of the $\varphi(M)$ coprime residue classes is represented once per block. For $M = 210$ the total is always 48 (verified across all 100 periods up to 21000, and the same is true at $M = 2310$). Figure 10.1 shows the stacked counts: every bar reaches 48.

What changes from block to block is the split. By the prime number theorem in arithmetic progressions, the primes in a block near height kM number

$$P(k) \sim \frac{M}{\log\left((k+\frac{1}{2})M\right)}, \tag{10.2}$$

which decreases as k grows; and since the total is pinned at $\varphi(M)$, the coprime composites $S(k) = \varphi(M) - P(k) - U(k)$ must increase to compensate, rising toward $\varphi(M)$. The primes that vacate the spokes are replaced by composites; the wheel fills in. The smooth crossover $P = S$ occurs where $M/\log(kM) = \varphi/2 = 24$, near $k \approx 30$, though single-period counts fluctuate around this trend.

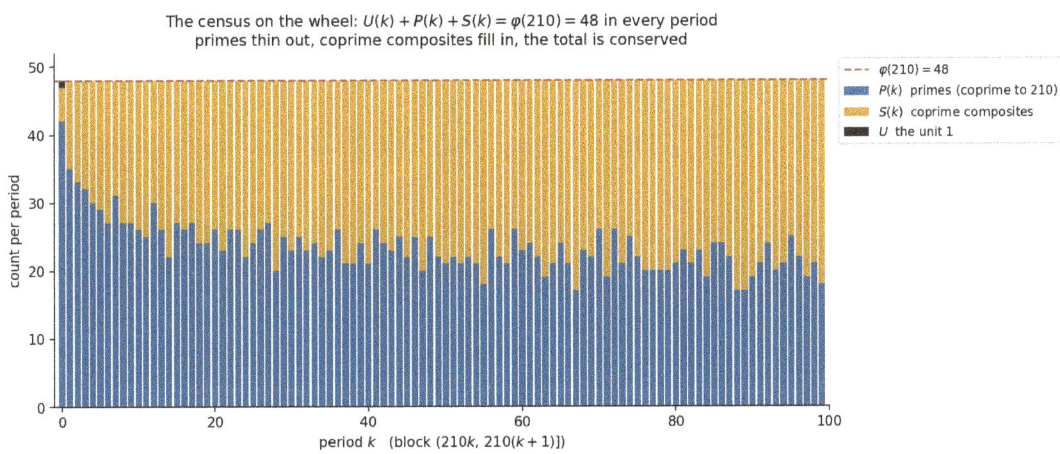

Figure 10.1: The census at $M = 210$. In each period the unit, the primes $P(k)$ (blue), and the coprime composites $S(k)$ (orange) sum to $\varphi(210) = 48$ (red line): the conservation law (10.1). As one moves out, $P(k)$ thins by the prime number theorem in progressions and $S(k)$ fills the vacated spokes; the total never moves.

10.3 Mertens' theorems and the thinning

The density of the wheel itself—the fraction of integers that are coprime to M—thins as M grows, and the rate is classical. Mertens' third theorem states

$$\prod_{p \le X} \left(1 - \frac{1}{p}\right) \sim \frac{e^{-\gamma}}{\log X}, \tag{10.3}$$

with γ the Euler–Mascheroni constant. Equivalently the product times $\log X$ approaches the constant $e^{-\gamma} = 0.5615\ldots$; Fig. 10.2 shows it doing so, approaching from below and settling by $X \sim 10^3$. Since the kth primorial M_k has $\log M_k \sim p_k$ (Chebyshev), the wheel's density obeys

$$\frac{\varphi(M_k)}{M_k} = \prod_{p \le p_k} \left(1 - \frac{1}{p}\right) \sim \frac{e^{-\gamma}}{\log p_k}, \tag{10.4}$$

the slow decay already pictured for the first few primorials in the opening chapter. (Mertens' second theorem, $\sum_{p \le X} 1/p = \log\log X + B$ with $B = 0.2615\ldots$, is the additive companion and governs the finer corrections.)

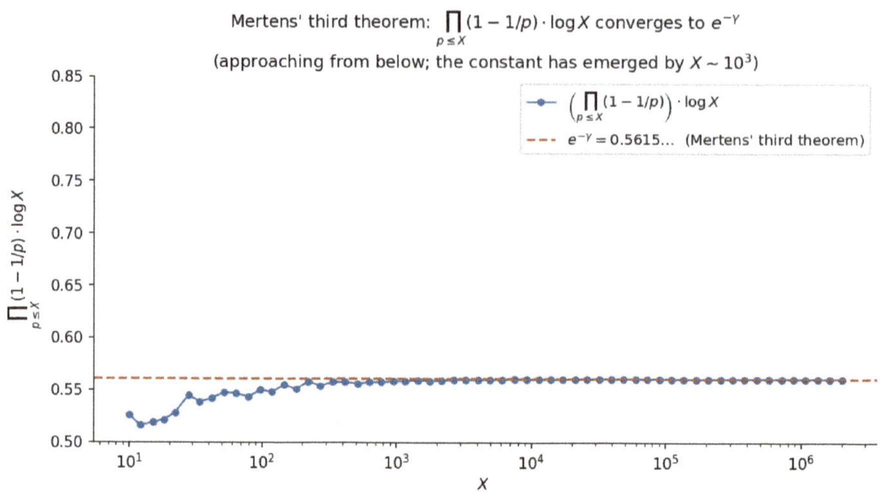

Figure 10.2: Mertens' third theorem (10.3) in the form $\prod_{p \le X}(1 - 1/p) \cdot \log X \to e^{-\gamma}$. The product approaches the constant from below and has effectively reached it by $X \sim 10^3$.

10.4 A bridge to the twin constant

The same density connects to the twins chapter through one short identity. For each prime $p \ge 3$,

$$\left(1 - \frac{1}{(p-1)^2}\right)\left(1 - \frac{1}{p}\right) = \frac{p-2}{p-1}, \tag{10.5}$$

a one-line algebraic fact. Taking the product over $3 \le p \le X$ turns it into

$$\prod_{3 \le p \le X} \frac{p-2}{p-1} = C_2 \prod_{3 \le p \le X} \left(1 - \frac{1}{p}\right)(1 + o(1)), \qquad C_2 = \prod_{p \ge 3}\left(1 - \frac{1}{(p-1)^2}\right) = 0.6601\ldots, \tag{10.6}$$

where C_2 is the Hardy–Littlewood twin-prime constant (the two sides agree to ratio 1.0000 already by $X \approx 2 \times 10^4$). Combining Eq. (10.6) with Mertens' third theorem gives the twin-admissible density of the wheel, $\prod_{3 \le p \le X}(p-2)/(p-1) \sim 2C_2 e^{-\gamma}/\log X$. This product is exactly the singular-series factor that sets the rate of twin primes in the twins chapter: the density that thins the wheel here is the same one that, refined by C_2, predicts how often the twin slots are filled.

10.5 What this is and is not

All of this is classical and self-contained. The conservation law (10.1) is elementary; the decay of $P(k)$ is the prime number theorem in progressions; the density rate is Mertens (1874); the twin-constant identity is a one-line computation. None of it depends on the Riemann Hypothesis or on any unproven conjecture—it uses only the prime number theorem, Mertens' theorems, and the classical zero-free region.

The census and the density developed here are simply facts about the wheel—true whether or not anything is built on top of them, and dependent only on the prime number theorem and Mertens' theorems. That is why they belong in the tour.

Reproducing the figures

```
import numpy as np
from math import gcd

M = 210; nper = 100; N = M*nper
s = np.ones(N+1, bool); s[:2] = False
for i in range(2, int(N**0.5)+1):
 if s[i]: s[i*i::i] = False
phi = sum(1 for r in range(M) if gcd(r, M) == 1) # 48

for k in range(nper): # conservation law
 U = P = S = 0
 for n in range(k*M+1, (k+1)*M+1):
 if gcd(n, M) != 1: continue
 if n == 1: U += 1
 elif s[n]: P += 1
 else: S += 1
 assert U + P + S == phi # holds every period

# Mertens' third: prod_{p<=X}(1-1/p) * log X -> e^{-gamma}
import numpy as np
pr = np.nonzero(s)[0]
prod = np.cumprod(1 - 1/pr.astype(float))
# prod[-1]*log(N) approaches exp(-0.5772156649) = 0.5615

# Lemma: prod_{3<=p<=X}(p-2)/(p-1) = C2 * prod(1-1/p) * (1+o(1))
C2 = 0.6601618158468695
odd = pr[(pr >= 3) & (pr <= N)]
lhs = np.prod((odd-2)/(odd-1)); rhs = C2*np.prod(1 - 1/odd.astype(float))
print(lhs/rhs) # ~ 1.0000
```

Chapter 11

The Repeating Decimals

Every residue r coprime to 210 gives a fraction $r/210$ whose decimal expansion repeats with a single digit of delay and a period of six. This short, self-contained chapter explains why: $210 = 10 \times 21$, the one delayed digit comes from the factor $10 = 2 \cdot 5$ shared with the base, and the period six is the multiplicative order of 10 modulo 21. The 48 coprime residues therefore carry only twelve distinct six-digit blocks, and those twelve fall into two *cyclic families—six rotations each of a single "cyclic number," both of which are the famous $1/7 = 0.\overline{142857}$ in disguise. We show that the family of a residue is nothing more elaborate than its value modulo 3; that the wheel's mirror $r \leftrightarrow 210 - r$ acts on the decimals as the digit-wise nines-complement while flipping the family; and that walking up the periods and walking across the residues are two perpendicular motions—one additive, one multiplicative. Everything here is classical (the theory of repeating decimals and cyclic numbers), and everything is a base-10 lens on the genuinely base-independent object underneath: the multiplicative group $(\mathbb{Z}/210\mathbb{Z})^\times$. No new theorem is claimed. This chapter is the decimal counterpart to the opening chapter and the two-mirrors chapter, and it points, for the part that is actually deep, to the chapters on how the primes distribute across these classes.*

11.1 Scope

This is an expository chapter. Its purpose is to make one piece of the mod-210 wheel vivid in base 10: the repeating decimals attached to the coprime residues. The opening chapter introduces the 48 units of $(\mathbb{Z}/210\mathbb{Z})^\times$ and the reflection $r \leftrightarrow 210 - r$; the two-mirrors chapter reads that reflection as the analytic functional equation. Here we add nothing to the arithmetic. We simply photograph the same residues through the lens of their decimal expansions and observe that the photograph is unusually orderly—twelve blocks, two families, a clean mirror law—and then we are honest about what that order is and is not. The order is real and classical; the decimals themselves are an artifact of writing numbers in base ten. The object they shadow—the multiplicative group and its behaviour on the primes—is what the rest of the book is about.

11.2 Why one digit of delay and a period of six

Write the modulus as

$$210 = 2 \cdot 3 \cdot 5 \cdot 7 = \underbrace{(2 \cdot 5)}_{10} \times \underbrace{(3 \cdot 7)}_{21} = 10 \times 21. \tag{11.1}$$

The factor 10 is exactly the part of the modulus that shares its primes with the base. For any reduced fraction the decimal expansion consists of a non-repeating *pre-period* followed by a repeating *period*, and the two lengths are governed separately: the pre-period length is $\max\big(v_2(210), v_5(210)\big) = \max(1, 1) = 1$, and the period length is the multiplicative order of 10 modulo the part coprime to the base, namely 21—the dependence of the period on this order being Gauss's result. Computing that order,

$$10^1 \equiv 10, \quad 10^2 \equiv 16, \quad 10^3 \equiv 13, \quad 10^4 \equiv 4, \quad 10^5 \equiv 19, \quad 10^6 \equiv 1 \pmod{21}, \tag{11.2}$$

so $\mathrm{ord}_{21}(10) = 6$. Hence every $r/210$ with $\gcd(r, 210) = 1$ looks like

$$\frac{r}{210} = 0.\,d\,\overline{b_1 b_2 b_3 b_4 b_5 b_6}, \tag{11.3}$$

one delayed digit d and then a six-digit block on infinite repeat. For example $\frac{11}{210} = 0.0\overline{523809}$.

11.3 Twelve blocks, two cyclic families

Because the repeating part is controlled entirely by the order of 10 modulo 21, the six-digit block depends only on $r \bmod 21$. There are $\varphi(21) = \varphi(3)\varphi(7) = 2 \cdot 6 = 12$ residues coprime to 21, so there are exactly *twelve* distinct blocks. Each is shared by $48/12 = 4$ of the residues coprime to 210 (those reducing to the same class mod 21). Table 11.1 lists all twelve with their residues.

The twelve residues mod 21 are organized by the cyclic subgroup $\langle 10 \rangle$. Since $\mathrm{ord}_{21}(10) = 6$, that subgroup has order 6 and index 2 in the group of order 12; the group therefore splits into *two* cosets of six elements each. Each coset is one orbit under multiplication by 10—and multiplication by 10 is precisely a one-step cyclic rotation of the decimal block. So each family is a single "cyclic number" read from six different starting points (Fig. 11.1):

$$\text{Family A}: \quad 047619 \to 476190 \to 761904 \to 619047 \to 190476 \to 904761, \tag{11.4}$$
$$\text{Family B}: \quad 095238 \to 952380 \to 523809 \to 238095 \to 380952 \to 809523. \tag{11.5}$$

Both are the most famous cyclic number of all in disguise. Since $21 = 3 \cdot 7$,

$$\frac{1}{21} = \frac{1}{3} \cdot \frac{1}{7} = 0.\overline{047619}, \qquad 047619 = \frac{142857}{3}, \qquad 095238 = 2 \times 047619, \tag{11.6}$$

where 142857 is the cyclic number of $1/7 = 0.\overline{142857}$. The whole decimal life of the wheel is one sixth of the seven-cycle, taken once (Family A) and twice (Family B).

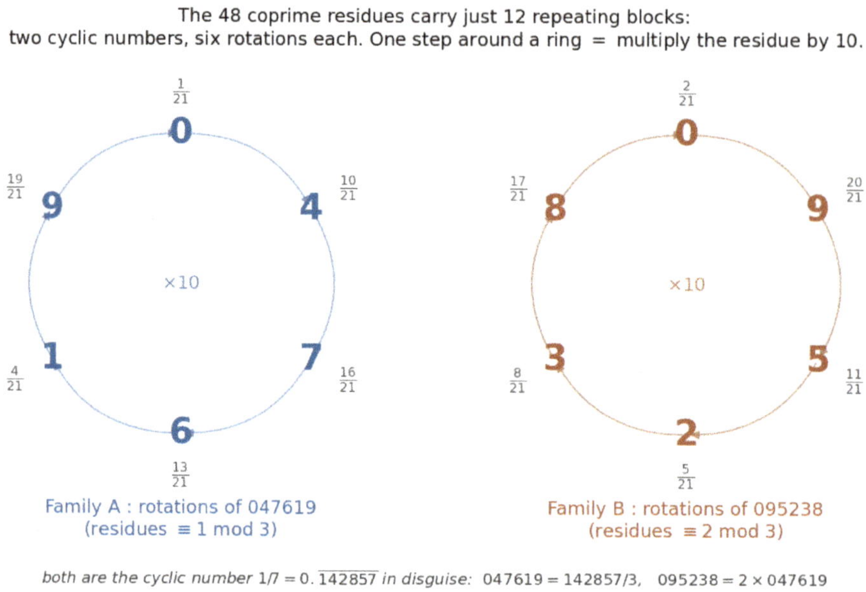

Figure 11.1: The 48 coprime residues carry only twelve repeating blocks, and those twelve are two cyclic numbers read from six starting points. Each digit on a ring is a rotation; one step clockwise multiplies the residue (shown as $r/21$) by 10. Family A is one copy of the 1/7 cycle, Family B is two copies. The split is exactly residue mod 3 (Sec. 11.4).

Table 11.1: The twelve repeating blocks of the mod-210 wheel. Each block is the six-digit period of $r/210$ for the four listed residues; "family" is the coset of $\langle 10 \rangle$, which coincides with residue mod 3; "mirror block" is the period of $(210 - r)/210$ and is the digit-wise nines-complement (Sec. 11.5).

Block	Family	Residues mod 210	Mirror block
047619	A	1, 43, 127, 169	952380
190476	A	67, 109, 151, 193	809523
476190	A	31, 73, 157, 199	523809
619047	A	13, 97, 139, 181	380952
761904	A	37, 79, 121, 163	238095
904761	A	19, 61, 103, 187	095238
095238	B	23, 107, 149, 191	904761
238095	B	47, 89, 131, 173	761904
380952	B	29, 71, 113, 197	619047
523809	B	11, 53, 137, 179	476190
809523	B	17, 59, 101, 143	190476
952380	B	41, 83, 167, 209	047619

11.4 The family is the mod-3 clock

The two families look like a fact about decimals, but they are a fact about residues mod 3. The family of r is its coset of $\langle 10 \rangle$ in $(\mathbb{Z}/21\mathbb{Z})^{\times}$. Multiplication by 10 moves within a coset, and

$$10 \equiv 1 \pmod{3}, \tag{11.7}$$

so multiplying by 10 leaves r mod 3 unchanged. The two cosets are therefore separated by exactly one invariant—the value modulo 3:

$$\text{Family A} \iff r \equiv 1 \pmod{3}, \qquad \text{Family B} \iff r \equiv 2 \pmod{3}. \tag{11.8}$$

This is the same mod-3 clock that runs through *Twin Primes on the Mod-210 Wheel.* There, the lower member of every twin pair is $\equiv 2 \pmod{3}$ and the upper is $\equiv 1 \pmod{3}$ (the integer between them carries the factor of 3), so every twin corridor reads "$B \to A$." The decimal families and the twin admissibility are two faces of the single elementary fact that the units mod 3 are $\{1, 2\}$.

11.5 The mirror is the nines-complement

The wheel's reflection $r \leftrightarrow 210 - r$ has an immediate decimal meaning. Since

$$\frac{r}{210} + \frac{210 - r}{210} = 1 = 0.\overline{9}, \tag{11.9}$$

the two repeating blocks must add, digit by digit, to all nines: their sum is 999999 (Fig. 11.2). The mirror also flips the family. Negation by -1 sends a coset of $\langle 10 \rangle$ to the other coset, because $-1 \equiv 2 \pmod{3}$ lies in Family B; so reflection always sends $A \leftrightarrow B$. In Table 11.1 every mirror block sits in the opposite family from its partner, and the two blocks are nines-complements. This is the decimal shadow of exactly the reflection that *The Two Mirrors* matches to the $s \leftrightarrow 1 - s$ symmetry of the completed zeta function.

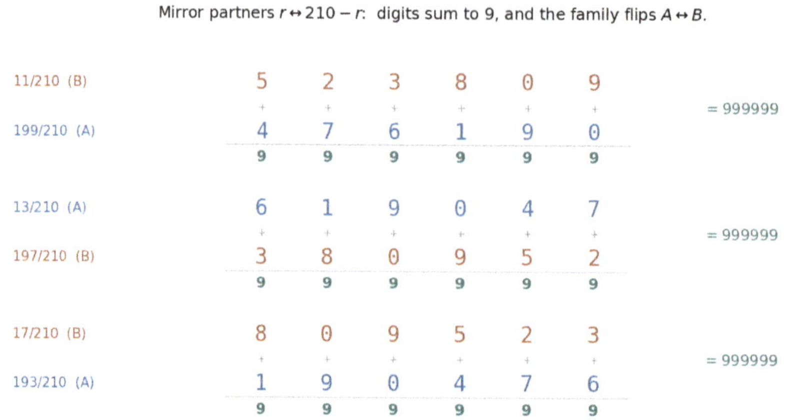

Figure 11.2: Mirror partners r and $210-r$. Their repeating blocks are digit-wise nines-complements (each column sums to 9, the whole block to 999999), and the family flips $A \leftrightarrow B$.

11.6 Two perpendicular axes

A number coprime to 210 has two coordinates, and the decimal displays them on perpendicular axes (Fig. 11.3). Writing

$$n = 210\,k + r, \qquad \frac{n}{210} = k + \frac{r}{210}, \tag{11.10}$$

the integer part is the *period* k and the fractional part is fixed by the *residue* r. These move independently:

- **Period axis (additive).** Increase k with r fixed: the integer in front climbs $0, 1, 2, \ldots$ while the repeating block stays frozen. This is place value; it is the elementary coordinate.
- **Residue axis (multiplicative).** Replace r by $10r \bmod 21$ within a period: the integer stays put while the block rotates one step. This is motion inside the group $(\mathbb{Z}/210\mathbb{Z})^{\times}$; it is the cyclic structure of Sec. 3.

Walking up the wheel's periods advances the front and holds the tail; walking around its residues rotates the tail and holds the front. The two are genuinely orthogonal: one is counting, the other is the multiplicative group at work.

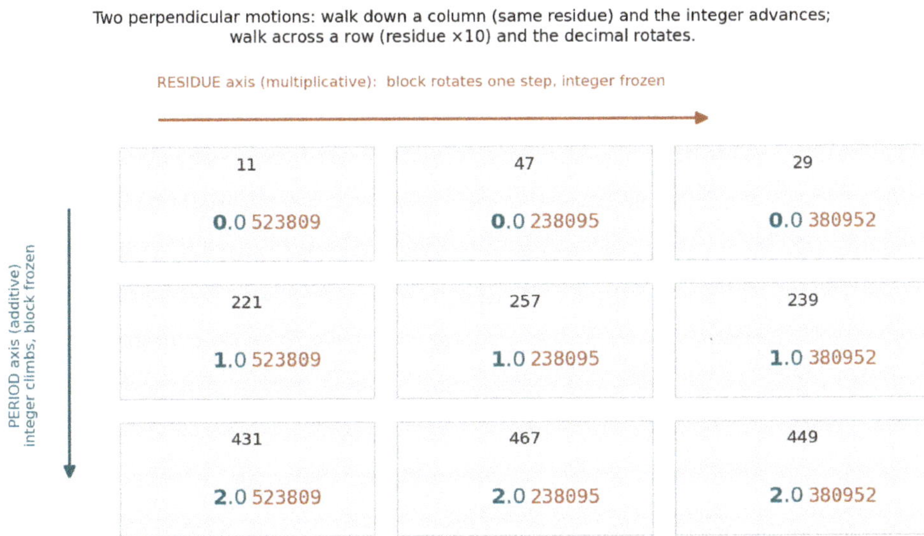

Figure 11.3: The two perpendicular motions. Down a column the residue is fixed, the integer climbs $0, 1, 2$ and the block is frozen. Across a row the residue is multiplied by 10, the integer is frozen and the block rotates ($523809 \to 238095 \to 380952$). Period is the additive axis; residue is the multiplicative axis.

11.7 What this is, and what it is not

What it is. A correct, classical description of the repeating decimals of $r/210$: one delayed digit, period six, twelve blocks, two cyclic families that are the $1/7$ cycle scaled, a family invariant equal to residue mod 3, and a mirror law that is the digit-wise nines-complement with a family flip. All of it follows from the theory of decimal expansions and the order of 10 modulo 21—material that traces back to Gauss, who worked out the periods of fractions and their dependence on the order of the base in Section VI of the *Disquisitiones Arithmeticae*, and which has been a staple of elementary number theory and the study of cyclic numbers ever since. It is a clean and, I think, beautiful way to organize the residues—and through them the primes and the coprime composites—on the wheel.

What it is not. It is not base-independent, and it is not new mathematics. The very existence of a six-digit block, and the particular digits in it, are artifacts of writing numbers in base 10: change the base b and the period becomes the order of b modulo the part of 210 coprime to b, the blocks regroup, and the pretty 142857 connection dissolves. In base 60—the base of a clock—the 3 is absorbed into the terminating part, leaving the 7 alone to drive a period-*three* repeat, and the families stop being residue mod 3 and become the quadratic-residue character mod 7: the same forty-eight numbers re-sorted by a different prime entirely (Fig. 11.4). The object that does *not* depend on the base—the thing the decimals are a shadow of—is the multiplicative group $(\mathbb{Z}/210\mathbb{Z})^\times \cong \mathbb{Z}/2\times\mathbb{Z}/4\times\mathbb{Z}/6$ together with its mod-3 quotient and its reflection.

Those are real and they recur throughout this book. Nothing in this chapter is a theorem that was not already known, and nothing here bears on factorization or on the twin prime conjecture: the residue, the period, and the decimal are all functions of n, and a function of n cannot reveal more than n itself contains. The genuinely deep questions are not about the digits but about how the primes *distribute* across the twelve blocks and the 48 classes, and those belong to the chapters on equidistribution, the prime race, and the connection of the wheel to the zeros. This chapter is the lens; those chapters are the landscape.

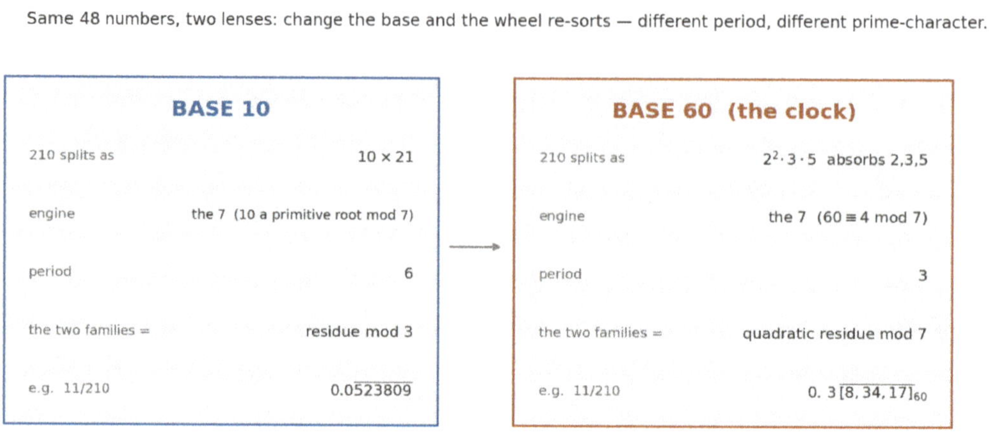

Figure 11.4: The same forty-eight numbers under two bases. In base 10 the repeating engine is the 7 (where 10 is a primitive root), the period is six, and the families are residue mod 3. In base 60—a clock's base—the 3 is absorbed into the terminating head, leaving the 7 alone to drive a period-three repeat, and the families become the quadratic-residue character mod 7. The period and the family-character are properties of the lens, not of the numbers; only the underlying group is invariant.

Reproducible computation

The blocks, families, mirror partners, and the order of 10 are reproduced by the following self-contained Python.

```
from math import gcd

def block(r): # six-digit repeating block of r/210
 x, s = r % 21, "" # period is governed by the part coprime to 10, i.e. 21
 for _ in range(6):
 x *= 10; s += str(x // 21); x %= 21
 return s

units = [r for r in range(1, 210) if gcd(r, 210) == 1] # 48 of them

groups = {}
```

```
for r in units:
 groups.setdefault(block(r), []).append(r)

assert len(groups) == 12 and all(len(v) == 4 for v in groups.values())

for b in sorted(groups):
 r0 = groups[b][0]
 fam = 'A' if r0 % 3 == 1 else 'B' # family = residue mod 3
 mir = block(210 - r0) # mirror block
 assert int(b) + int(mir) == 999999 # nines-complement
 print(b, fam, sorted(groups[b]), mir)

# order of 10 modulo 21 is 6
o, x = 0, 1
while True:
 x = (x * 10) % 21; o += 1
 if x == 1: break
assert o == 6
```

Chapter 12

Twin Primes

This is the capstone of the tour, and it returns to the question the wheel was built around: twin primes. A pair $(p, p+2)$ can both be prime only if both residues are coprime to 210, and exactly 15 of the 48 spokes begin such an admissible slot. These 15 slots are perfectly mirror-symmetric under $r \mapsto 210 - r$. The primes, however, are not. Among the residues themselves, six of the seven mirror pairs hold two genuine twins, but one is broken: $(41, 43)$ is a twin while its mirror $(167, 169)$ is not, because $169 = 13^2$. A single square spoils one half of a symmetric pair. The Hardy–Littlewood heuristic predicts how the twins are distributed across these slots and how many there are—a prediction the data confirm to within half a percent at 10^7—but whether the twins are infinite at all is the twin prime conjecture, still open. This chapter is expository: a well-tested heuristic and an exact structural observation, resting on a question no one can yet answer.

12.1 Scope

This chapter is expository. The Hardy–Littlewood heuristic it describes is classical; the slot structure and its mirror symmetry are elementary; the numerical agreement is reproduced in the appendix. It proves nothing, and it is careful in Sec. 12.5 about the one thing that matters most here: the infinitude of twin primes is conjectural. This chapter draws on the opening chapter for the wheel and on the two-mirrors chapter for the reflection, and it closes the tour.

12.2 The twin slots

A twin prime is a pair $(p, p+2)$ with both members prime. Past 7, each member must be coprime to 210, so a twin can occupy only those residues r for which both r and $r+2$ are coprime to 210. Counting these prime by prime,

$$\underbrace{1}_{p=2} \cdot \underbrace{1}_{p=3} \cdot \underbrace{3}_{p=5} \cdot \underbrace{5}_{p=7} = 15, \tag{12.1}$$

where each factor is the number of residues mod p that avoid both 0 and -2. So there are exactly 15 *admissible twin slots* on the wheel (Fig. 12.1). This is the wheel made to do the work of the notion of an *admissible pattern*: the pair $(0, 2)$ is admissible because, modulo every prime, it misses at least one residue (it never fills all of them, as the inadmissible triple $(0, 2, 4)$ does modulo 3), and the 15 slots are simply where that admissibility lands on the 210-cycle.

The single factor 1 at $p = 3$ deserves a moment, because it is one elementary fact wearing three masks across this book. It equals 1 because the lower member of a twin must avoid both 0 and $-2 \equiv 1$ modulo 3, leaving only $r \equiv 2 \pmod 3$; the upper member $r + 2$ is then $\equiv 1 \pmod 3$, and the multiple of 3 falls between them. So every twin slot reads $2 \to 1$ modulo 3. That same mod-3 split is exactly what the repeating-decimals chapter calls the two cyclic *families* of the wheel when it is written in base 10, and exactly the conductor-3 character $\chi_{3.2}$ of the conductor-lattice chapter. Twin admissibility, the decimal families, and $\chi_{3.2}$ are a single elementary fact about the prime 3, seen from three directions.

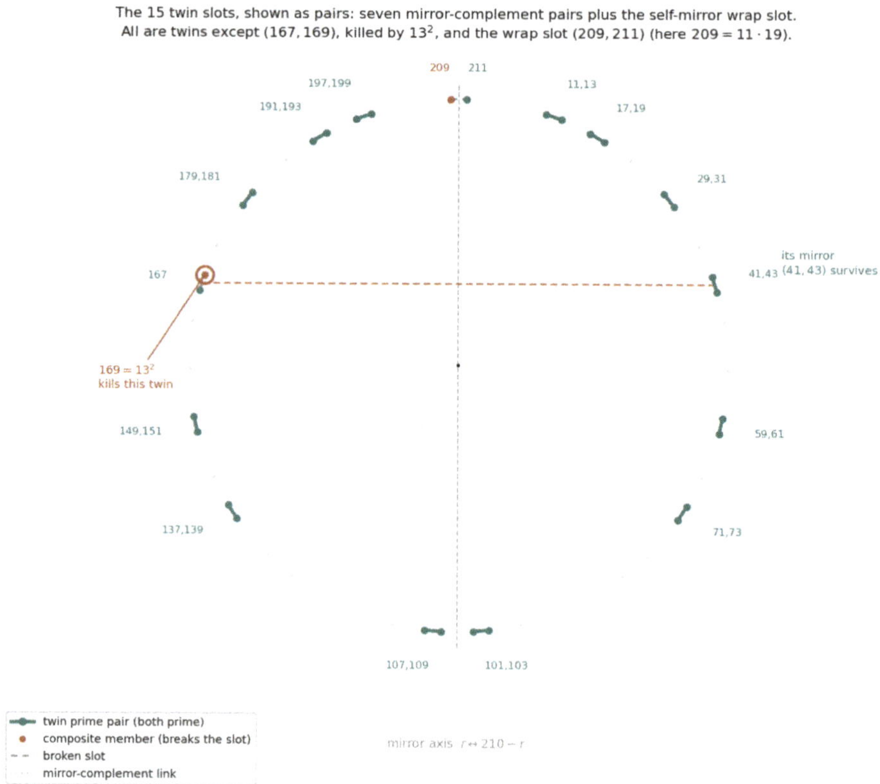

Figure 12.1: The 15 admissible twin slots on the wheel, each drawn as its two members $(r, r+2)$ joined by a bond; faint dotted links connect the mirror-complement pairs across the axis. Each member is colored by what it is—green if prime, red if composite—and a bond is green for a genuine twin, grey for a broken slot. The slots are mirror-symmetric, but the primes are not: $(41, 43)$ (right) is a twin while its mirror $(167, 169)$ (left) is broken at the single composite $169 = 13^2$, with 167 itself still prime. The self-mirror wrap slot at the top is broken at $209 = 11 \cdot 19$, with 211 prime.

12.3 The mirror, broken

The reflection $r \mapsto 210 - r$ of the opening and two-mirrors chapters carries a twin slot to a twin slot: it sends $(r, r+2)$ to $(208 - r, 210 - r)$, again a pair of the right shape. The 15 slots therefore arrange themselves into 7 mirror pairs together with one slot fixed by the reflection—the wrap-around slot $(209, 211)$, sitting on the axis at the top of Fig. 12.1. (This spanning slot is a genuine admissible slot and is counted with the rest: its first representative $(209, 211)$ is broken since $209 = 11 \cdot 19$, but the class holds true twins higher up, the first being $(419, 421)$.) As a pattern of *slots*, the twins are exactly symmetric.

The primes break the symmetry. Looking at the residues themselves, six of the seven mirror pairs hold two genuine twins apiece, but the pair $\{(41, 43), (167, 169)\}$ does not: 41 and 43 are both prime, while across the axis 167 is prime and $169 = 13^2$ is not. The reflection promises a twin on each side; the square 13^2 falls into one slot and quietly removes it, leaving its mirror standing. It is the only interior pair so broken in the first period, and the cause is sharp—13 is the first prime beyond the wheel's own $2, 3, 5, 7$ whose square lands in a twin slot ($11^2 = 121$ misses, having no coprime neighbor two away).

The break is a statement about the small cases, and it is worth being exact about its scope. The *slots* are symmetric; the *asymptotic densities* are symmetric too, since the twins equidistribute among all 15 admissible slots as $x \to \infty$, a mirror slot receiving the same share as its partner. What is *not* symmetric is the finite, concrete arithmetic: which of the literal residues happen to be prime. That is the recurring lesson of the whole book in one picture—the wheel is regular and its symmetries are exact, while the primes scattered on it are not, and the square sitting in $(167, 169)$ is the irregularity made visible.

12.4 The Hardy–Littlewood heuristic

How many twins are there, and how are they shared among the slots? Hardy and Littlewood conjectured in 1923 that the number of twin primes up to x is

$$\pi_2(x) \sim 2C_2 \int_2^x \frac{dt}{(\ln t)^2}, \qquad C_2 = \prod_{p \geq 3} \left(1 - \frac{1}{(p-1)^2}\right) \approx 0.6601618, \tag{12.2}$$

where C_2 is the twin prime constant. The factor in front is the *singular series* for the pair $(0, 2)$: a product of local densities, one per prime, measuring how the admissibility of the pattern translates into a global rate. The wheel is where those local densities live—the 15 surviving slots out of 210 are precisely the $p = 2, 3, 5, 7$ factors of the singular series, made concrete.

Equation (12.2) is only a heuristic, but it is an extraordinarily accurate one. Figure 12.2 compares it with the actual twin count up to 10^7: the prediction tracks the data across four decades, and at $x = 10^7$ there are 58,980 twin pairs against a predicted 58,754, a ratio of 1.004.

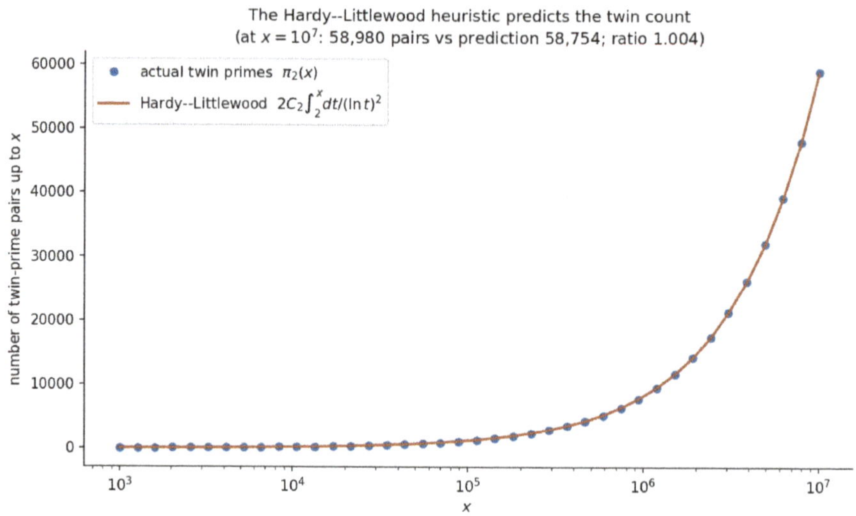

Figure 12.2: The actual count of twin primes $\pi_2(x)$ (points) against the Hardy–Littlewood prediction (12.2) (curve). The heuristic matches the data to within half a percent at 10^7. It predicts the rate with great precision; it does not prove the count is unbounded.

12.5 What this is and is not

The slot structure and the broken mirror are exact and elementary; they are facts. The Hardy–Littlewood formula is a heuristic—a conjecture, not a theorem—and although it agrees with computation to remarkable precision, it is unproven. And the most basic question stands behind both: it is not known whether there are infinitely many twin primes at all. The wheel guarantees 15 admissible slots; it does not guarantee that any single slot is filled infinitely often.

The progress that exists is real but falls short of the target. Zhang proved in 2013 that infinitely many prime pairs differ by at most a fixed bound, and the work of Maynard, Tao, and the Polymath project lowered that bound to 246. A gap of 2—an actual twin—infinitely often is not among the things proved. So the capstone of this tour ends where the whole subject does: at the primes, and at a wall.

Reproducing the figures

```
import numpy as np
from math import gcd
from sympy import isprime

# the 15 admissible twin slots mod 210, and which are genuine twins
slots = [r for r in range(210)
 if gcd(r,210)==1 and gcd((r+2)%210,210)==1]
print(len(slots)) # 15
for r in slots:
```

```
 print(r, r+2, isprime(r) and isprime(r+2)) # (167,169),(209,211) False

# Hardy-Littlewood vs actual twin count up to 1e7
N = 10_000_000
s = np.ones(N+2, bool); s[:2] = False
for i in range(2, int(N**0.5)+1):
 if s[i]: s[i*i::i] = False
twins = np.nonzero(s[:N-1] & s[2:N+1])[0]
from scipy.integrate import quad
C2 = 0.6601618158468695
HL = 2*C2*quad(lambda t: 1/np.log(t)**2, 2, N)[0]
print(len(twins), round(HL), len(twins)/HL) # 58980 58754 1.0038
```

Part V

Closing

Chapter 13

The Walls

13.1 Why 210?

A natural temptation, having spent a whole book on a single number, is to grant 210 a standing of its own—to feel that $2 \cdot 3 \cdot 5 \cdot 7$ must be doing something the integers themselves care about. It is worth being clear, at the close, that it is not.

The wheel has been a lens, not a law. A law would be a rule the primes obey, something fundamental about them; a lens is only a thing we look through. The number 210 is the second kind of object. The primes do not obey it or notice it. It has been the clearest window we have onto structure that belongs to all the integers, and it earned that role by being small: 210 is the smallest modulus that resolves the first four primes at once, and so the smallest wheel on which the whole early sieve is visible in a single turn—the forty-eight spokes, the mirror across the diameter, the fifteen admissible twin slots, the metronome diagonals of 11, 13, and 17. Every one of these is fixed by the arithmetic of $2 \cdot 3 \cdot 5 \cdot 7$. The lattice is ruled by arithmetic at every point; nothing on it is arbitrary, and that complete determinism is the source of the wheel's clarity. It is also the source of a tempting misreading.

The misreading is to take the tidy patterns of the first turn for laws. They are not laws; they are what the arithmetic of this particular small modulus produces. On the first turn four of the five coprime composites reflect onto primes (only 209 reflects onto the unit 1, not a prime), and the symmetry looks all but total—but that is the arithmetic of 210 speaking about 210, not about the integers, and it does not survive the climb. By the second turn the mirror already pairs two composites with each other, $253 = 11 \cdot 23$ against $377 = 13 \cdot 29$; a primorial or two further up, the clean reflections of the first wheel are gone. Nothing has broken, and nothing was ever arbitrary. The same arithmetic that forces the tidy alignment at 210 forces a different, less tidy one at 2310, and a messier one again at 30030—each fully determined, each particular to its own modulus. What looked like a privilege of 210 was the legibility of a small number, and legibility is not law.

So the structures the wheel made visible—the characters, the explicit formula, the mirror, the singular series, the primon gas—are properties of the integers, reachable from any modulus; 210 only gave the clearest first view of them. And the questions they run into cannot be answered by choosing a better modulus, because they are the same on every wheel. They are the walls of the subject, and the rest of this chapter walks up to them.

13.2 Three walls, one frontier

This book has been, from the first spoke, a work of description. It located the primes on the wheel, measured how evenly they spread, drew the faint bias in their race, found their shadow in the zeros, and counted the slots a twin could occupy. Description, however, is not proof, and three times the description ran all the way up to a fact it could not establish. Those three places are not separate dead ends. They are one frontier seen from three sides.

13.2.1 Whether the primes are forced

The coin-flip model of the opening chapter made the primes look like chance at density $1/\log n$, and indeed they pass almost every test of randomness one can put to them while being completely determined. The explicit formula, in the diffraction and spectrum chapters, showed that the gap between *behaves like* and *is* has an exact address: the zeros of the zeta function. If every nontrivial zero lies on the critical line, the primes' apparent randomness is reined in as tightly as the coin-flip suggests; the Riemann Hypothesis is precisely the assertion that they do. We have checked the first many trillions of zeros and found every one on the line, and we cannot prove the next will be. The spectrum chapter pointed at where a proof might live—a self-adjoint operator whose eigenvalues are the zeros, which would pin them to the line at a stroke—but no such operator is known. The prime race stands on the same wall: its small, stubborn bias is explained, but only by assuming the zeros behave, the generalized form of the same unproven statement.

13.2.2 Whether the twins run out

The wheel guarantees fifteen admissible twin slots and not one more; it guarantees nothing about whether those slots stay filled forever. The Hardy–Littlewood heuristic of the twin-slot chapter predicts that they do—infinitely many twins, at a density the singular series fixes from the very same fifteen-of-210 count—and the prediction tracks the data beautifully. But a heuristic is not a proof, and admissibility is only a necessary condition: the wheel says where a twin *can* sit, never that one *must*. The breakthrough work on bounded gaps showed that *some* gap below a few hundred recurs infinitely often—an extraordinary result, and still a finite distance from the gap of 2 that the slots describe. This is the additive wall: the wheel's multiplicative bookkeeping, for all its precision, cannot by itself reach a statement about pairs.

13.2.3 Whether multiplication can be undone

The two sides of the coin, back in the opening chapter, were never peers. Forming a subprime is a moment's multiplication; recovering the primes it was built from is factoring, for which no efficient method is known. The wheel drew that asymmetry honestly—the duality perfect, the direction not—but it could not prove it permanent. No one has shown that factoring *must* be hard; it is not even known to be as hard as the famous hard problems, and a quantum machine running Shor's algorithm would undo it efficiently in a way no classical method is known to. So even this wall is, strictly, unproven to be a wall. What is certain is only that the easy direction and the hard direction look nothing alike, and that a great deal of the modern world is built on the gap between them holding.

13.2.4 One frontier

Set side by side, the three walls are one thing seen from three angles. The primes are deterministic and yet pseudorandom, and each wall is a place where the precise description the wheel supplies—a density, a set of slots, an easy multiplication—meets a deep fact it cannot prove: that the randomness is genuinely bounded, that the slots are genuinely filled forever, that the multiplication is genuinely irreversible. The Riemann Hypothesis is this frontier on its multiplicative, spectral side; the twin-prime conjecture is the same frontier on its additive side; the hardness of factoring is the same frontier on its algorithmic side. The wheel gave the clearest available view of all three. It cannot pass any of them, and no better-chosen modulus can either.

13.3 Where the wheel leaves us

We set out to see real analytic number theory through a single object, at human scale, and that is what the wheel delivered: the Dirichlet characters as its harmonics, the zeros as its spectrum, the explicit formula as honest interference, the singular series as fifteen surviving slots, the primes as the modes of a gas whose partition function is ζ. None of it was metaphor borrowed to dress up something thinner; it was the standard machinery of the subject, made concrete enough to point at.

The other half of the promise was that the tour would end here—in front of the walls, rather than pretending to pass them. It is worth saying why that is not a disappointment. Knowing exactly where a wall stands, knowing why it stands, and knowing that no clever choice of lens will move it, is not a lesser thing than a proof. It is what honest understanding of an unfinished subject actually looks like. The wheel was a good place to stand. It set us down at the edge of what is known, looking straight at what is not.

The forty-eight units and their mirror partners

The $\varphi(210) = 48$ residues coprime to 210, listed as the 24 mirror pairs $r \leftrightarrow 210 - r$.

$1 \leftrightarrow 209$	$11 \leftrightarrow 199$	$13 \leftrightarrow 197$	$17 \leftrightarrow 193$
$19 \leftrightarrow 191$	$23 \leftrightarrow 187$	$29 \leftrightarrow 181$	$31 \leftrightarrow 179$
$37 \leftrightarrow 173$	$41 \leftrightarrow 169$	$43 \leftrightarrow 167$	$47 \leftrightarrow 163$
$53 \leftrightarrow 157$	$59 \leftrightarrow 151$	$61 \leftrightarrow 149$	$67 \leftrightarrow 143$
$71 \leftrightarrow 139$	$73 \leftrightarrow 137$	$79 \leftrightarrow 131$	$83 \leftrightarrow 127$
$89 \leftrightarrow 121$	$97 \leftrightarrow 113$	$101 \leftrightarrow 109$	$103 \leftrightarrow 107$

BOOK II

A Spectrum Through the Primes

Preface

This is the second of three books about the same set of facts. The first, *A Wheel Through the Primes*, built its picture of the primes from the wheel of 210: the residue classes coprime to $2 \cdot 3 \cdot 5 \cdot 7$, the race among them, the conductor lattice of their characters, the explicit formula tying the primes to the zeros of the Riemann zeta function, and the three walls at which honest exposition stops — the Riemann Hypothesis, the twin primes, and the hardness of factoring. This book assumes that one. It does not re-derive the wheel or re-introduce the explicit formula; it takes the headline ideas of the first book — including the Hilbert–Pólya dream, the Berry–Keating model, and Connes's trace formula, all met there in outline — and goes one level deeper into the single most seductive face of the subject: the idea that the zeros are the *spectrum* of something, and that to understand the primes one should find the instrument whose notes they are.

The stance is the same as the first book's, and it is worth stating plainly. Everything credited to the established literature is the work of the people named; the metaphors are marked as metaphors and the mechanisms as mechanisms; and where a calculation is the book's own, it is presented with the numbers and the means to reproduce it. The book does not pass the three walls, and it does not pretend to. It ends at them on purpose, because the most useful thing a tour can do at a wall is show exactly where it stands and why it is hard, rather than tell a story that walks through it.

A few conventions carry over. The word *subprime* means, throughout, a composite number coprime to 210 whose smallest prime factor exceeds 7. Each chapter opens with a short SCOPE and closes with a section called *What this is and is not*, separating the established from the merely measured from the frankly conjectural; where a chapter has a figure, a closing note prints the short program that produced it, and the longer computations are reproduced by the verification programs that accompany this book. The voice is plain and declarative on purpose; the subject supplies enough difficulty without ornament.

The book is in four parts. *The statistics* looks at the zeros as a population and asks what kind of spectrum they resemble. *The operator dream* takes the Hilbert–Pólya idea seriously enough to try to build the operator, and watches the attempts fail in instructive ways. *Hearing and reading* asks what the primes can tell us about the zeros — what can be detected, what cannot be resolved, and which apparent laws do not survive being pushed. *The walls* walks up to the three obstacles again, through the spectral lens, and finds them standing where the first book left them.

Acknowledgements

The author gratefully acknowledges the use of Anthropic's Claude as a computational and drafting assistant during the preparation of this work. Any mathematical results or interpretations presented here remain entirely the responsibility of the author. — J.M.

Part I

The statistics

Chapter 1

The Symmetry of a Family

Book One looked at the zeros of one zeta function among themselves and found the statistics of the eigenvalues of a large random Hermitian matrix. This chapter makes the move Book One did not: from a single spectrum to a whole family of them. The governing fact, due to Katz and Sarnak, is that the local statistics deep in the spectrum are the same for every family — the sine-kernel hole already met in Book One — but each family also carries a symmetry type, *unitary or orthogonal or symplectic, and that type is written in one place only: the way the very lowest zeros sit against the central point. A symplectic family pushes its lowest zeros away from the centre; an orthogonal family pulls them in; a unitary family is indifferent. And there is exactly one arithmetic world in which all of this is a theorem rather than a hope — curves over a finite field, where the zeros really are the eigenvalues of a real operator and the Riemann Hypothesis is Weil's theorem. Over the rational numbers the same statistics hold to every digit anyone has computed, and stay conjectural. This chapter is expository.*

1.1 Scope

This chapter is expository and reports established mathematics. It assumes the previous book's spectral chapter — unfolding, level repulsion, Montgomery's pair correlation, and the Hilbert–Pólya idea — and goes one step past where that chapter stopped. There, the object was the single sequence of zeros of ζ. Here the object is a *family* of L-functions, and the new content is threefold: that the members of a family are statistically independent of one another, that families sort into three symmetry types according to the behaviour of their lowest zeros, and that this sorting is a genuine theorem in one setting and a well-supported conjecture in ours. The chapter introduces the independence of distinct L-functions seen on the wheel's seven, the Katz–Sarnak symmetry types, the one-level density that detects them, the function-field world where they are proved, and the precise sense in which the rational world falls short. It proves nothing new. The one thing this circle of ideas would most like to hand over — an operator over the rationals to play the role the Frobenius operator plays over a finite field — is exactly what the next chapter is unable to find, and Section 1.7 is careful about that line.

1.2 From one spectrum to a family

A single zeta function gives you a single spectrum: one critical line, one ladder of zeros climbing it. Book One showed that ladder, once unfolded to unit spacing, keeps the statistics of a random Hermitian matrix. But a statistical law wants more than one sample. The eigenvalues of *one* large matrix already display the law; the law itself, though, is a statement about the *ensemble* — about what happens on average as the matrix is drawn again and again. The arithmetic analogue of drawing the matrix again is changing the L-function.

So the natural object is not ζ alone but a family of L-functions that resemble it: the Dirichlet L-functions of a fixed modulus; the quadratic L-functions $L(s, \chi_d)$ as d runs over fundamental discriminants; the L-functions of the quadratic twists of a fixed elliptic curve. Each member has its own critical line and its own lowest zeros. Averaging a statistic across the family is the arithmetic version of averaging over the ensemble of matrices. The seven real characters of the wheel are the smallest such family sitting in plain sight — seven quadratic L-functions, one per spoke-symmetry of Book One. Seven is far too few to see anything statistical, a point we return to; but they fix the idea, and they will turn out to belong to a definite one of the three types.

1.3 The bulk is one law for everyone

The first thing to say about families is a negative: deep inside the spectrum, they are all the same. The two-point statistic of the high zeros — Montgomery's pair correlation from Book One,

$$R_2(u) \;=\; 1 - \left(\frac{\sin \pi u}{\pi u}\right)^2, \tag{1.1}$$

with its deep hole at small separation — is the *same* function for every family, and it is the same as the bulk two-point law of the unitary, orthogonal, and symplectic random-matrix ensembles alike. This is the content of *universality*: the local spacing of the high zeros does not know which family it came from. Panel A of Figure 1.1 is this one law. If the bulk were all we could measure, the three symmetry types would be indistinguishable.

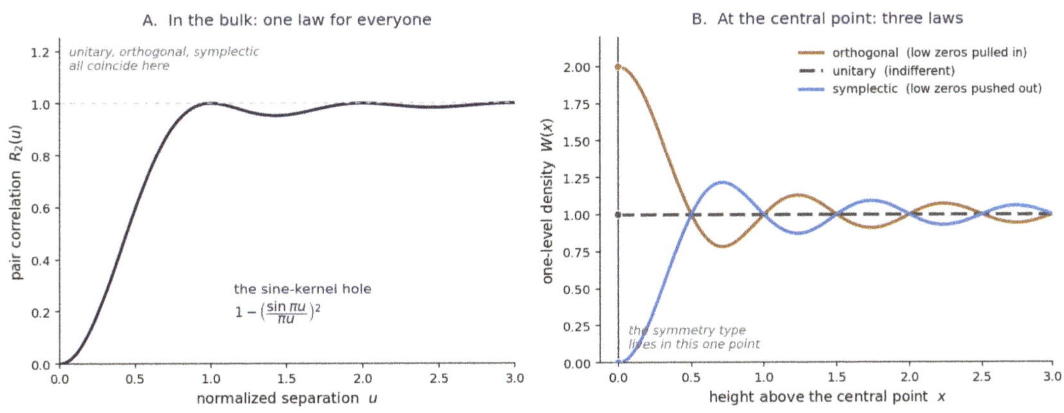

Figure 1.1: Where a family hides its symmetry. **A:** the bulk two-point statistic, Eq. (1.1), is the same sine-kernel hole for every L-function in every family, and the same for the unitary, orthogonal, and symplectic ensembles — the bulk cannot tell them apart. **B:** the one-level density $W(x)$, which measures how the lowest zeros sit at height x above the central point. Here the three types separate, and only at the central point itself: a symplectic family has $W(0) = 0$ (lowest zeros repelled), a unitary family $W(0) = 1$ (indifferent), an orthogonal family $W(0) = 2$ (lowest zeros attracted). The symmetry type lives in that single point.

1.4 The families are independent draws

Universality says every family obeys the same bulk law. It does not say the families have anything to do with one another, and they do not. Take the seven real L-functions of the wheel and measure, for each one alone, the average ratio of consecutive gaps between its zeros — a single number, needing no unfolding, that a unitary spectrum drives to 0.603 and an uncorrelated one to 0.386. Each of the seven returns about 0.62 — at the unitary end, a shade above the asymptotic 0.603 and far from 0.386: each spectrum, on its own, repels exactly as Eq. (1.1) demands.

Now pour all seven sets of zeros into one list and measure again. The number collapses to about 0.39 — the value for points with no correlation (Figure 1.2). The repulsion each family carried alone is gone from the merged sequence: its gaps are those of a structureless list, the sine-kernel hole of Eq. (1.1) filled flat. Rigid spectra can superimpose into a structureless one only when they are mutually independent: any dependence among the seven — coincident zeros, correlated ordinates, or a rigidity shared across the families — would show as a departure from the Poisson value, in one direction or the other, and none appears. This is the statistical independence of distinct L-functions that Rudnick and Sarnak established, seen directly on the wheel's seven. Like the bulk law, it is a statement about how the zeros are distributed and not about where they lie, and it makes no claim about the Riemann Hypothesis.

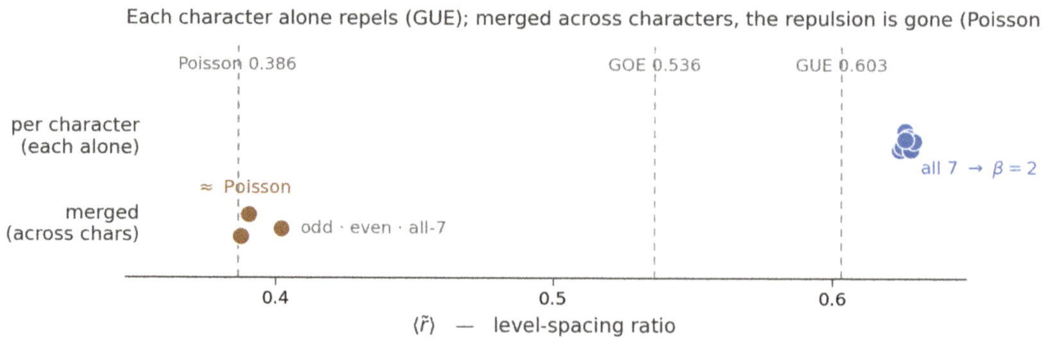

Figure 1.2: Independence by collapse. For each of the seven real L-functions of the wheel, measured alone, the level-spacing ratio sits at the unitary end (blue, ≈ 0.62, just above the $\beta = 2$ value 0.603): the zeros repel, as in Eq. (1.1). Pooling the seven — as the odd four, the even three, and all seven — sends the ratio to the Poisson value (orange, the three points near 0.39): the repulsion is gone. Rigid spectra merge to a structureless one only when they are independent, so the collapse is the independence of the families made visible. Reference values: Poisson 0.386, GOE 0.536, GUE 0.603.

1.5 The symmetry lives at the central point

What separates the families is not how their zeros sit among themselves high up, but how the *lowest* zeros sit relative to one distinguished height: the central point $s = \frac{1}{2}$, where the functional equation pivots. The statistic that sees this is the *one-level density* — loosely, the averaged density of low zeros at height x above the centre, measured across the family. Its scaling form is one of three closed expressions, shown in Panel B:

$$W_{\text{unitary}}(x) = 1, \tag{1.2}$$

$$W_{\text{symplectic}}(x) = 1 - \frac{\sin 2\pi x}{2\pi x}, \tag{1.3}$$

$$W_{\text{orthogonal}}(x) = 1 + \frac{\sin 2\pi x}{2\pi x}. \tag{1.4}$$

At the centre, $x \to 0$, the kernel $\sin(2\pi x)/(2\pi x) \to 1$, so the three densities take the values 0, 1, and 2. That is the whole distinction, and it is a statement about a single point. A *symplectic* family clears a gap at the centre — its lowest zeros are pushed up, away from $s = \frac{1}{2}$. An *orthogonal* family does the opposite, crowding its lowest zeros down toward the centre. (The density $W_{\text{orthogonal}}$ of Eq. (1.4) is the even orthogonal SO(even) density; the odd orthogonal SO(odd) case instead carries a forced zero at the centre.) A *unitary* family does neither; its low zeros are as indifferent to the centre as to any other height.

Which family is which is decided by the arithmetic symmetry each member carries. The full set of Dirichlet characters to a modulus is a *unitary* family. The quadratic characters — the real ones, the kind the wheel is built from — form a *symplectic* family. The quadratic twists of an elliptic curve are *orthogonal*, and split further by the sign of the functional equation into an even part and an odd part. So the wheel's seven real characters are, in principle, members of a symplectic family: their lowest zeros should sit a touch higher than a unitary family's would. The qualification "in principle" is doing real work. To *see* the curve of Eq. (1.3) you need the lowest zeros of thousands of quadratic L-functions, smoothed into a density. Seven characters give you seven numbers and no curve at all. (The independence of the previous section escaped this shortage: it pooled the *zeros* of the seven, which run to the thousands, while a symmetry type is a statistic across *family members* — and of those, seven is far too few.) This is the first, quiet appearance of a problem that will dominate the second half of the book: the law is real, and the data in hand is too thin to resolve it.

1.6 The one world where it is a theorem

Everything so far is conjecture over the rational numbers. There is a different arithmetic world where the same statements are theorems, and the difference between the two worlds is the heart of this book.

Replace the rational numbers by the functions on a curve over a finite field $\mathbb{F}_q$. The L-functions attached to such a curve are not mysterious infinite products that might or might not have their zeros on a line; they are *polynomials*. By Weil, and in full generality by Deligne in 1974, the reciprocal roots of these polynomials are $q^{1/2}$ times numbers of absolute value one — the eigenvalues of the Frobenius operator acting on the cohomology of the curve. In this world the Riemann Hypothesis is not open. It is Weil's theorem. The zeros lie on the critical line not because a real operator has a real spectrum — Frobenius is not self-adjoint — but because of *purity*: the Weil–Deligne weight argument forces the Frobenius eigenvalues to absolute value $q^{1/2}$, which is exactly the Riemann Hypothesis for the curve. An operator whose eigenvalues are the zeros genuinely exists here, realizing the Hilbert–Pólya dream over a finite field rather than over $\mathbb{Q}$; what places the zeros on the line, though, is purity, not self-adjointness.

Katz and Sarnak then proved, in 1999, that as the field grows the zeros of families of these L-functions follow exactly the symmetry-type laws of Panel B. And the symmetry type is not an analogy there: it is the *monodromy group* of the family, the group through which the Frobenius eigenvalues turn as the curve varies. "Unitary," "orthogonal," "symplectic" name actual structure groups, and Deligne's equidistribution theorem makes the random-matrix average an equality rather than a hope. Over function fields, in short, the spectrum is an operator's spectrum, the symmetry is a group, and the statistics are theorems.

Back over $\mathbb{Q}$, the best available results are real but bounded. Özlük and Snyder proved in 1999 (later extended by Rubinstein) that the quadratic family has the symplectic one-level density of Eq. (1.3) — but only when tested against functions whose Fourier transform is supported in a limited window. That is the very same restriction Montgomery met in Book One, where the pair correlation could be confirmed only for test functions of limited support. Inside the window it is a theorem; outside it, the symmetry type is conjecture backed by overwhelming numerics.

1.7 What this is and is not

What is established is of two kinds. The bulk universality — the sine-kernel hole of Eq. (1.1), shared by every family — is on the firmest footing the subject offers. And over function fields the entire symmetry-type theory is a theorem, because there the zeros are the eigenvalues of the Frobenius operator and the Riemann Hypothesis is Weil's. Over the rationals, the symplectic one-level density of the quadratic family is a theorem for test functions of restricted support, by Özlük–Snyder (and Rubinstein).

What is not established is the rest of it over $\mathbb{Q}$: the symmetry types beyond the restricted window, for every family, remain conjectural. The gap is sharp, and it is worth naming exactly, because the rest of the book lives in it. The function-field case is a theorem and the rational case is not, and the difference between them is not a difference of evidence — the numerical agreement over $\mathbb{Q}$ is as good as numerical agreement ever gets. It is a difference in the *kind of object* available. Over a finite field there is an operator, Frobenius on cohomology, whose eigenvalues are the zeros; the symmetry types fall out of its monodromy as a matter of proof. Over $\mathbb{Q}$ there is no such operator, and so the same statistics can only be observed, not derived.

Book One said the zeros of ζ behave *as though* they were a spectrum. This chapter sharpens that into a contrast. In one arithmetic world the zeros simply *are* a spectrum, and one can watch the symmetry types emerge from the operator. In ours, we have the identical statistics and no operator to hang them on. Whether such an operator exists over $\mathbb{Q}$ — and what would have to be true of it — is the subject of the next chapter.

Reproducing the figures

```
import numpy as np

# Panel A: the bulk two-point law, shared by all symmetry types
def sinc1(u):                      # sin(pi u)/(pi u), = 1 at u = 0
    s = np.ones_like(u, dtype=float); nz = u != 0
    s[nz] = np.sin(np.pi*u[nz])/(np.pi*u[nz]); return s
u  = np.linspace(1e-4, 3, 800)
R2 = 1 - sinc1(u)**2               # Montgomery pair correlation

# Panel B: one-level densities; the kernel is sin(2 pi x)/(2 pi x)
def sinc2(x):                      # = 1 at x = 0
    s = np.ones_like(x, dtype=float); nz = x != 0
    s[nz] = np.sin(2*np.pi*x[nz])/(2*np.pi*x[nz]); return s
x      = np.linspace(0, 3, 800)
W_unit = np.ones_like(x)             # unitary     : W(0) = 1
W_symp = 1 - sinc2(x)                # symplectic  : W(0) = 0
W_orth = 1 + sinc2(x)                # orthogonal  : W(0) = 2

# central-point values that carry the whole distinction
print(1 - sinc2(np.array([0.0])))    # [0.] symplectic: lowest zeros repelled
print(1 + sinc2(np.array([0.0])))    # [2.] orthogonal: lowest zeros attracted

import numpy as np
```

```
# Independence figure: the level-spacing ratio r~ of a sorted ordinate list.
#   no unfolding needed; ~0.603 for GUE (beta=2), ~0.386 for uncorrelated points.
def rtilde(g):
    g = np.sort(np.asarray(g, float)); s = np.diff(g)
    r = np.minimum(s[1:], s[:-1]) / np.maximum(s[1:], s[:-1])
    return float(r.mean())

# each of the seven real L-functions, alone, gives ~0.62 (GUE);
# the seven ordinate lists poured into one give ~0.39 (Poisson) -- the collapse.
# the per-character and merged ordinates are produced by the verification
# programs accompanying this book.
```

Part II

The operator dream

Chapter 2

What Would Count as an Operator

The previous chapter ended on a single missing object: over a finite field the zeros are the eigenvalues of the Frobenius operator and the Riemann Hypothesis is a theorem, while over the rationals we have the same statistics and no operator to hang them on. This chapter states exactly what such an operator would have to be, and then watches the two most natural attempts to build one fail — in opposite ways. A candidate must pass a two-part test: it must be defined without already knowing the zeros, *and its eigenvalues must* be *the zeros. The first attempt builds an object whose spectrum is exactly the zeros, by writing the zeros into it by hand; it has the right spectrum and fails the first half of the test. The second builds the most natural operator the primes offer, intrinsic and honest, and finds that its spectrum is the logarithms of the integers, not the zeros; it passes the first half and fails the second. The two failures are not bad luck. They are the two walls of a corridor, and the corridor is narrow.*

2.1 Scope

This chapter is partly expository and partly a pair of worked computations. It assumes the previous book's primon-gas chapter, whose bosonic gas of prime "modes" returns here as a genuine operator, and it assumes the Hilbert–Pólya idea as that book stated it. The new content is a sharpening and two examples. The sharpening is the two-part test that any Hilbert–Pólya operator must pass. The examples are two operators built explicitly and examined honestly: one with the zeros as its spectrum and no independent definition, one with an impeccable independent definition and the wrong spectrum. Both computations are exact and checkable; neither produces a Hilbert–Pólya operator, and Section 2.6 says why in the sharpest terms the chapter can manage. The genuinely promising candidate — Berry and Keating's — is held over to the next chapter.

2.2 The dream, stated exactly

The Hilbert–Pólya idea, as Book One gave it, is that the ordinates γ_n of the nontrivial zeros are the eigenvalues of some self-adjoint operator H. If that were so, the γ_n would be real because the spectrum of a self-adjoint operator is real, and the zeros $\frac{1}{2} + i\gamma_n$ would be pinned to the critical line not by a delicate analytic balance but by the reality of a quantum spectrum. The Riemann Hypothesis would follow at a stroke (Figure 2.1).

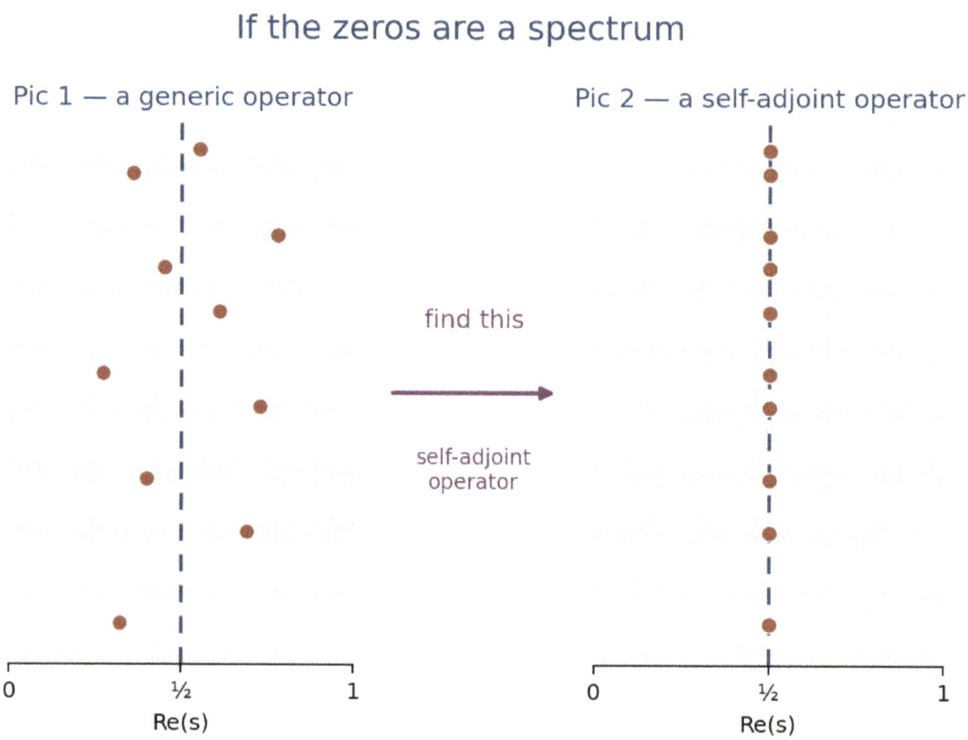

Figure 2.1: The shape of the dream. Were the zero ordinates γ_n the spectrum of a self-adjoint operator, that spectrum would be real, and every zero would lie on the line $\mathrm{Re}(s) = \frac{1}{2}$ (Pic 2). Pic 1 shows what a *generic* operator would permit; the arrow is the unsolved part—no such operator has been constructed. Schematic.

To turn this from a wish into a target, we ask what would actually count as a solution. Two conditions, and a candidate must meet both.

(I) **Independence.** The operator H must be defined without already knowing the zeros. It should be built from arithmetic — from the primes, from a space of functions, from a geometry — by a recipe that never mentions the γ_n. An operator you can only write down after someone hands you the zeros explains nothing, because the explaining was done by whoever found the zeros.

(II) **Correct spectrum.** The eigenvalues of H must be the γ_n themselves — not a sequence that resembles them, not a sequence with the same statistics, but the actual ordinates.

The two conditions sound almost the same on first reading, and they are almost opposite in practice. Condition (II) alone is easy; condition (I) alone is easy; the difficulty is that no known construction meets both at once. The one place they have been met together is the function-field world of the previous chapter, where the Frobenius operator is defined intrinsically from the geometry of a curve — it never mentions the zeros (I) — and its eigenvalues *are* the zeros of the curve's L-function (II). That is the whole reason the Riemann Hypothesis is a theorem over a finite field and an open problem over the rationals. With the bar set, we try to clear it.

2.3 The first failure: a spectrum made to order

Take the zeros as given — the first several hundred ordinates γ_n, computed to high precision — and ask for an object whose spectrum is exactly those numbers. There is an embarrassingly direct way to get one. Form the function

$$K(\xi) = \sum_n \cos(\gamma_n \xi), \tag{2.1}$$

a finite sum of cosines, one per ordinate. Its frequency content — the thing a physicist would call its spectrum — is a set of spikes sitting exactly at the γ_n, because that is what each cosine contributes and nothing else is present. Panel A of Figure 2.2 is this spectrum: the zeros, reproduced perfectly.

One can do this with more care and get more structure. Pair each ordinate with the midpoint $m_n = \frac{1}{2}(\gamma_n + \gamma_{n+1})$ to its successor and form the windowed difference

$$K(\xi) = \frac{1}{|I|} \sum_{n \in I} \big[\cos(\gamma_n \xi) - \cos(m_n \xi)\big], \tag{2.2}$$

a trigonometric polynomial of some hundreds of cosine terms — in one natural windowing, equally many at the ordinates and at the midpoints. This object has a genuine closed form, a genuine asymptotic behaviour, and even a genuine combinatorial topology: which zeros pair with which midpoints depends on the arithmetic of the underlying wheel, so the kernel carries real structure that took real work to describe. None of that rescues it. Its spectrum is a signed sum of spikes at the ordinates and the midpoints, and the ordinates are in the spectrum for one reason only: we wrote them in. The object satisfies condition (II) and violates condition (I) completely. It is not a window onto the zeros; it is a photograph of them, and a photograph predicts nothing about its subject.

This is the first lesson of the operator dream, and it is worth stating bluntly because the literature is full of subtler versions of the same move: *having the right spectrum is no achievement if the spectrum was the input.* Structure is not the issue — this kernel has plenty of structure. Independence is the issue.

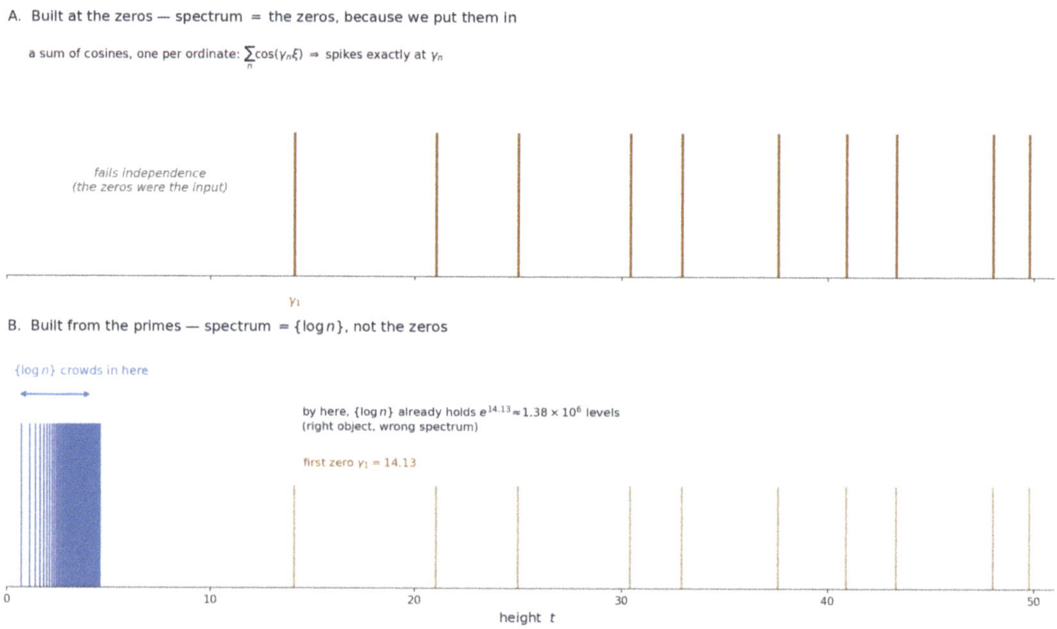

Figure 2.2: Two natural operators, two ways to miss. **A:** an object built as a sum of cosines at the ordinates, Eq. (2.1). Its spectrum is the zeros exactly — because the zeros were written into it. It has the right spectrum and no independent definition. **B:** the primon-gas Hamiltonian of Book One, built only from the primes. Its spectrum is $\{\log n\}$, the logarithms of the integers, which thin out near the origin and crowd together at the top of the range, never taking the shape of the zeros (faint, for reference, at 14.13 and beyond). By the height of the first zero, this spectrum already holds about 1.38 million levels. It has an impeccable independent definition and the wrong spectrum.

2.4 The second failure: the right object, the wrong spectrum

Now do the honest thing and refuse to look at the zeros at all. Build an operator from the primes alone. The natural one is the Hamiltonian of Book One's primon gas: a bosonic Fock space with one mode for each prime, and

$$\widehat{H} = \sum_p (\log p)\, \hat{n}_p, \tag{2.3}$$

where $\hat{n}_p$ counts the quanta in the mode of the prime p and each such quantum costs energy $\log p$. This is as intrinsic as an operator can be. It mentions no zeros; it is written down from the primes and the single rule that a prime p carries energy $\log p$. It passes condition (I) without an argument.

Its spectrum is immediate and beautiful. A state with k_p quanta in each prime mode has energy $\sum_p k_p \log p = \log(\prod_p p^{k_p}) = \log n$, where n is the integer with that factorization. By unique factorization every positive integer arises exactly once, so the spectrum is

$$\operatorname{spec}\widehat{H} = \{\log n : n = 1, 2, 3, \dots\}, \tag{2.4}$$

the logarithms of the integers, each appearing once. This is the object whose partition function $\sum_n e^{-s\log n} = \sum_n n^{-s}$ is the zeta function itself — which is the sharp irony of the construction. The zeta function is the primon gas's partition function, and the primon gas's spectrum is $\{\log n\}$. Zeta enters through its Dirichlet series, not through its zeros, and the operator never sees a single γ_n.

The mismatch is total, and Panel B makes it visible. The logarithms of the integers thin out near the origin and crowd together at the top of the range — $\log 2 \approx 0.69$, $\log 3 \approx 1.10$, and the gaps shrink like $1/n$, so the spectrum becomes a near-continuum — while the zeros begin only at $\gamma_1 = 14.13$ and stay discrete, repelling one another. The cleanest way to see how far apart these two sequences are is to count. The number of integers with $\log n \le t$ is $\lfloor e^t \rfloor$, growing exponentially in t. The number of zeros with $\gamma_n \le t$ grows like $\frac{t}{2\pi}\log\frac{t}{2\pi}$, barely faster than linearly. By the height of the very first zero, $t = 14.13$, the primon spectrum already contains $e^{14.13} \approx 1.38$ million levels, and the zeros contain one. No reweighting, no change of variable, repairs a discrepancy of that order. The operator is the right kind of thing and has entirely the wrong spectrum. It passes (I) and fails (II).

A word on names, because they cause real confusion. The Hamiltonian Eq. (2.3) is sometimes called a Berry–Keating Hamiltonian. It is not the one Berry and Keating proposed. Theirs is a different operator, built on a single classical degree of freedom rather than a gas of prime modes, and it is the subject of the next chapter. Equation (2.3) is the free primon gas of Book One, and its spectrum is $\{\log n\}$. Keeping the two apart is the difference between an operator that fails the second test outright and one that might, conceivably, pass it.

2.5 Why the two failures are one

The two operators fail opposite halves of the test, and that opposition is the real content of the chapter. Push on either one and you are pushed toward the other's failure. Try to repair the cosine kernel of Eq. (2.1) by giving it an intrinsic definition — some rule that produces those cosines without being handed the ordinates — and you are asking for exactly the operator nobody has: condition (I) is the whole problem. Try to repair the primon gas of Eq. (2.3) by bending its spectrum from $\{\log n\}$ toward the zeros, and every natural bending — a different weight on each prime, a chemical potential, an interaction between modes — moves the spectrum to some other arithmetic sequence built from the primes, never to the zeros, because the zeros are not a simple functional of the primes in that way; they are the primes' Fourier-dual, the subject of Book One's diffraction chapter, and Fourier duals do not appear as the energy levels of the thing they are dual to.

So the corridor is narrow on purpose. An intrinsic construction from the primes wants to give you $\{\log n\}$ or a relative of it; a construction with the zeros as spectrum wants the zeros as input. The function-field world escapes the corridor because there the dual variable is itself geometric — Frobenius acts on a cohomology that the zeros live in — so intrinsic definition and correct spectrum coincide. Over the rationals no one has found the room where they coincide. That room, if it exists, is what the next two chapters go looking for.

2.6 What this is and is not

What is established here is two exact facts. The cosine kernel of Eqs. (2.1) and (2.2) has spectrum equal to the ordinates by construction; this is a triviality dressed as a computation, and naming it as such is the point. The primon-gas Hamiltonian of Eq. (2.3) has spectrum $\{\log n\}$; this is a clean theorem about a clean operator, proved in one line from unique factorization. Both are mathematics, and neither is in doubt.

What is not established — by anyone — is a Hilbert–Pólya operator over the rationals. The two constructions here are not failed attempts at one; they are demonstrations of the two ways the attempt characteristically fails. And it is important to be honest about the status of the test itself. There is no theorem that says no such operator exists. The two-part test is not an impossibility proof; it is a statement of the bar, and a map of the terrain just below it. Many serious programs — Berry and Keating's, Connes', de Branges' — are attempts to find the narrow room where conditions (I) and (II) hold together, and the chapters that follow take each at its true status: real ideas, real partial results, no operator yet. What this chapter shows is only this. The easy ways to get the zeros as a spectrum cheat, and the honest operator the primes hand you has the logarithms of the integers for a spectrum, which is the wrong sequence by a factor of a million levels before the first zero is reached.

Reproducing the figure

```
import numpy as np
from mpmath import zetazero
g = np.sort(np.array([float(zetazero(k).imag) for k in range(1, 31)]))

# Panel A: spectrum of the cosine kernel sum_n cos(gamma_n xi)
#   = spikes at the ordinates themselves (they were the input)
zeros_in_window = g[g <= 52]                    # gamma_1 .. ~gamma_10

# Panel B: spectrum of the primon-gas Hamiltonian  H = sum_p (log p) n_p
#   eigenvalue of an occupation state {k_p} is sum k_p log p = log n
#   so spec(H) = { log n : n = 1, 2, 3, ... }
n    = np.arange(2, 60)
logn = np.log(n)                                # all land in [0.69, ~4.07]

# how far apart the two spectra are: count levels below height t
#   primon:  #{ log n <= t } = floor(e**t)
#   zeros :  #{ gamma_n <= t } ~ (t/2pi) log(t/2pi)
import math
print(math.floor(math.exp(g[0])))              # 1376046  levels below gamma_1 = 14.13
```

Chapter 3

Primes as Orbits

The last chapter left a corridor with two walls: an intrinsic operator from the primes gives the logarithms of the integers, and an operator with the zeros as spectrum has to be handed the zeros. This chapter walks the one candidate that gets furthest down the corridor. Berry and Keating's classical Hamiltonian $H = xp$ is built from nothing but a position and a momentum — it never mentions the zeros, so it passes the independence test cleanly — and its regularized semiclassical level count reproduces the average density of the Riemann zeros exactly. *That is the deepest partial success in the subject. But the average is all it gives. The individual zeros live in the fluctuation around the average, and the fluctuation is supplied not by xp but by the primes, entering the explicit formula in precisely the shape of a sum over periodic orbits — one orbit per prime, of period* $\log p$. *The explicit formula, read this way, is a trace formula. The catch is that we have the trace formula and the orbits but not the dynamical system they belong to, and the reason that system is so hard to find is the same reason its spectrum looks like the zeros in the first place: it would have to be chaotic.*

3.1 Scope

This chapter is expository, building on two chapters of Book One — the diffraction chapter that read the zeros as the Fourier-dual of the primes, and the primon-gas chapter — and on the two-part test of the previous chapter. The new content is the Berry–Keating proposal at its true weight: what $H = xp$ achieves, which is real and exact, and what it does not, which is everything that distinguishes one zero from the next. The chapter then rereads Book One's explicit formula as a trace formula in the sense of quantum chaos, with the primes as periodic orbits, and explains why that reading is a genuine structural match and still not a proof. It closes on the reason the whole programme is hard, which is the reason it is attractive: the spectrum behaves like that of a chaotic system, and chaos resists exactly the kind of forward computation a proof would need. Connes' and de Branges' programmes, which try to supply the missing system by other means, are named here and taken up later.

3.2 An operator that passes the first test

Berry and Keating proposed, as the classical mechanics underneath the Riemann zeros, the simplest Hamiltonian that is not a harmonic oscillator:

$$H = x\,p, \tag{3.1}$$

a single particle on a line, with energy equal to position times momentum. The trajectories are hyperbolae $xp = E$ in the phase plane. Nothing in this definition knows anything about ζ; it is written from a position and a momentum and a product. So $H = xp$ passes the independence test of the previous chapter without discussion — and it is the first candidate in this book to do so while also getting something quantitative right.

Here is what it gets right. To count the quantum levels of a system below energy E, the semiclassical rule is to measure the area of the phase-space region with $H \leq E$ and divide by 2π (in units where Planck's constant is one). For $H = xp$ that region is unbounded — a hyperbola encloses infinite area — so one must regularize, by refusing to resolve phase-space cells smaller than a single quantum: $|x| \geq \ell_x$, $|p| \geq \ell_p$. With that cutoff the area below E is finite, and the count of levels below height T comes out as

$$N_{\text{smooth}}(T) \;=\; \frac{T}{2\pi}\log\frac{T}{2\pi} - \frac{T}{2\pi} + \frac{7}{8} + \cdots \tag{3.2}$$

This is not merely the right order of magnitude. It is, term for term, the smooth part of the Riemann–von Mangoldt counting function for the zeros of ζ — the same $N_{\text{smooth}}(T)$ Book One used to unfold them, down to the constant $\frac{7}{8}$. Panel A of Figure 3.1 shows the curve of Eq. (3.2) laid over the actual staircase of zeros: at $T = 100$ the formula gives 29.00 and there are exactly 29 zeros below that height. A Hamiltonian written from a position and a momentum, regularized by a single quantum cell, counts the Riemann zeros on average to the last term. That is a real and surprising fact, and it is the reason xp is taken seriously.

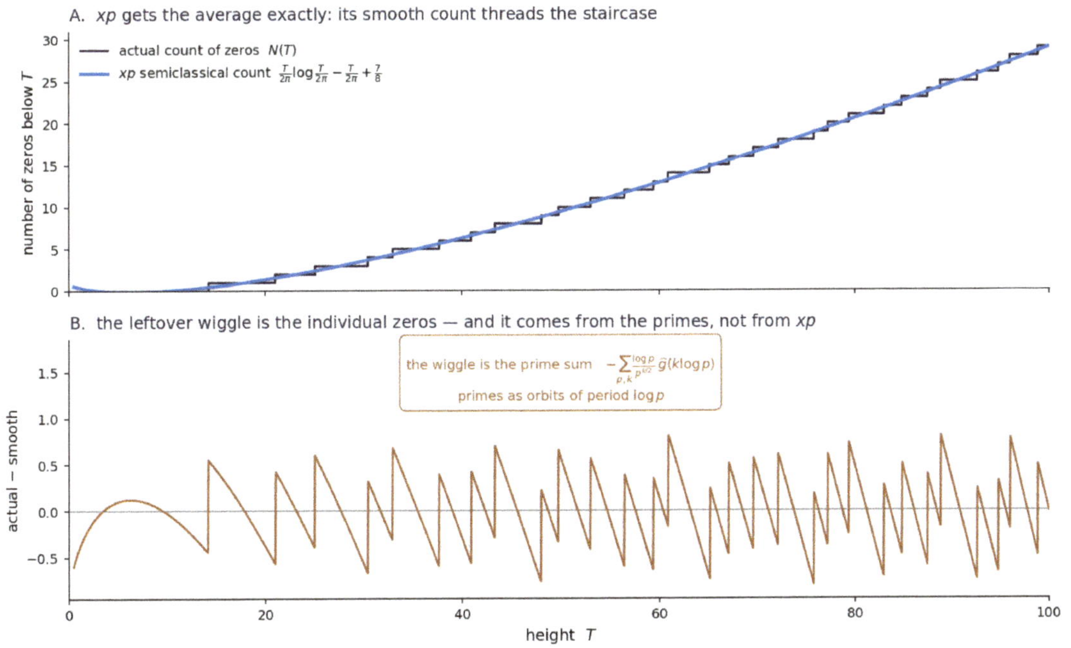

Figure 3.1: What xp gives and what it misses. **A:** the regularized semiclassical count of $H = xp$, Eq. (3.2) (smooth curve), threads the staircase of actual zeros (steps); it reproduces the average density of zeros term for term. **B:** the leftover, actual count minus smooth count. This fluctuation — a unit jump at each zero, sloping between — is where the individual zeros live, and xp does not produce it. In the explicit formula it is exactly the sum over primes, each prime acting as a periodic orbit of period $\log p$.

3.3 The average and the fluctuation

Now the limitation, which is visible the moment one subtracts. Panel B is the difference between the actual count of zeros and the smooth count of Eq. (3.2): a sawtooth that jumps up by one at each zero and slopes gently down between them. The smooth curve is the *average* behaviour; the sawtooth is the *fluctuation*, and every fact that distinguishes one zero from another — where exactly γ_1 sits, how far it is to γ_2 — is contained in the fluctuation, not in the average.

$H = xp$ produces the average and nothing else. It is a free system, in the sense that its regularized phase space has no internal structure beyond the cutoff; its level density is smooth by construction. The fluctuation has to come from somewhere else, and in the Riemann problem there is only one other place it can come from: the primes. This is the content of Book One's diffraction chapter restated as a budget. The zeros are an average part, set by a smooth count that xp reproduces, plus a fluctuating part, set by the primes. xp owns the first line of the budget and contributes nothing to the second.

3.4 The explicit formula as a trace formula

The way the primes supply the fluctuation has a shape, and the shape is the point of the chapter. Book One's explicit formula relates a sum over the zeros to a sum over the primes; in one standard form, for a suitable test function g and its transform $\widehat{g}$,

$$\sum_n g(\gamma_n) = \underbrace{\text{(smooth archimedean term)}}_{\text{the average}} - \underbrace{\sum_p \sum_{k\geq 1} \frac{\log p}{p^{k/2}} \widehat{g}(k \log p)}_{\text{the fluctuation}}. \tag{3.3}$$

Read the two sides in the language of quantum chaos. On the left is a sum over the *energy levels* of a system — the zeros. On the right, the first term is the smooth density of states, the very term xp supplies. The second term is a sum over the primes, and each prime contributes a wave $\widehat{g}(k \log p)$ of "length" $k \log p$, weighted by $(\log p)\, p^{-k/2}$. That is exactly the form of Gutzwiller's trace formula, the central identity of quantum chaos, in which the fluctuating density of states of a chaotic system is a sum over its classical *periodic orbits*: each orbit contributes a wave whose phase is its length and whose amplitude is set by how unstable it is. The dictionary is clean. A primitive periodic orbit is a prime p; its period is $\log p$; traversing it k times gives the repetition $k \log p$; its amplitude $(\log p)\, p^{-k/2}$ is the period times a stability factor $p^{-k/2}$. The explicit formula, in other words, *is* a trace formula — and unlike almost everything else in this book, it is a theorem, stated by Riemann, first proved rigorously by von Mangoldt, and cast in general form by Weil, not a hope.

The primon gas of Book One makes the period $\log p$ concrete rather than metaphorical. Evolve that gas in time: the mode of the prime p rotates by the phase $p^{it} = e^{i(\log p)t}$, so in the gas each prime is literally an oscillator of frequency $\log p$. The "orbit of period $\log p$" is not an analogy borrowed from billiards; it is the natural motion of the prime mode that Book One already built. The primes are oscillators, the zeros are the spectrum that their interference would produce, and Eq. (3.3) is the bookkeeping that ties the two together.

3.5 Why it is a near-miss and not a proof

Everything above is real, and none of it is a proof of anything. The reason is worth stating precisely, because xp and the trace formula are the place where the operator dream comes closest, and seeing exactly how close is seeing exactly where the wall is.

In an honest application of Gutzwiller's formula one starts with a dynamical system, finds its periodic orbits and their periods and stabilities *from the dynamics*, and the trace formula then follows as a theorem about that system. The Riemann case runs the logic backwards. We have the trace formula, Eq. (3.3), as a theorem. We can read off the orbits and their periods — they are the primes, with periods $\log p$. What is missing is the system: no one has exhibited a dynamical system, classical or quantum, whose periodic orbits are the primes with these periods and these amplitudes. The Hilbert–Pólya operator would *be* that system, so "find the system whose orbits are the primes" and "find the operator whose spectrum is the zeros" are the same unsolved problem wearing two costumes. The trace formula is a true identity between two sums you would like to compute; it is not a machine that takes the primes in and hands the zeros out.

Three specific features show that the missing system, if it exists, is not an ordinary one. First, the periodic-orbit sum in Eq. (3.3) carries an overall *minus* sign: the primes enter as subtractions, so the zeros behave less like the emission lines of a spectrum and more like absorption lines — gaps where energy is missing. (It was Connes who turned this sign into a precise spectral statement, the zeros as an absorption spectrum on a large symmetry space; we return to it.) Second, xp is integrable rather than chaotic (and in fact breaks time-reversal symmetry), so although its phase-space volume reproduces the average counting function, it cannot generate the chaotic fluctuations the individual zeros require. Third, xp without its regularization has a continuous spectrum — the whole real line — not the discrete ladder of the γ_n; the discreteness has to be installed by a boundary condition or a cutoff, and every known way of installing it either fails to give the zeros or smuggles in a choice that amounts to knowing them. On the two-part test, then, xp lands in a third position, different from either failure of the previous chapter: it passes independence, and it passes the *average* half of the correct-spectrum condition exactly, while missing the fluctuating half that contains the actual zeros. It is the most honest near-miss the subject has.

3.6 Why chaos is the wall

There is a temptation to read all of this as a near-miss that one more clever idea will close. The last thing this chapter wants to leave is the reason that is unlikely, and it is the same fact that makes the trace-formula picture beautiful: the system, if it exists, is chaotic.

A chaotic system is precisely one whose long trajectories are unpredictable and whose periodic orbits proliferate exponentially, and the trace formula for such a system is a delicate object. Its periodic-orbit sum does not converge in the ordinary sense; it must be handled by analytic continuation, by careful resummation, by using the very spectrum it is meant to produce to make sense of itself. This is visible already in Eq. (3.3): the prime sum on the right converges only conditionally, and giving it a value requires information equivalent to knowing where the zeros are. The identity is genuinely two-directional — it constrains primes by zeros as much as zeros by primes — which is why it cannot be run as a one-way computation. You cannot feed in the primes and grind out the zeros, because the grinding only makes sense once the zeros are in hand.

So chaos sits on both sides of the ledger. It is the best available explanation for why the zeros look the way they do: a chaotic spectrum with broken time-reversal symmetry would have exactly the level repulsion and pair correlation Book One measured. And it is the reason the zeros cannot be computed from below: a chaotic system has no closed form for its individual levels, its trace formula is conditionally convergent and self-referential, and the dynamical system that would anchor the whole picture has never been found. The same word explains the statistics and forbids the shortcut.

That is what it means for this to be a wall and not a gap, and it is why the chapters that follow turn from building the instrument to asking what the instrument could ever measure.

3.7 What this is and is not

What is established is twofold and solid. The regularized semiclassical count of $H = xp$ equals the smooth Riemann–von Mangoldt count of the zeros, Eq. (3.2), term for term; this is Berry and Keating's computation and it is correct. And the explicit formula, Eq. (3.3), is a theorem with the exact structure of a Gutzwiller trace formula, primes for orbits and zeros for levels. Both facts are real mathematics.

What is not established is a Hilbert–Pólya operator, and xp is not one. It supplies the average density of the zeros and not the fluctuation; the fluctuation is the primes, entering as a periodic-orbit sum whose underlying dynamical system is unknown; and the obstructions to finding that system — the absorption-spectrum sign, the chaotic-fluctuation requirement, the missing discreteness — are not technical loose ends but the fingerprints of a chaotic system that no one has constructed. The trace formula is a true and beautiful identity, and it is not a derivation of the zeros from the primes, because chaos makes the identity self-referential. Book One showed that the zeros keep the statistics of a chaotic spectrum. This chapter shows that the explicit formula is the trace formula such a spectrum would obey — and that having the trace formula without the system is exactly as far as the operator dream has come.

Reproducing the figure

```
import numpy as np
from mpmath import zetazero
g = np.sort(np.array([float(zetazero(k).imag) for k in range(1, 41)]))
T = np.linspace(0.5, 100, 2000)

# smooth (Berry-Keating / Riemann-von Mangoldt) count = xp's semiclassical density
def Nsmooth(t):
    return t/(2*np.pi)*np.log(t/(2*np.pi)) - t/(2*np.pi) + 7/8

# Panel A: smooth curve vs actual staircase  (number of gamma_n <= T)
count = np.searchsorted(g, T, side="right")
# Panel B: the fluctuation that xp does NOT supply -- the prime/periodic-orbit term
fluctuation = count - Nsmooth(T)

print(round(float(Nsmooth(100)), 3), int((g <= 100).sum()))   # 29.002  29
```

Chapter 4

An Algebra in the Primes

The natural operator from the primes had the wrong spectrum, and the natural classical Hamiltonian had only the average. This chapter follows the impulse to go bigger — to build not a single operator but a whole algebra of them from the prime data, and to hope that somewhere in its structure the zeros are hiding. The algebra is real and it is beautiful. The prime-mode bilinears close into the rank-two symplectic algebra $\mathfrak{sp}(4, \mathbb{R})$*; its natural representation splits into Dirac's two famous singletons; and because* $\mathfrak{sp}(4, \mathbb{R})$ *is the same Lie algebra as the symmetry algebra of a small curved spacetime, one can dress the entire construction as a little universe living inside the primes, complete with a field, a stress tensor, and a gesture at the cosmological constant. None of that dress is mathematics, and the chapter is careful to say which is which. The test that matters is unmoved: the algebra's own Hamiltonian still has the logarithms of the integers for a spectrum, not the zeros. The chapter ends by scoring every candidate of Part II against the two-part test, and by placing the two programs — Connes' and de Branges' — that try to clear it for real.*

4.1 Scope

This chapter is part construction, part exposition, part warning. The construction is the operator algebra generated by the prime-mode bilinears of the previous chapters, carried up to its full closed form. The exposition is the representation theory that algebra turns out to carry, which is classical and which we credit to its discoverers. The warning concerns what that representation theory does and does not mean: it is genuine structure, and it is also the most inviting place in the whole subject to mistake structure for progress, because the algebra coincides with one that physics has every reason to find interesting. The chapter keeps the mathematics and the physical costume strictly apart. It then steps back to the two serious functional-analytic programmes aimed at the operator dream, Connes' and de Branges', and reports each at its true status. As before, nothing here proves anything about the zeros, and the closing section says so plainly.

4.2 The algebra the primes generate

A word first on what is being built, in plain terms. An *algebra* of operators is a collection closed under the operations one can perform on them—here, taking commutators, the operator measure of how far two operations fail to commute. To say the collection *closes* is to say one never escapes it: combine any two members and the result is a third already in the set. Building one from the primes means asking for the smallest such set the two prime modes generate, and the answer turns out to be a named object carrying a great deal of structure—none of which mentions the zeros, which is the whole point.

Recall the two distinguished modes of the previous chapters: the prime-side spectral function and its complement, the operators we have been calling $\widehat{S}_{\psi}$ and $\widehat{S}_{\psi^{\dagger}}$, built as weighted sums over the primes in two complementary families of residues. Each is a single creation-and-annihilation object on the bosonic Fock space of prime modes. From two such modes one can form Hermitian bilinears — products of one creation or annihilation operator with another — and the question is what algebra those bilinears generate under commutation.

The answer is clean. The independent Hermitian bilinears of a two-mode system number ten, and under commutation they close: the commutator of any two is again a combination of the ten. The algebra they form is $\mathfrak{sp}(4,\mathbb{R})$, the real symplectic Lie algebra of rank two and dimension ten. This is not an analogy or an approximate match; the bilinears satisfy the defining commutation relations exactly, and one can write down a Cartan–Weyl basis — two commuting generators and the four root directions with their conjugates — entirely in terms of the prime-mode operators. The four observables of the earlier chapters sit inside this algebra as four of its ten elements; the remaining six are the squeezing and rotation generators that mix the two modes. The primes, in short, generate a copy of $\mathfrak{sp}(4,\mathbb{R})$, by hand, with no reference to the zeros at all.

4.3 Two famous representations

An algebra is one thing; how it acts is another. That "how it acts" has a name—a *representation*, a way of realizing the abstract generators as concrete operators on a concrete space of states—and the space the prime modes supply carries a famous one. The two-mode Fock space on which these bilinears act carries the *oscillator representation* of $\mathfrak{sp}(4,\mathbb{R})$ — the same representation, built from quantized oscillators, that Segal, Shale, and Weil constructed in general. It is reducible, and it splits in the simplest possible way: by the parity of the number of quanta. The even-quantum states form one irreducible piece, the odd-quantum states the other. In the prime construction these are exactly the symmetric and antisymmetric combinations of the two modes, $\widehat{S}_{\psi}+\widehat{S}_{\psi^{\dagger}}$ and $\widehat{S}_{\psi}-\widehat{S}_{\psi^{\dagger}}$, that the earlier chapters met as the even and odd sectors.

These two irreducible pieces have names, and they are famous ones. They are the *singletons* that Dirac singled out in 1963 and that Flato and Fronsdal later called the

Rac and the *Di*: two representations so small that they have no analogue in ordinary quantum mechanics, each describing a single irreducible tower of states. The Rac is the scalar singleton, with lowest weight $\Delta = \frac{1}{2}$; the Di is the spinor singleton, with lowest weight $\Delta = 1$. The even and odd sectors of the prime two-mode gas are a copy of the Rac and the Di. This is correct, and it is the kind of fact that makes the subject feel as though it is about to open onto something enormous.

4.4 The isomorphism, and the costume

The reason it feels that way is a genuine coincidence of Lie theory. $\mathfrak{sp}(4,\mathbb{R})$ is isomorphic, as a Lie algebra, to $\mathfrak{so}(2,3)$ — one of the small "accidental" isomorphisms among the low-rank classical algebras. And $\mathfrak{so}(2,3)$ is not just any algebra: it is the symmetry algebra of four-dimensional anti-de Sitter space, the curved spacetime of the same name, and equivalently the conformal algebra of its three-dimensional boundary. The Rac and the Di, in that reading, are fields living on the boundary of a small universe. So the prime two-mode gas can be dressed, step by faithful step, as a physical system: a spacetime, a boundary, a Weyl field on that boundary, a Lagrangian for the field, a stress-energy tensor, an energy density — and, with one more assumption about counting degrees of freedom on a cosmological horizon, even a number of the rough size of the observed cosmological constant.

Every arrow in that chain is a real piece of representation theory or differential geometry. And the chain is a costume. The mathematics is this: a ten-dimensional Lie algebra, generated by prime-mode bilinears, acting on a Fock space through a representation that splits into two irreducibles. The spacetime, the field, the horizon, and the cosmological constant are physical names laid over that mathematics; they are suggested by the isomorphism $\mathfrak{sp}(4,\mathbb{R}) \cong \mathfrak{so}(2,3)$ and they are not consequences of the primes. The honest version of the claim keeps only what is derived — the algebra, its representations, the proportionalities among the prime-mode quantities — and refuses the identification with physical space, physical mass, or physical vacuum energy. It is worth saying that this restraint is not a reluctance imposed from outside; the construction's own internal accounting reaches the same verdict, marking the mass as dimensionless and not a particle mass, and the geometry as not identified with the spacetime we live in. A beautiful isomorphism is an invitation to a metaphor, not a derivation of a world.

Producing a striking number. The algebra is generous with numbers, and some of them are striking. Build two real observables from the prime two-mode quantities — a quadratic and a quartic combination of the mode amplitudes, each averaged with the windowed difference of Part III, evaluated at the zeros and again at the midpoints between them — and measure how they move together. At the canonical split modulo 210, with primes counted to $X = 10^5$, the two mode energies come out as $E_+ = 0.16513$ and $E_- = 0.15251$, and the covariance of the two observables is

$$\mathrm{Cov}(Q_2, Q_4) \;=\; -0.82040.$$

A correlation of -0.82 is not a faint signal. It is a near-total anti-alignment, the kind of number that looks like a conservation law: as one observable rises the other falls, almost exactly, as though something were being held fixed between the two modes.

Dressing it as relativity. Here the costume goes on, and for a reader trained in physics it goes on almost by itself. Stack the two mode amplitudes into a single two-component object, $\psi = (a, b)^\top$, and read it as a Weyl spinor — the basic object of relativistic field theory. Contract it with the Pauli matrices $\sigma^\mu = (\mathbb{1}, \sigma^x, \sigma^y, \sigma^z)$ to form a current $J^\mu = \psi^\dagger \sigma^\mu \psi$. Because the two modes are carried on disjoint residues the spatial components J^x and J^y vanish identically, and what survives is

$$J^0 = E_+ + E_-, \qquad J^z = E_+ - E_-.$$

Now take the Minkowski length of this current, contracting with the metric $\eta = \mathrm{diag}(+, -, -, -)$:

$$m^2 \;:=\; J^\mu J_\mu \;=\; (J^0)^2 - (J^z)^2 \;=\; 4\,E_+E_-.$$

Call m a rest mass, J^0 an energy and J^z a momentum, and the relation $E^2 = m^2 + p^2$ — the relativistic dispersion relation, the most recognizable equation in the subject — holds to machine precision across every modulus and cutoff one cares to test. Push further and the mass acquires a closed form: in the limit of large primorial moduli the density factor ties, through Mertens' theorem, to the twin-prime constant C_2, giving

$$m_\infty \;\sim\; \frac{4\,C_2\,e^{-\gamma}\,S(\infty)}{\log p_k} \;=\; \frac{0.770\ldots}{\log p_k}, \qquad S(\infty) = \frac{4\,\mathrm{Cin}(\pi) - \mathrm{Cin}(2\pi)}{8} = 0.519\ldots$$

Here Cin denotes the entire cosine integral, $\mathrm{Cin}(x) = \int_0^x \frac{1-\cos t}{t}\,dt$. A spinor, a conserved current, a Lorentz-invariant mass, a dispersion relation, and a clean closed-form mass with one of the famous constants of prime theory sitting inside it. It is, I will admit plainly, the most convincing thing in either book. When I first assembled it I spent the better part of a day believing the primes had handed me special relativity.

Taking the costume off. They had not. Every equation above is true, and not one of them is physics. The dispersion relation holds to machine precision because it is an identity, not a measurement: $(J^0)^2 - (J^z)^2 = (E_+ + E_-)^2 - (E_+ - E_-)^2 = 4E_+E_-$ is the difference-of-squares rule, true of any two numbers whatever. The Minkowski minus sign that makes it look Lorentzian was put in by hand, the moment we chose to contract with $\eta = \mathrm{diag}(+, -, -, -)$ rather than with the ordinary dot product; the "spinor" is two real energy sums; the spatial components vanished by construction, not by any dynamics. The mass is, in a word, kinematically thin: it is built from the mode energies alone and carries no information whatever about the zeros — the very objects the chapter set out to reach. The one genuinely arithmetic ingredient, the closed form, is honest but bare: the specific number 0.770 is fixed by the choice of window and the choice of residue split, both conventions of the construction and neither a physical input. And the striking -0.82 does not even hold still — recomputed at $X = 10^4$ and

$X = 10^6$ it reads -0.76 and -0.70. It is the value the finite computation happens to sit at in this regime, set by which residues are small primes, not a constant of anything.

So the relativity was a translation, not a discovery. The isomorphism $\mathfrak{sp}(4, \mathbb{R}) \cong \mathfrak{so}(2, 3)$ is real, the Weyl-spinor contraction is legitimate notation, and the dressing is internally flawless — which is precisely what makes it dangerous. A Lie algebra appearing in the primes is not a spacetime arriving; a minus sign one has inserted oneself is not a discovery that the world is Lorentzian; and a number that drifts with the cutoff is not a law. This is the warning the chapter is really for, and it is sharpest for the reader most fluent in the costume: the more relativity you know, the more readily this assembles itself, and the more carefully you must ask which of its symbols you put in and which the primes actually gave you.

4.5 Does the algebra reach the zeros?

Put the costume aside and ask the only question that matters. This algebra is defined intrinsically from the primes, so it passes the independence half of the two-part test without effort. Does it pass the other half — is the spectrum of its natural Hamiltonian the zeros?

It is not. The Hamiltonian that generates time evolution on this Fock space is the same one from the operator chapter, the primon-gas Hamiltonian, and its spectrum is the logarithms of the integers, $\{\log n\}$, exactly as before. Enlarging the single operator to a ten-dimensional algebra, and recognizing two celebrated representations inside it, changes nothing about that spectrum. The algebra is richer; the spectrum is identical; and the spectrum is the wrong one. This is the sharpest form of the lesson the operator chapter began. Structure is not the obstacle and never was. One can have as much structure as one likes — a closed Lie algebra, a famous pair of representations, an entire borrowed spacetime — and still not have an operator whose eigenvalues are the zeros. Reaching the zeros is a different kind of achievement from building structure, and no amount of the second is any of the first.

Figure 4.1 scores the candidates of Part II against the two conditions. The made-to-order kernel has the spectrum and no independence; the primon gas, with or without its $\mathfrak{sp}(4, \mathbb{R})$ dress, has independence and the wrong spectrum; Berry and Keating's xp has independence and the average but not the fluctuation. The one candidate in the corner that passes both — Frobenius, whose eigenvalues are the zeros and which is defined without them — lives only over a finite field, where the previous chapters located the single world in which the Riemann Hypothesis is a theorem. Over the rational numbers the corner is empty.

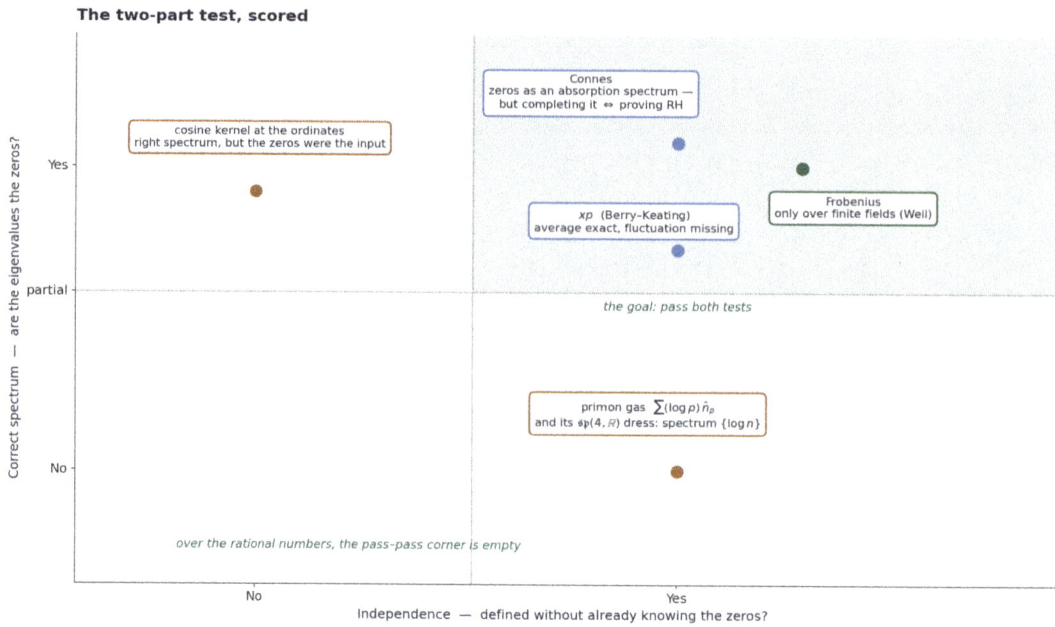

Figure 4.1: Part II against the two-part test. Horizontal: is the operator defined without already knowing the zeros? Vertical: are its eigenvalues the zeros? The made-to-order cosine kernel has the spectrum but no independence (upper left). The primon gas, and its $\mathfrak{sp}(4, \mathbb{R})$ enlargement, have independence and the spectrum $\{\log n\}$ (lower right). Berry–Keating's xp is independent and gets the average density exactly but not the fluctuating part (right, partial). Connes' construction is independent and exhibits the zeros as an absorption spectrum, but completing it is equivalent to proving the Riemann Hypothesis. The only *unconditional* occupant of the pass–pass corner is Frobenius — and only over finite fields. Over $\mathbb{Q}$ that corner is empty.

4.6 The two programmes that aim at the corner

The hand-built algebra is a toy, useful for seeing exactly how structure and spectrum come apart. Two serious programmes aim at the empty corner with the full weight of modern analysis, and both deserve their true status rather than a slogan.

Connes' programme places the problem in noncommutative geometry. On a large symmetry space attached to the rational numbers — the adele class space — there is a natural action whose spectral content can be computed, and in it the zeros of ζ appear not as emission lines but as an *absorption spectrum*: as the frequencies missing from an otherwise continuous background. This is the precise form of the minus sign noticed in the trace-formula chapter, where the primes entered as subtractions. Connes' theorem is that the Riemann Hypothesis is equivalent to the positivity of a particular functional — the Weil form of the explicit formula, read as a trace. That is a real and deep reformulation. It is also, exactly, a reformulation: the construction is independent of the zeros, and the zeros do appear in its spectrum, but establishing that the absorption spectrum is complete and sits on the critical line is equivalent to the Riemann Hypothesis itself. Connes' work moves the problem into a setting where it can be stated as a single positivity, and the positivity is as hard as the thing it would prove. On the scorecard it sits at the very corner, reachable only by an argument that would already be the proof.

De Branges' programme is functional-analytic. His spaces of entire functions — a theory powerful enough that he used it to settle the Bieberbach conjecture — come with structure functions whose positivity, in the right generality, would force the zeros of ζ and of the Dirichlet L-functions onto the critical line. The difficulty has been specific and documented: the positivity conditions de Branges required were shown by Conrey and Li not to hold in the form the argument needs, so the proposed proofs are not accepted. The machinery itself is sound and valuable; the application to the Riemann Hypothesis is unproven, and particular versions of it have been refuted. It is neither a success nor a dead end — it is an open programme with a known obstruction, which is the honest thing to call it.

4.7 What this is and is not

What is established is a chain of correct mathematics. The prime-mode bilinears close into $\mathfrak{sp}(4,\mathbb{R})$; the two-mode oscillator representation splits into the Rac and the Di singletons; $\mathfrak{sp}(4,\mathbb{R})$ is isomorphic to $\mathfrak{so}(2,3)$. Each of these is a theorem, and together they are a genuinely elegant structure sitting inside the primes. Connes' equivalence of the Riemann Hypothesis with a trace positivity, and the de Branges theory of entire-function spaces, are likewise real mathematics of a high order.

What is not established is anything about the location of the zeros. The $\mathfrak{sp}(4,\mathbb{R})$ algebra, for all its structure and its borrowed spacetime, has a Hamiltonian whose spectrum is $\{\log n\}$; it passes independence and fails the spectrum, and the famous representations do not change that. The spacetime, the field, and the cosmological constant that the isomorphism invites are a costume over the representation theory and carry no inference about the primes. Connes' programme converts the problem into a positivity that is exactly as hard as the Riemann Hypothesis; de Branges' converts it into a positivity that is known, in the proposed form, to fail. The operator dream ends Part II where the scorecard leaves it: the corner that would settle everything is occupied only over finite fields, and over the rationals the most beautiful algebra we can build from the primes still does not reach the zeros. Part III stops trying to build the instrument and asks instead what such an instrument, if we had it, could ever measure.

On the figure

Figure 4.1 is a schematic, not a computation: it places each operator of Part II by how it fares on the two conditions argued through Chapters 2–4. The one numerical anchor behind it is the spectrum of the primon-gas Hamiltonian, $\{\log n\}$, established in the operator chapter from unique factorization, and shared without change by the $\mathfrak{sp}(4, \mathbb{R})$ enlargement of this chapter.

Reproducing the number

The mode energies and the covariance are the book's own computation; the program below produces them at the canonical split modulo 210 with primes counted to $X = 10^5$ (the thousand zeta zeros take a few minutes to generate via `mpmath`). Changing X reproduces the drift the previous section describes — the -0.82 does not hold still.

```
import numpy as np
from math import gcd, log, sqrt
from mpmath import zetazero, mp

M, X = 210, 10**5

# substrate: the twin-corridor split of the wheel
Zstar = [r for r in range(1, M) if gcd(r, M) == 1]          # 48 reduced residues
Zset  = set(Zstar)
T     = [r for r in Zstar if (r+2) % M in Zset or (r-2) % M in Zset]
S_psi = {r for r in T if r < M/2}        # psi sector       (15 residues)
S_pd  = {r for r in T if r > M/2}        # psi-dagger sector (15 residues)

# primes coprime to M, with the sin^2 taper
def primes_to(N):
    s = np.ones(N+1, bool); s[:2] = False
    for i in range(2, int(N**0.5)+1):
        if s[i]: s[i*i::i] = False
    return np.where(s)[0]
primes = np.array([p for p in primes_to(X) if gcd(int(p), M) == 1])
logp  = np.log(primes.astype(float))
w     = np.sin(np.pi * (logp/log(X)) / 2)**2
c     = w / np.sqrt(primes.astype(float))
res   = primes % M
inpsi = np.array([int(r) in S_psi for r in res], float)
inpd  = np.array([int(r) in S_pd  for r in res], float)
Ep = float(np.sum((c*inpsi)**2))      # E_+
Em = float(np.sum((c*inpd )**2))      # E_-

# first 1000 zeta zeros; zero-centres and midpoint-centres
mp.dps = 15
g = np.array([float(zetazero(n).imag) for n in range(1, 1001)])
K, step = 12, 23
W = (1000 - 120)//step + 1
zc = np.array([g[j*step+k]                        for j in range(W) for k in range(K)])
mc = np.array([0.5*(g[j*step+k]+g[j*step+k+1]) for j in range(W) for k in range(K)])
n  = len(zc)

# windowed-difference covariance of Q1..Q4
norm = 4.0/sqrt(Ep*Em)
cpsi = (c*inpsi).astype(complex); cpd = (c*inpd).astype(complex)
def cov(centres):
    ph = np.exp(1j*np.outer(centres, logp))
```

```
    sp, sd = ph@cpsi, ph@cpd; qp, qd = np.conj(sp), np.conj(sd)
    a, b = norm/2, norm/2j
    AB = np.array([[a*sp,a*sd],[b*sp,-b*sd],[a*qp,a*qd],[-b*qp,-b*qd]])
    CD = np.array([[a*qp,a*qd],[-b*qp,b*qd],[a*sp,a*sd],[b*sp,b*sd]])
    Am = AB[:,0:1,:]*CD[None,:,0,:]*Em + AB[:,1:2,:]*CD[None,:,1,:]*Ep
    return (0.5*np.real(Am+np.conj(Am.transpose(1,0,2)))).sum(2)
C = (cov(zc) - cov(mc))/n

print("E_+ =", round(Ep,5), "  E_- =", round(Em,5))
print("Cov(Q2,Q4) =", round(C[1,3],5))
```

Part III

Hearing and reading

Chapter 5

Hearing the Zeros

Part II could not build the instrument. This part asks what the instrument could measure if we had it, and the answer divides cleanly in two. From the primes alone one can detect *the zeros: a coherence built out of the prime data lights up, at the very floor of statistical significance, whenever it is evaluated within one mean spacing of a zero ordinate, and goes dark elsewhere. The zeros announce themselves; you can hear that they are there. What one cannot do is* resolve *them — separate two nearby zeros, or read off an ordinate to better than the data allows — and this is not for want of trying. The sharpest modern line-finding methods, ESPRIT and atomic-norm super-resolution, applied honestly to the prime-side signal, do no better than the crude periodogram, and in one case slightly worse. The reason is a single limit that ties this chapter to the two before it: the resolution is set by the aperture, the aperture is set by the largest prime in hand, and the fine structure of the zeros sits below it. You can hear the zeros. You cannot read them.*

5.1 Scope

This chapter is part exposition and part a set of honest measurements. It assumes Book One's diffraction picture — primes and zeros as Fourier duals — and the resolution limit hinted at in this book's first chapter, where seeing a symmetry type required not seven L-functions but thousands. The new content is the clean separation of two operations that are easy to confuse: detecting that a spectral line is present, and resolving where it is. The chapter reports one positive result, that a prime-side coherence detects the zeros under a strict null control, and four negative results, that the standard super-resolution methods do not resolve them. It makes no claim about the Riemann Hypothesis: the zeros are computed independently and used only as ground truth, to score detection and resolution against. Section 5.6 states the boundary the chapter draws, which is the resolution wall in the language of signal processing.

5.2 Two questions, easily confused

When a signal contains a pure tone, two different things can be asked of it. The first is *detection*: is there a line near this frequency at all? The second is *resolution*: exactly where is it, and if there are two lines close together, can they be told apart? In ordinary listening the two run together, but they are governed by different limits. Detection needs only that the line carry enough energy to rise above the noise. Resolution needs that the observation be long enough — the aperture wide enough — for two nearby lines to separate. A short recording of a chord tells you a chord is there (detection) without letting you name the two closest notes (resolution). The whole of this chapter is the claim that, for the zeros read off the primes, these two come apart: detection succeeds and resolution fails.

5.3 You can hear them

Build, from the primes in two complementary families of residues, the two prime-side signals of the earlier chapters, and form a coherence between them — a measure of how strongly the two move together — as a function of the height t at which they are evaluated. Now ask whether that coherence knows about the zeros. It does, and sharply.

Restricted to heights within one mean spacing of a zero ordinate, the coherence is large and statistically unmistakable. Tested against a null model that randomly flips the signs of the contributing terms — a Rademacher null, run for twenty thousand replications — the observed coherence is more extreme than every single replicate, pinning its significance at the Monte Carlo floor, a p-value of 5×10^{-5}. Restricted instead to heights more than one mean spacing away from any zero, the same coherence collapses to the null: it is indistinguishable from the sign-flipped surrogates. The signal is null-controlled and grid-replicated, so the effect is not an artifact of one window or one character. In plain terms: the primes, combined this way and listened to at the heights of the zeros, light up; listened to anywhere else, they are silent.

This is the explicit formula of Book One made audible. Primes and zeros are Fourier duals, so a prime-side quantity evaluated at the dual frequencies — the zero ordinates — interferes constructively, and the coherence is the sound of that interference. It is a genuine detection of the zeros from the prime side, and it is exactly as much as the prime side will give.

5.4 You cannot read them

Detection invites a hope: if the coherence rises near a zero, perhaps a sharp enough method can turn the rise into a precise location, or split two zeros that lie close together. The hope is natural, and over a careful study it fails four times.

First, separating the prime-side signal into orthogonal channels — demixing, the obvious move to clean up the data — does not buy resolution. Across thirty-six configurations, sweeping characters, window lengths, and model orders, the standard subspace method ESPRIT applied to the *undemixed* signal resolved 52.5% of close pairs, while ESPRIT on the demixed channel resolved only 42.5%. Demixing slightly *hurt.* It is a tool for telling the contributions of different L-functions apart, not for telling two nearby zeros apart.

Second, exploiting the wheel's mirror symmetry, which halves the effective bank of corridors, reduces the cross-talk between opposite-parity characters by a mere 3.7 percentage points and yields no resolution advantage at all.

Third, the most celebrated modern super-resolution method — atomic-norm minimization, the convex approach of Tang, Bhaskar, Shah, and Recht that provably beats the classical limit under the right conditions — failed to converge to a usable answer on the prime-side problem, returning degenerate spectra even on a synthetic test signal where it is known to work. The honest reading, and the one the study records, is that we cannot tell a fundamental obstruction from a solver shortfall; either way, no resolution was gained.

Fourth, stacking many channels and reweighting them by prime-gap information — gap-weighted multichannel ESPRIT — does not resolve close pairs either, and the reason is the one that matters most. The channels carry different amplitudes, but they all share the same finite-window basis, and so they share the same systematic errors. Adding channels that all look through the same window cannot widen the window.

5.5 The aperture

That last reason is the whole chapter. Resolution is set by the aperture — the span over which the signal is observed — and for a spectrum read off the primes, the aperture is fixed by the largest prime in hand: it is roughly the logarithm of that prime. Two lines separated by less than the inverse aperture merge into a single bump; no amount of cleverness applied to the same finite data widens the window that produced it.

Figure 5.1 makes the limit visible on two zeros placed close together. Observed through the short aperture the primes actually provide, the two produce a single rounded peak: the coherence detects that there is spectral content here, and cannot say there are two lines rather than one. Observed through an aperture wide enough to separate them, the two stand apart as clean peaks — but that aperture would require primes out to astronomical size, far beyond any computation. The super-resolution methods of the previous section are precisely the attempts to fake the wide aperture from the narrow data, and on this signal they do not succeed.

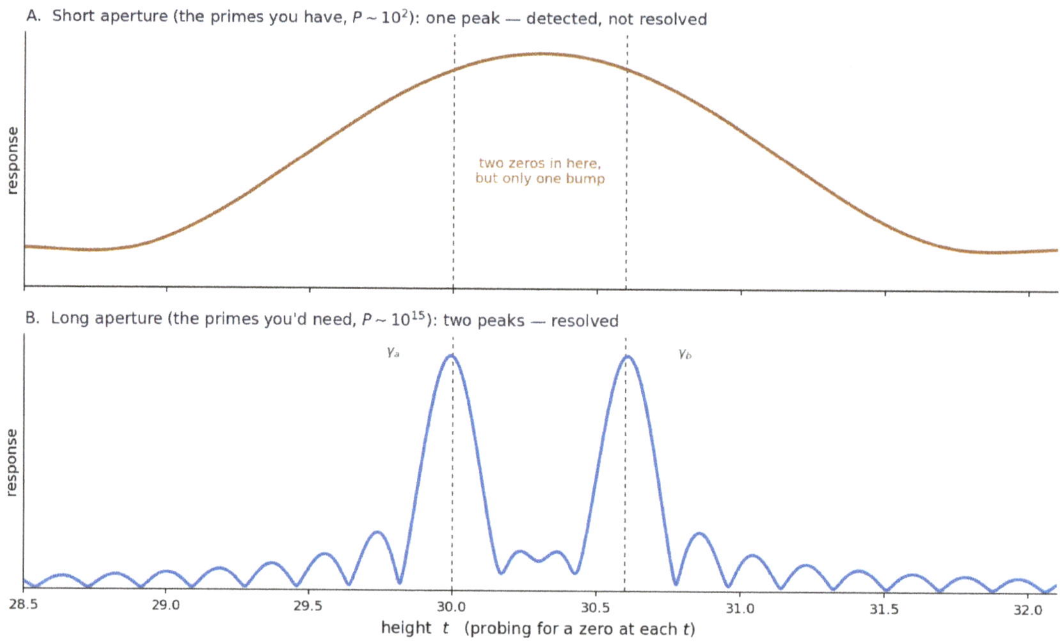

Figure 5.1: Detection without resolution. Two nearby zeros, probed by a prime-side coherence. **A:** through the short aperture the available primes provide, the two merge into one bump — their presence is detected, their separation is not resolved. **B:** through an aperture wide enough to resolve them, they appear as two peaks — but that aperture needs primes of astronomical size. The dashed lines mark the two true ordinates. Resolution is an aperture-limited quantity, and the aperture is the logarithm of the largest prime in hand.

This is the same wall met twice before, now in the language of signal processing. The first chapter needed thousands of L-functions, not seven, to see a symmetry type, because a density cannot be read from too few samples. The trace-formula chapter found a chaotic spectrum whose individual levels have no closed form and cannot be extrapolated. Here the same limit appears as an aperture: the average, coherent behaviour of the zeros is available to detection, and their individual fine positions sit below the resolution floor set by the finite prime data. The three descriptions — too few samples, no closed form, too narrow an aperture — are one fact wearing three coats.

5.6 What this is and is not

What is established is a detection and four non-resolutions, all measured. The prime-side coherence detects the zeros within one mean spacing, at the floor of a twenty-thousand-replicate null; this is a real, reproducible, null-controlled effect. The four resolution attempts — demixing, mirror symmetrization, atomic-norm minimization, gap-weighted multichannel ESPRIT — are honest negative results, reported as such, including the one where method and solver cannot be cleanly separated. None of this touches the Riemann Hypothesis. The zero ordinates were computed independently to high precision and used only as the ground truth against which detection and resolution were scored; nothing here certifies a zero on the critical line or bounds one away from it.

What this is, finally, is the line between hearing and reading. You can confirm, from the primes alone, that a zero sits at a given height — the coherence rises and the null rejects chance. You cannot, from the same primes, resolve two zeros that lie within an aperture of each other, or predict where the next one falls. The same chaos that gives detection a coherent signal to latch onto is what denies resolution a shortcut: a chaotic spectrum is detectable in the average and unpredictable in detail. The instrument Part II could not build would, even if built, face this second wall. The next chapter turns to what happens when one forgets the wall and mistakes a detected pattern for a law.

Reproducing the figure

```
import numpy as np

# two zeros placed close together near height 30
w1, w2 = 30.0, 30.6

# matched-filter (periodogram) response of a finite-aperture observation:
#   aperture L = log(largest prime).  short = the primes you have;
#   long  = the primes you would need to separate the pair.
def response(L, probe):
    tau = np.linspace(0, L, 6000); dt = tau[1] - tau[0]
    s = np.cos(w1*tau) + np.cos(w2*tau)
    return np.array([abs((s*np.exp(-1j*w*tau)).sum()*dt) for w in probe])

probe = np.linspace(28.5, 32.1, 1400)
short = response(5.0,  probe)   # largest prime ~ e^5  ~ 1.5e2  -> one bump
long  = response(34.0, probe)   # largest prime ~ e^34 ~ 6e14   -> two peaks
```

The figure isolates the aperture limit. The resolution percentages and the complement-symmetry figure quoted earlier in the chapter come from a longer sweep across characters, window lengths, and model orders; that sweep, and the symmetry test, are reproduced by the verification programs accompanying this book.

Chapter 6

Laws That Died

The previous chapter showed that the primes detect the zeros: patterns are there to be found. This chapter is about the trap that opens the moment a pattern is found, which is to mistake it for a law. A pattern is whatever the finite data happens to show; a law is what survives being pushed. The difference is measured by pushing, and most apparent spectral laws do not survive. This chapter is a catalogue of ones that did not: a clean cross-modulus relation that vanished when the prime cutoff was raised; a rule that held to a prime cutoff of a billion and then broke the instant the modulus was lifted to the next wheel; a ranking that inverted depending on which of two observables was used to read it; a contribution ruled out because it came up a thousandfold too small; a celebrated pattern from elsewhere in the subject that simply was not present when looked for; and a near-perfect correlation that turned out to be an artifact of which residues happen to be small primes. None of these is a failure to be hidden. The running, numbered tally of them is the most honest instrument the subject has, and keeping it is what separates the work from numerology.

6.1 Scope

This chapter is a set of negative results, each reported as a measurement. It assumes nothing beyond the observables of the earlier chapters and the wheel of Book One. Its content is six patterns that looked like laws and were killed by one or another standard test — raising the prime cutoff, lifting the modulus, changing the observable, checking the magnitude, looking for a known pattern, asking whether a striking number is structure or artifact — together with the discipline of recording such deaths in a numbered list. The chapter proves nothing new; that is the point. It is the inventory of what the methods of this book throw up that does not hold, and Section 6.9 draws the lesson for the three walls: they are hard in part because the false laws on the way to them are so plausible.

6.2 A relation that died when the cutoff rose

The simplest death comes from raising the prime cutoff. Working at primes up to 10^5, one finds a clean numerical relationship between observables computed at two different moduli — a cross-modulus shape similarity, agreeing to within a few percent, exactly the kind of tidy fact one would want to promote to a structural law. Raise the cutoff to 5×10^5 and the relationship is gone. It was a feature of the smaller sample, not of the integers, and the honest record marks it as retracted rather than burying it. The first test any candidate law must pass is that it not depend on how far up the primes you happened to count, and this one failed it.

6.3 A rule that survived a billion and died at the next wheel

The second death is sharper, because the pattern survived a far more demanding version of the first test and died on a different axis entirely. It is also, on the wheel of 210, genuinely beautiful, which is what makes its death worth telling.

Recall from Book One that the seven real characters of the wheel are indexed by their conductors, and that every conductor is a product of a subset of the three odd primes $\{3, 5, 7\}$: conductor 7, conductor $15 = 3 \cdot 5$, conductor $105 = 3 \cdot 5 \cdot 7$, and so on through all seven non-empty subsets. The conductors form a Boolean algebra — the lattice of subsets of $\{3, 5, 7\}$ — and one can ask whether the *sign* of the spectral observable is a Boolean function of which primes a conductor contains. It is, and a strikingly simple one. Writing x_p for "the conductor contains p," the observable is positive exactly when

$$x_5 \oplus x_7 = 1, \tag{6.1}$$

that is, when the conductor contains exactly one of 5 and 7. The prime 3 plays no role at all; it only pairs the characters within each cell of the rule. Figure 6.1 is the rule as a truth table, with each of the seven characters in its place.

This is exactly the kind of clean statement one wants to promote to a law, and it was tested as one. The rule was written down and time-stamped *before* the decisive run, then checked at a prime cutoff of 10^9 — several thousand zeros per character, hours of computation each — and it held, seven characters out of seven. One character, of conductor 21, sat on a knife-edge: its observable was the smallest of all seven and had changed sign at lower cutoffs, so the sign-only prediction was caught correctly, but by a hair. Another, of conductor 35, was the lone character never to change sign across every cutoff from 10^4 to 10^9. A pre-registered parity law, clean to a billion, is the sort of evidence that ordinarily settles a question.

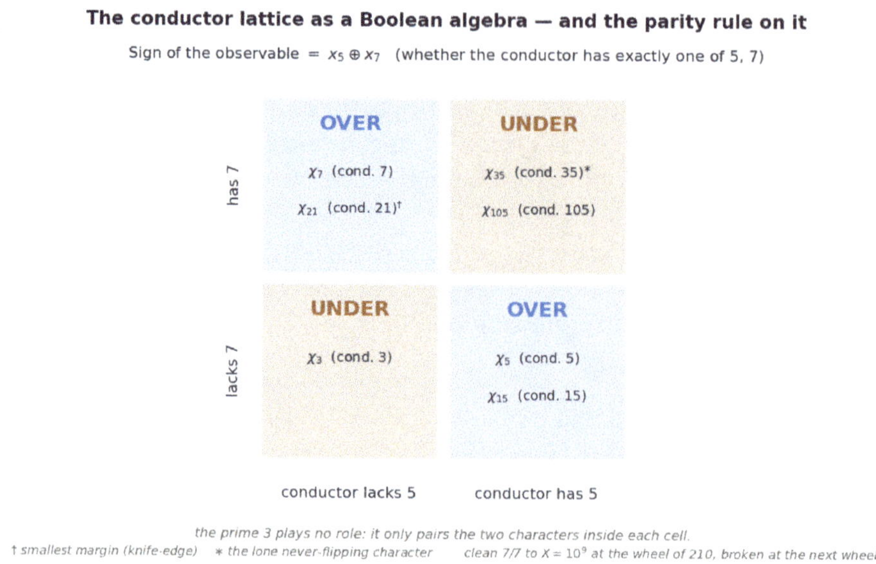

Figure 6.1: The parity rule on the conductor lattice of the wheel of 210. The seven characters are the seven non-empty subsets of $\{3, 5, 7\}$; the sign of the observable is positive (OVER) exactly when the conductor contains exactly one of 5 and 7, Eq. (6.1). The prime 3 is irrelevant: it only pairs the two characters inside each cell. Clean across all seven characters to a prime cutoff of 10^9 — and broken at the next wheel.

Then lift the modulus. Carry the same rule, transcribed literally, from the wheel of 210 to the next wheel, $2310 = 2 \cdot 3 \cdot 5 \cdot 7 \cdot 11$, and it breaks. It does not weaken or blur; it fails outright, on several of the original characters and on the eight new ones the larger wheel introduces. A brute-force search over the fifteen characters of the new wheel for the best simple Boolean replacement turns up a different, weaker rule — positive unless the conductor contains 7 — which fits thirteen of fifteen, with two stubborn exceptions, and which is significant only after correcting for the handful of simple rules one tried, not against the full space of possibilities. The clean parity law of the wheel of 210 has become, at the wheel of 2310, a rough rule of thumb. (Stranger still, the wheel of 1155, which is 2310 with the factor of two removed, behaves almost identically, the two agreeing to within a few parts in ten thousand — so the failure tracks the odd part of the modulus, not the size.)

The lesson is that the prime cutoff and the modulus are two independent axes of stress, and surviving one is no guarantee of surviving the other. A pattern can be true for every prime up to a billion on one wheel and false on the next. Holding to 10^9 proved only that the rule was not an artifact of small primes; it said nothing about whether the rule was an artifact of the wheel 210 itself, and it was.

6.4 A verdict that inverts

The third death needs no new data at all — only a second way of reading the same data. Throughout these chapters a "canonical" arrangement of the residues has been compared against a thousand random arrangements, and rated by where it falls in the random distribution: a high percentile means the canonical arrangement is special, a percentile near the middle means it is ordinary. The verdict depends, it turns out, on which observable does the rating.

Figure 6.2 shows two observables — a signed bilinear quantity and its absolute-value cousin — rating the same canonical split. On the wheel of 210 they agree: both place it high, at the 99.9th and 92.8th percentiles, both calling it special. On the wheel of 2310 they invert. The signed observable rates the canonical split at the 100.0th percentile, above all thousand random arrangements; the absolute-value observable rates the identical split, on the identical data, at the 43.7th percentile — below the random average. One observable says this arrangement is more extreme than anything chance produces; the other says it is slightly less special than a coin flip. A property that flips from "extraordinary" to "below average" depending on how you choose to measure it is not a property of the arrangement. It is a property of the measurement, and a candidate law that survives only under one of two equally natural observables has already failed.

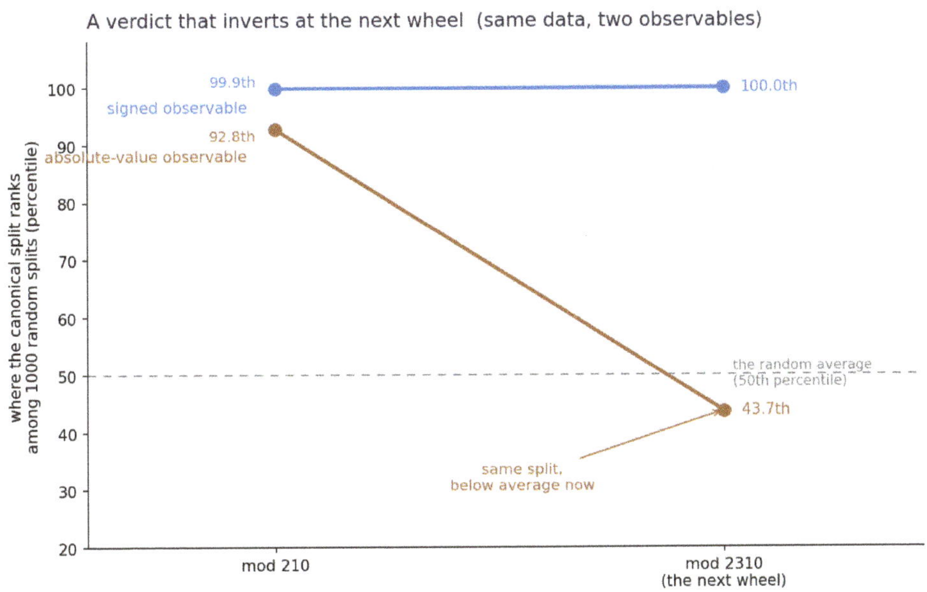

Figure 6.2: The same canonical split, rated by two observables, on two wheels. On the wheel of 210 both rate it special (top left). On the wheel of 2310 the signed observable still rates it at the very top (100.0th percentile) while the absolute-value observable drops it below the random average (43.7th). Same data, opposite verdict: the ranking is a fact about the observable, not about the split.

6.5 An idea killed by its size

The fourth death is the cleanest, because it is not subtle. One of the most natural ways to try to explain a residual gap in the data is to attribute it to the archimedean factor — the gamma-function piece and root number that complete an L-function at the infinite place. It is a real contribution and an obvious suspect. So one computes it, for all seven real characters of the wheel, and compares its size to the gap it is meant to explain. It is more than three orders of magnitude too small. There is no fitting to be done and no parameter to adjust; the candidate explanation is simply the wrong size by a factor of thousands, and it joins the running list of tested-and-falsified hypotheses — a magnitude failure, a different kind of death from the change-of-wheel collapses tallied later in the chapter. Not every death is a subtle reversal at scale. Some are a number that is off by a thousandfold, and the discipline is to check the magnitude before believing the story.

6.6 A pattern that was not there

The fifth death is of a pattern that belongs to someone else. In 2022 a striking correlation was discovered by He, Lee, Oliver, and Pozdnyakov, using machine learning, in the data of elliptic-curve L-functions: averaged over a family and plotted against the prime p, for curves in a fixed conductor window, the coefficients trace out a faint, coherent, oscillating shape, since named a murmuration. It is a real and now much-studied phenomenon, and its discovery is exactly the kind of family-wide structure the methods of this book are built to look for. So one looks, in the families natural to the wheel. The murmuration is not there. The honest record states it plainly — no such correlation is found, and no claim is made that the wheel's families murmurate. Looking for a beautiful pattern known to exist elsewhere, and reporting its absence here, is as much a result as finding one, and a good deal rarer in practice.

6.7 A number that was an artifact

The sixth death was promised two chapters ago. The hand-built algebra of the operator part threw up a striking number: a correlation of about -0.82 between two of its observables, a near-total anti-alignment that looks for all the world like a conservation law binding the two modes. Pushed, it dissolves. The two channels of that algebra are built from two complementary families of residues, and the family on one side consists of residues that are themselves small primes — 11, 13, and their kin — while the family on the other side contains no small prime at all below one hundred. The strong anti-alignment is a consequence of that lopsided arithmetic: one channel is carried by the smallest, heaviest-weighted primes and the other is not. Change which residues fall on which side — lift the modulus, or redraw the families — and the -0.82 moves. It was never a law coupling two abstract observables; it was a shadow of which numbers happen to be small primes. A striking constant is the most seductive kind of pattern, and the test it must pass is whether it stays put when the arithmetic underneath it is changed.

6.8 Keeping a body count

Six deaths, six different tests: the prime cutoff, the modulus lift, the choice of observable, the magnitude, the search for a known pattern, the change of arithmetic. What ties them together is not that the methods are weak — they are the same methods that detected the zeros in the previous chapter — but that finite data over a small modulus will always show patterns, and the chaotic richness of the zeros guarantees that many of those patterns will look meaningful. The only defense is to try, systematically, to kill every pattern one finds: to push the cutoff, lift the wheel, swap the observable, check the size, compare against a null, and ask whether a number is structure or coincidence. The findings that survive all of it are few, and the ones that do not deserve to be written down, in order, with their cause of death.

That is why the most honest instrument in this subject is a numbered list of tested-and-falsified hypotheses, kept openly, with retractions marked in the text rather than quietly deleted. The list is real and it is specific. The exclusive-or law of this chapter was, when it broke, the *ninth* entry in one such running tally — a tally of patterns clean on the wheel of 210 that failed to survive a change of wheel, joining a factor-two hypothesis, a seam-pair relation, a character-concentration effect, the residue pairing of the next section, and several others before it. A program that records its dead ends in order is doing the one thing that separates this work from numerology, which is distinguishing what it has shown from what it has merely seen. The instinct to promote a clean coincidence to a law is exactly the instinct that the three walls punish, and the habit of killing one's own patterns is the habit that keeps the walls in view.

6.9 What this is and is not

What is established is that each of these six patterns failed a specific, standard test, and that the failures are recorded as measurements: the cross-modulus relation did not survive a fivefold increase in the prime cutoff; the exclusive-or rule held to a cutoff of 10^9 on one wheel and broke on the next; the canonical ranking inverted between two observables on the same data; the archimedean contribution was too small by three orders of magnitude; the murmuration was absent from the wheel's families; the -0.82 correlation was an artifact of small primes. These are real results, and reporting them is not a confession but the method working as intended.

What is not established is any law. Not one of these patterns is a theorem, and the chapter is deliberately a catalogue of near-laws rather than laws. The lesson it leaves for the rest of the book is about the walls. The Riemann Hypothesis, the twin-prime conjecture, and the hardness of factoring are not surrounded by a desert in which nothing happens; they are surrounded by a thicket of plausible patterns, clean coincidences, and striking numbers, most of which die when pushed. That thicket is part of why the walls are hard: it is genuinely difficult to tell, from inside a finite computation, the pattern that is the edge of a theorem from the pattern that is the edge of nothing. The remaining chapters walk up to the three walls themselves, having learned to distrust the view.

On the figure

Figure 6.2 plots reported percentile ranks, not a fresh computation: the canonical split's standing among 1000 random splits, under the signed observable (99.9th at mod 210, 100.0th at mod 2310) and the absolute-value observable (92.8th at mod 210, 43.7th at mod 2310), all at prime cutoff $X = 10^5$. These are the measured values discussed in the text; the chart simply draws the inversion between the two wheels.

Part IV

The walls

Chapter 7

A Family of Walls

Book One ended at three walls. This part walks up to them again through the spectral lens, and the first thing the lens shows is that the Riemann wall is not one wall. The spectral signal of the wheel — the coherence at the zeros, the cancellation in the observables, the prime race of Book One — is built not from ζ alone but from the whole family of Dirichlet L-functions the wheel carries, and to control it one would need the Riemann Hypothesis for every one of them at once. That is the Generalized Riemann Hypothesis, strictly stronger than the original and exactly as open. The chapter makes the dependence concrete. A handful of subprime residues do most of the cancelling, dominated by $209 = 11 \cdot 19$; and 209 matters because the characters of conductor 11 and 19 carry the two L-functions whose zeros push hardest against ζ's. Worse for any hope of a shortcut, which L-functions oppose ζ is not even fixed: lift the wheel from 210 to 2310 and the clean split dissolves. The spectral approach does not reduce the Riemann wall to something smaller. It reveals it as a coupled, wheel-dependent family of walls.

7.1 Scope

This chapter is compressed and assumes Book One's closing chapter, where the three walls were named, and its prime-race chapter, where the race was found to wander at the mercy of the zeros of the wheel's L-functions. It does not re-teach the Riemann Hypothesis; it views the wall through the observables of this book. The content is a single concrete dependence, traced in three steps: which residues carry the spectral cancellation, which L-functions among the wheel's family move against ζ, and how both reorganize when the modulus is lifted. The conclusion is that the spectral structure of the wheel is hostage to the Generalized Riemann Hypothesis, and Section 7.8 is careful that this is a door into the wall, not a passage through it. Nothing here proves, approaches, or weakens any form of the hypothesis.

7.2 The wall is not just zeta

Book One's prime race already stood at this wall without quite naming it. The race among the residue classes wandered, rather than settling, because its fluctuations are governed by the nontrivial zeros — and not only the zeros of ζ, but the zeros of all the Dirichlet L-functions attached to the wheel's characters. The same is true of every spectral quantity in this book. The coherence that detected the zeros, the cancellations that filled the observables, the rankings that inverted in the last part: all of them are sums over the primes sorted by residue, and the behaviour of such sums is controlled, through the explicit formula, by where the zeros of the corresponding L-functions sit.

So the wall that matters for the wheel is not the Riemann Hypothesis alone. It is the *Generalized* Riemann Hypothesis: the statement that *every* Dirichlet L-function has its nontrivial zeros on the critical line. The ordinary hypothesis is the single case of the trivial character; the generalized one asks the same of the whole family at once. It is strictly stronger, and it is just as unproved. Reading the wheel spectrally does not let one trade the Riemann wall for an easier obstacle; it commits one to the harder one.

7.3 Which numbers do the cancelling

The dependence can be made concrete by asking which residues actually carry the spectral cancellation. Among the residues coprime to 210, a few are composite — subprime, in the vocabulary of these books: composite, but with no factor below eleven. The smallest of them are $121 = 11^2$, $143 = 11 \cdot 13$, $169 = 13^2$, $187 = 11 \cdot 17$, and $209 = 11 \cdot 19$. When one measures how much each residue class contributes to the cancellation in the relevant observable, the work is done almost entirely by two of these: 169 and 209. Removing the primes in class 169 alone shifts the cancellation by some seventy per cent, and per prime the class 209 carries about three times the weight of 169.

A natural first guess is that this is about size — that the residue with the largest prime factor should dominate. It is not. The reason 209 outweighs 169 is not that 19 exceeds 13 but that of the wheel's characters, the ones that matter are the characters of conductor 11 and conductor 19, and $209 = 11 \cdot 19$ is built from both of them, while $169 = 13^2$ is built from the single, weaker conductor 13. The driver is not the magnitude of the prime factors. It is which characters' L-functions the residue is made of.

7.4 Which L-functions move against zeta

That sends the question one level deeper, to the L-functions themselves. Here *the wheel's characters* means the small set of low-conductor quadratic characters $\chi_{-3}, \chi_{-4}, \chi_{-7}, \chi_{-11}, \chi_{-19}$ — a spread of internal and external conductors, 3 and 7 dividing 210 and 4, 11, 19 not — and not the seven real Dirichlet characters of Chapter 6. These five expose the internal/external split rather than exhausting the wheel: the remaining external conductors, 13 and 17 behind the subprimes 169 and 187, are left off the figure to keep it readable — 13 being the single, weaker conductor the previous section identified behind 169. Measure the correlation between each of the wheel's L-functions and ζ, across the same windowed grid of zero heights used throughout, and a clean pattern appears at the wheel of 210 (Figure 7.1, left). The characters whose conductor divides 210 — the internal ones, conductors 3 and 7 — move *with* ζ, correlations +0.70 and +0.58. The characters whose conductor does not divide 210 — the external ones, conductors 4, 11, 19 — move *against* it. And the two strongest opponents, at −0.82 and −0.77, are exactly the characters of conductor 11 and 19: the two factors of 209. The residue 209 dominates the cancellation, then, for a reason that runs all the way down to the zeros: it is the product of the two characters whose L-function zeros push hardest in the direction opposite to ζ's. The cancellation is L-function content, and the content is the joint behaviour of a particular pair of zero sequences.

This −0.82 is not the −0.82 of Chapter 4. That one—the covariance $\mathrm{Cov}(Q_2, Q_4)$ of the operator algebra—was a hand-built artifact that drifts to −0.70 as the cutoff grows; this one is a genuine correlation between an L-function and ζ, and it collapses instead to −0.49 when the modulus is lifted to 2310. Two unrelated quantities, each regime-dependent in its own way, happen to read −0.82 at the canonical wheel: the agreement is a coincidence of value, not a hidden identity—of a piece with the near-laws this chapter has been taking apart.

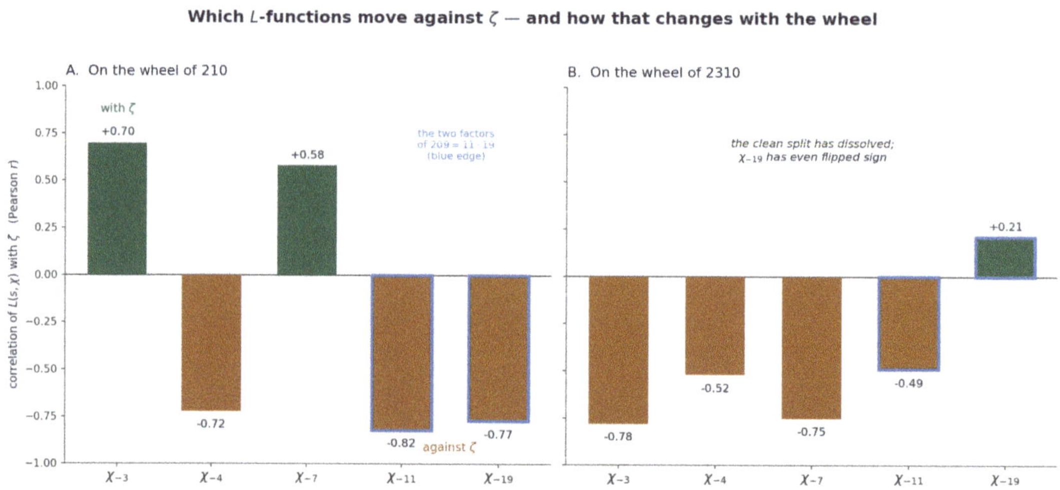

Figure 7.1: The wheel's L-functions, correlated with ζ. **Left (wheel of 210):** a clean split — internal characters (conductors $3, 7$, dividing 210) move with ζ; external characters (conductors $4, 11, 19$) move against it; the two factors of $209 = 11 \cdot 19$, outlined, are the strongest opponents. **Right (wheel of 2310):** the same correlations recomputed after lifting the modulus. The clean split has dissolved — two internal characters have gone negative and χ_{-19} has flipped to positive. Which L-functions oppose ζ is not fixed; it depends on which primes divide the wheel.

7.5 The split dissolves at the next wheel

If that pattern were stable, it would at least be a fixed target: a definite short list of L-functions to worry about. It is not stable. Lift the modulus from 210 to 2310 — adjoin the prime 11 to the wheel — and recompute the same correlations, and the clean split falls apart (Figure 7.1, right). The internal characters that moved with ζ now move against it; the conductor 11, newly internal, weakens; and χ_{-19}, the strong opponent on the wheel of 210, flips to a mild positive correlation. The organizing principle of the left panel — internal with ζ, external against — has simply dissolved.

This is the modulus-lift lesson of the previous part, arriving now at the quantities that control the Riemann wall. It is not only that a given correlation changes value; it is that the *structure* of which L-functions oppose ζ is itself a function of the wheel. There is no wheel-independent fact of the form "these characters are the opponents." Change which primes divide the modulus and the membership of the opposition changes with it, because the internal–external distinction — and with it the whole pattern — is defined relative to the wheel.

7.6 How far the explicit formula gets

So far the chapter has been diagnostic: the signal depends on a whole family of L-functions whose membership shifts with the wheel. The constructive question is sharper. How much of the observable can one actually *predict* from analytic inputs — the explicit formula, the zeros, the characters, with nothing fitted to the answer — and where does the prediction stop? The answer draws the wall in a single picture.

The observable decomposes, over the Dirichlet characters of the wheel, into four pieces supported on disjoint parity classes; the decomposition rests on an exact parity rule among the characters and the vanishing of the mixed-parity terms. One of the four pieces is special: the principal-character piece, built from the trivial character, whose L-function is ζ itself (stripped of the Euler factors the wheel removes). That piece, and only that piece, can be written directly through a Riemann-style explicit formula — an amplitude expressed as a sum over the zeros of ζ against an explicit kernel — and evaluated from the first thousand zeros. Figure 7.2 is the result of doing so.

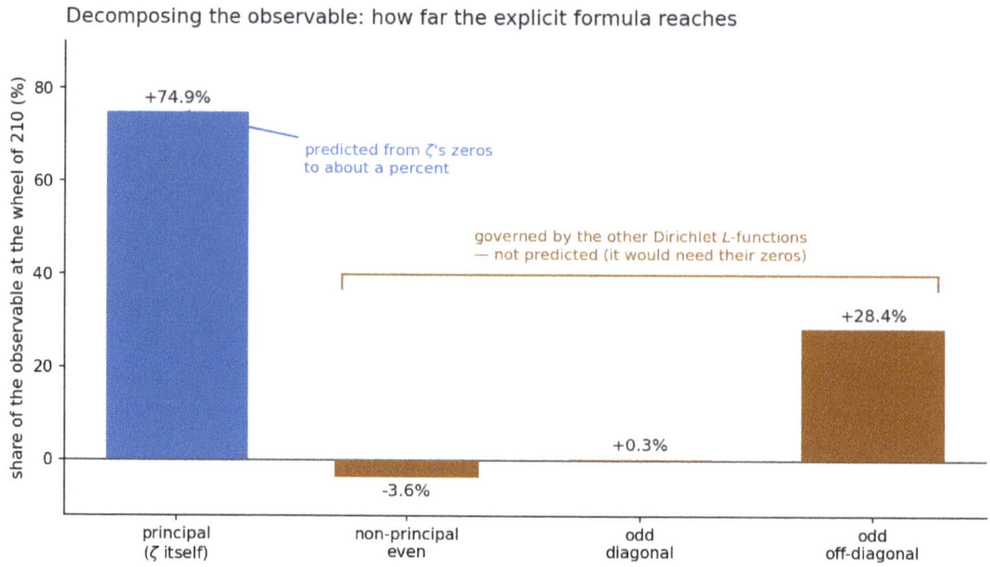

Figure 7.2: The observable at the wheel of 210, split into the four character-parity pieces. The principal piece — governed by ζ's own zeros — is about three-quarters of the whole and can be predicted from the explicit formula to about a percent. The remaining three pieces are governed by the non-principal Dirichlet L-functions of the wheel, and predicting them would require the zeros of that whole family; that prediction is not made.

At the wheel of 210 the principal piece is about three-quarters of the entire observable, and the prediction lands within roughly a percent of the measured value. The honest figure is "about a percent," not less: the pointwise prediction carries a conditional-convergence bias of several to many percent that does *not* shrink as more zeros are added, and it is the windowed difference — the same zero-minus-midpoint contrast of the detection chapter — that cancels most of that bias, so the sub-percent agreement on any one run rides on a cancellation rather than on genuine convergence. Across other wheels the principal piece is a smaller or larger share of the observable, from about a quarter to almost all of it, and the same method tracks it at the same percent level. That much is a real and non-trivial reach: most of the signal, at the canonical wheel, recovered from the zeros of ζ alone.

And then it stops, exactly where this chapter says it must. The remaining pieces — a quarter of the signal at the canonical wheel, dominated by a single off-diagonal contribution worth nearly thirty percent — are governed not by ζ but by the non-principal L-functions of the family. Predicting them would require locating and summing the zeros of those L-functions, the same family whose shifting allegiances filled the earlier sections. That prediction is left unmade, as the open problem it honestly is. The boundary between what the explicit formula delivers and what it cannot is the boundary between ζ and the rest of the family — which is to say, the boundary between the Riemann Hypothesis and its generalization. The method reaches the first wall and halts at the family.

7.7 Why this is the wall

Put the three steps together. The spectral signal of the wheel is carried by a handful of residues; those residues matter because of the L-functions they are built from; those L-functions matter because of where their zeros sit relative to ζ's; and the entire arrangement reorganizes when the wheel is lifted. To predict or control the signal one would need to know the joint placement of the zeros of all the wheel's L-functions at once — which is the Generalized Riemann Hypothesis — and one would need to know it afresh for each wheel, because the couplings are not invariant. The spectral lens does not shrink the Riemann wall to a single manageable question. It resolves the one wall into a coupled family of them, one per L-function, with the couplings themselves depending on the modulus.

That is the honest shape of the obstacle, and it is larger than the one Book One drew, not smaller. Book One's wall was the placement of the zeros of ζ. The wheel's spectral structure depends on the placement of the zeros of every L-function it carries, on how those placements correlate with one another, and on an arithmetic — which primes divide the wheel — that rearranges the correlations from one modulus to the next. The door this chapter opens leads into the wall and onto a larger room behind it. It is a clearer view of the obstruction, not a way around it.

7.8 What this is and is not

What is established is a chain of measurements. The spectral cancellation is dominated by the subprime residues 169 and 209, with 209 carrying roughly three times the per-prime weight; the wheel's L-functions correlate with ζ in the pattern shown, with the conductor-11 and conductor-19 characters the strongest opponents at -0.82 and -0.77; and that pattern dissolves when the modulus is lifted to 2310. And the principal-character piece of the observable — about three-quarters of it at the canonical wheel — is predictable from the explicit formula and ζ's zeros to roughly a percent, while the pieces carried by the other L-functions are not predicted at all. These are computed facts about the wheel's observables and the first thousand zeros, nothing more.

What is not established is anything about the Generalized Riemann Hypothesis. Nothing here bounds a zero, certifies one on the line, or moves the hypothesis an inch. The chapter shows only that the spectral structure of the wheel is hostage to GRH — that to control the signal one would need the zeros of an entire family of L-functions under simultaneous control, on each wheel separately — which makes the spectral approach a door into a harder wall rather than a shortcut past an easier one. The correlations are real and the dependence is real; the wall they lead to is the strongest form of the Riemann question, and it stands exactly where it stood. The next chapter turns to the second wall, the twin primes, where the spectral lens delivers something cleaner: not a passage through, but a single number on the far side, and an honest account of what it is.

On the figure

Figure 7.1 plots measured Pearson correlations between each L-function's windowed observable and ζ's, over the grid of the first 1000 zeros at prime cutoff $X = 10^5$. Wheel of 210: χ_{-3} +0.70, χ_{-4} −0.72, χ_{-7} +0.58, χ_{-11} −0.82, χ_{-19} −0.77. Wheel of 2310: χ_{-3} −0.78, χ_{-4} −0.52, χ_{-7} −0.75, χ_{-11} −0.49, χ_{-19} +0.21. The chart draws these values; it does not compute anything new.

Chapter 8

The Barrier and the Constant

The second wall is the twin primes, and the spectral lens does two clean things to it — neither of which is to pass it. On the near side it sharpens the obstruction to a finite, provable shape. The whole twin question localizes to fifteen corridors of the wheel; inside a corridor the relevant sieve reduces to two explicit matrices, B and M; and the centered form $B - M$ is negative semidefinite, *by an elementary identity. That is the parity barrier — the known reason sieve methods cannot, by themselves, deliver twin primes — written as the sign of a quadratic form one can compute and display. On the far side the lens produces a number. A natural invariant built from the two prime-side channels, followed up the primorials, has a closed asymptotic form in which the Hardy–Littlewood twin-prime constant $C_2 =$ 0.6601618... appears, by way of Mertens' theorem. The most concrete quantity the construction yields is the very constant that would count the twin primes if there are infinitely many of them. You can compute the constant. You cannot, by computing it, earn the conjecture it belongs to.*

8.1 Scope

This chapter is compressed and assumes Book One's twin-primes chapter, where the twin corridors of the wheel were introduced and the conjecture left standing at its wall. It views that wall through the linear-algebraic and spectral instruments of this book. Two results, both honest about their limits. The first is a finite identity that exhibits the parity barrier as the negative-semidefiniteness of an explicit matrix; it is elementary and exact, and it proves nothing about the existence of twins. The second is a closed-form computation in which the twin constant emerges from a two-channel invariant; it is a genuine calculation, and it does not prove the conjecture either. Section 8.6 keeps the two firmly on the near side of the wall. As with the previous chapter, the lens clarifies the obstruction; it does not remove it.

8.2 The problem, localized

Book One placed the twin primes on the wheel and observed that a twin pair $(n, n+2)$ can occupy only certain residue corridors. Sharpened, the observation is a reduction.

Among the integers greater than 7 and coprime to 210, the residues split into forty-eight survivors, and a twin pair can sit in only fifteen of them — the twin-admissible corridors, those residues r for which both r and $r+2$ are coprime to 210. The twin-prime conjecture is then exactly the statement that at least one of these fifteen corridors contains infinitely many twin primes. Nothing is lost in this reduction and nothing is gained; it simply puts the question where the wheel can see it, as a statement about fifteen arithmetic progressions.

The fifteen corridors are worth seeing all at once (Figure 8.1). Under the wheel's mirror symmetry $r \leftrightarrow 210 - r$ they fall into seven mirror pairs and a single self-mirror slot that wraps across the top. Thirteen of the fifteen hold genuine twin primes; the two that fail are the instructive ones. The slot $(167, 169)$ is broken because $169 = 13^2$, and the wrap slot $(209, 211)$ because $209 = 11 \cdot 19$ — in both cases by a *subprime*, a composite coprime to 210 whose least factor exceeds 7, the same rough numbers that did the cancelling in the previous chapter. The picture also shows the asymmetry the rest of the chapter is about: $(167, 169)$ and its mirror $(41, 43)$ are identical as corridors, related by the exact symmetry of the wheel, yet one holds a twin and the other does not. Admissibility is symmetric; realization is not, and nothing in the wheel decides which admissible slot is filled.

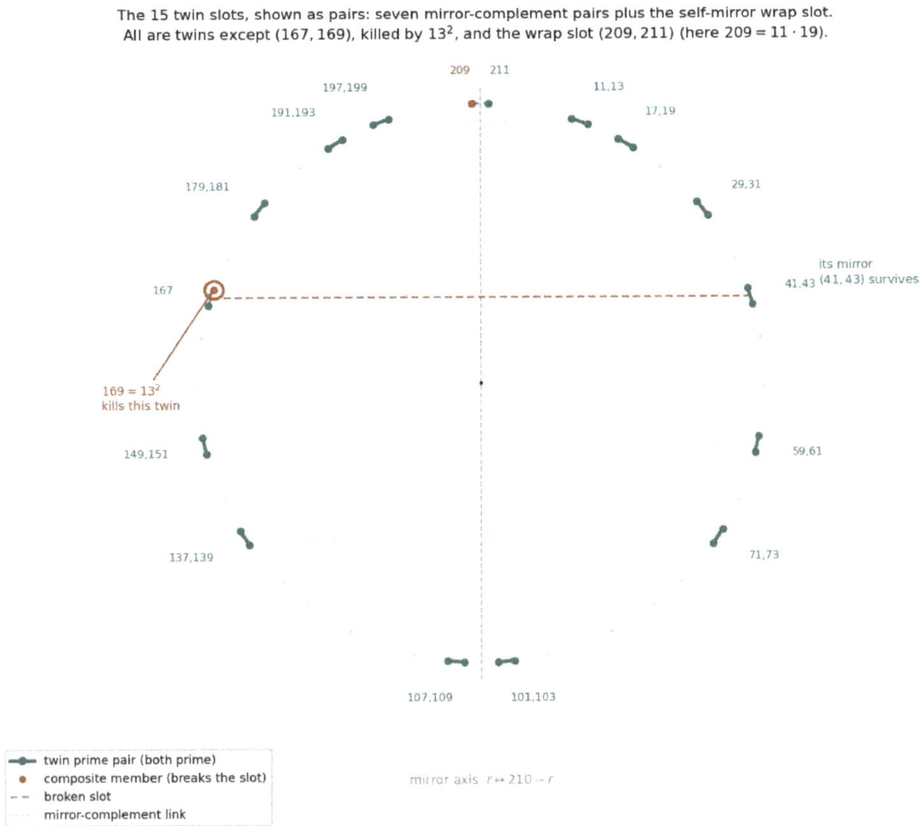

Figure 8.1: The fifteen twin-admissible corridors modulo 210. The mirror pair $(41, 43)$ and $(167, 169)$ is the chapter in miniature: identical as corridors under $r \leftrightarrow 210 - r$, related by the exact symmetry of the wheel, yet one realizes a twin and the other is broken — here by the subprime $169 = 13^2$, just as the wrap slot $(209, 211)$ is broken by $209 = 11 \cdot 19$. Admissibility is symmetric; realization is not.

8.3 Two kernels

Inside a single twin-admissible corridor, the sieve machinery for the shift pair $\{0, 2\}$ — the Selberg and Goldston–Pintz–Yıldırım variance and main terms — is governed by two matrices, indexed by squarefree moduli d coprime to 210. The first counts how often two sieve conditions overlap:

$$B(d_1, d_2) = \frac{\mathbf{1}_{\gcd(d_1,d_2)=1}}{d_1 d_2}, \tag{8.1}$$

which by the Chinese Remainder Theorem is the density of integers divisible by both d_1 and d_2 in the corridor — and which is *sparse*, carrying a coprimality indicator, precisely because 2 divides 210 while the d_i do not. The second is the smoother form

$$M(d_1, d_2) = \frac{1}{\operatorname{lcm}(d_1, d_2)} = \frac{\gcd(d_1, d_2)}{d_1 d_2}. \tag{8.2}$$

These two are the entire arithmetic content of the corridor sieve for twins; everything the method can do is built from how a chosen vector of sieve weights $\lambda = (\lambda_d)$ pairs against B and against M.

8.4 The centered identity and the barrier

The decisive comparison is between the two. Form the centered quadratic form $\lambda^\top(B-M)\lambda$ — the difference between what the overlap count gives and what the smoother form gives — and it has an exact, elementary closed expression. Using the two Möbius identities $\mathbf{1}_{\gcd=1} = \sum_{k|\gcd} \mu(k)$ and $\gcd = \sum_{k|\gcd} \varphi(k)$, one finds

$$\lambda^\top(B-M)\,\lambda \;=\; -\sum_{k\geq 2}(\varphi(k)-\mu(k))\,W_k(\lambda)^2, \qquad W_k(\lambda) = \sum_{k|d} \frac{\lambda_d}{d}. \tag{8.3}$$

Every coefficient $\varphi(k) - \mu(k)$ is positive for $k \geq 2$, and every $W_k(\lambda)^2$ is a square, so the entire right-hand side is a sum of nonnegative terms with a minus sign in front. The centered form is therefore negative or zero for *every* choice of sieve weights: the matrix $B - M$ is negative semidefinite. Figure 8.2A is its spectrum, computed on the squarefree moduli coprime to 210 — one eigenvalue exactly zero, every other strictly below.

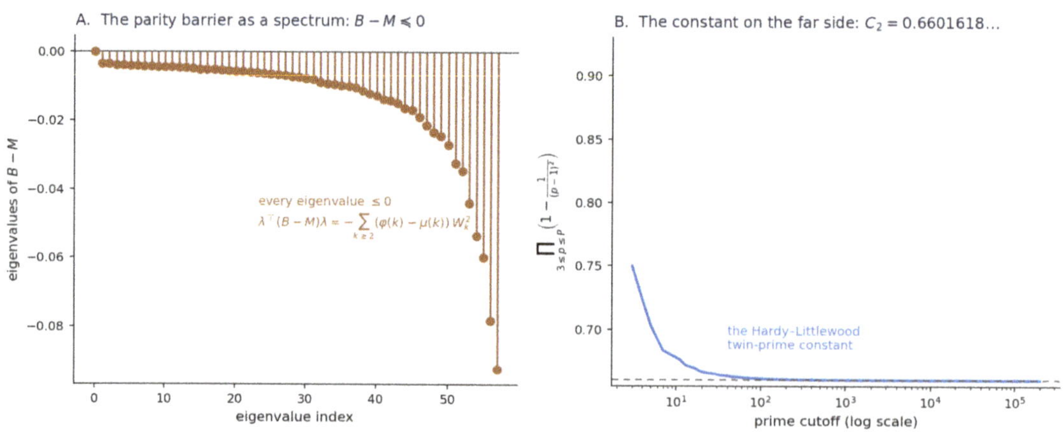

Figure 8.2: The twin wall, both sides. **A:** the spectrum of the centered matrix $B-M$, Eqs. (8.1)–(8.2), computed on squarefree moduli coprime to 210. Every eigenvalue is ≤ 0: the form is negative semidefinite, by the identity (8.3). This is the parity barrier as the sign of a quadratic form. **B:** on the far side, the Hardy–Littlewood twin-prime constant $C_2 = \prod_{p\geq 3}(1-1/(p-1)^2) = 0.6601618\ldots$, the number that would count the twin primes if there are infinitely many.

This negative-semidefiniteness is not a technicality; it is the parity barrier in finite dimensions. The barrier, in its classical form, is the fact that sieve methods cannot tell a prime from a product of two primes — cannot resolve the parity of the number of prime factors — and so cannot, on their own, force the count of primes (or twin primes) to be positive. Equation (8.3) is that obstruction made explicit and finite: the overlap form B never exceeds the smooth form M in the centered sense, so no choice of weights λ turns the sieve into a positive lower bound for twins. The best unconditional result, Chen's theorem, lives exactly at this edge: it gives infinitely many primes p for which $p+2$ is either a prime or a product of two primes, in each corridor — everything the method yields right up to the barrier, and not the final step across it, because that step is the one the parity barrier forbids.

8.5 The constant on the far side

Turn now to the far side of the wall and ask what number the spectral construction produces when pointed at the twins. Each of the two prime-side channels of the earlier chapters carries an energy — call them E_ψ and $E_{\psi^\dagger}$ — and the natural scalar the two-channel system forms is the product, $4\,E_\psi E_{\psi^\dagger}$. (The construction that produced these channels dressed this product as a relativistic mass; that dressing is a costume, and its own authors decline to attach it to any physical mass. The honest object is a product of two energies.) Follow that invariant up the primorials M_k, with p_k the largest prime dividing M_k, and it has a closed asymptotic form:

$$4\,E_\psi E_{\psi^\dagger}(M_k) \;\sim\; \frac{4\,C_2\,e^{-\gamma}\,S(\infty)}{\log p_k}, \tag{8.4}$$

where γ is the Euler–Mascheroni constant, $S(\infty)$ is a fixed number set by the windowing, and $C_2 = 0.6601618\ldots$ is the Hardy–Littlewood twin-prime constant of Figure 8.2B. The mechanism is Mertens' third theorem — the product $\prod_{p\le x}(1-1/p) \sim e^{-\gamma}/\log x$ — applied to the channel energies, with a leading-order cancellation that leaves C_2 standing.

The overall scale in (8.4) depends on the windowing convention and carries no meaning; what does not depend on any convention is which constant appears. Of all the numbers the construction could have produced for its most concrete, "physical-feeling" quantity, the one it produces is C_2 — the constant in the Hardy–Littlewood conjecture $\pi_2(x) \sim 2C_2\,x/(\log x)^2$, the number that would give the asymptotic count of twin primes if the conjecture holds. The spectral construction reaches, on the far side of the barrier, the exact constant that would weigh the twins. It reaches the number; it does not reach the theorem.

8.6 What this is and is not

What is established is two exact facts on opposite sides of the wall. On the near side, the centered identity (8.3) is elementary and proved, and the matrix $B-M$ is negative semidefinite — a finite, displayable form of the parity barrier, with Chen's theorem sitting at its edge. On the far side, the two-channel invariant has the closed asymptotic form (8.4), in which the Hardy–Littlewood constant C_2 provably appears by Mertens' theorem. Both are real mathematics; the figure computes both directly.

What is not established is the twin-prime conjecture, and nothing here moves it. The localization to fifteen corridors loses no difficulty; the negative-semidefiniteness is the obstruction, not a way around it; and the appearance of C_2 is the appearance of a number, not a proof of the infinitude it would measure. This is the honest shape of the second wall under the spectral lens. The barrier is the parity barrier, now visible as the sign of a small matrix one can write down; and the prize on the far side is a constant one can compute to as many digits as one likes, which is exactly not the same as proving there are infinitely many twins for it to count. Book One left the twins at their wall. This chapter shows the wall has a precise finite shadow and a precise number behind it, and stands exactly where it stood.

On the figure

Figure 8.2A is a direct computation: the matrices B and M of Eqs. (8.1)–(8.2) on the 58 squarefree moduli coprime to 210 below 260, and the eigenvalues of $B - M$, whose maximum is 0 to machine precision and whose remainder is negative — the negative-semidefiniteness of Eq. (8.3), checked rather than asserted. Panel B is the partial product $\prod_{3 \le p \le P}(1 - 1/(p-1)^2)$, which converges to $C_2 = 0.6601618\ldots$ (at $P = 2 \times 10^5$ it reads 0.6601621).
The corridor wheel (Figure 8.1) is a direct enumeration: the fifteen residues r with both r and $r + 2$ coprime to 210, arranged by the mirror $r \leftrightarrow 210 - r$, with the primality of each member checked outright — which is what exposes the two broken slots.

Chapter 9

What the Zeros Don't Know

The third wall is the hardness of factoring, and it stands apart from the other two. The Riemann wall and the twin wall are places where the spectral approach reaches and then stops; the factoring wall is a place where it never points. The zeros of ζ encode the global distribution *of the primes — how many there are below a height, how they thin, how they fluctuate — and factoring asks a* local *question about one specific number: which two primes multiply to give it. These are different kinds of knowledge, and the first does not contain the second. Even a full proof of the Riemann Hypothesis, which would pin the distribution exactly, would not hand anyone an efficient way to factor. And the quantum that does break factoring — Shor's algorithm — works by a mechanism that has nothing to do with the quantum chaos of the zeros that this book has been about. There are two different "quantums" in this subject, and the one that threatens the codes is not the one that might prove the hypothesis. This chapter is short, because the honest thing to report about the third wall is mostly that the lens does not address it.*

9.1 Scope

This chapter is brief and assumes Book One's closing chapter, where the hardness of factoring was named as the third wall. It does not develop the theory of factoring or of public-key cryptography; it asks only what the spectral approach of this book has to say about it, and the answer is: very little, for a reason worth stating clearly. The point is the distinction between two kinds of knowledge about the primes, and the distinction between two unrelated uses of the word "quantum." Section 9.5 draws the line, and with it closes the book's walk along the three walls.

9.2 Two kinds of knowledge about primes

Everything in this book and the last has been about the distribution of the primes: their density on the wheel, their race across the residue classes, the smooth count of them below a height and the fluctuation around it, the zeros that govern that fluctuation. This is global knowledge. It answers questions of the form "how many primes are there with such-and-such a property," and it answers them in the aggregate, as statements about counts and averages and error terms.

Factoring asks something else entirely. Given a single large number N, known to be the product of two primes, find them. This is local knowledge — a question about the internal multiplicative structure of one specific integer, not about the population of primes around it. The two kinds of knowledge are genuinely different, and the gap between them is the reason the spectral approach falls silent here. Knowing the distribution of the primes with perfect precision tells you how many primes lie near $\sqrt{N}$; it does not tell you which of them divides N.

9.3 Why the zeros do not factor

The point can be made sharp through the explicit formula, the bridge of Book One. That formula relates a sum over the primes to a sum over the zeros, and what it computes is the count of primes (with prime powers, suitably weighted) below a height x. The zeros control the error in that count. So the information the zeros carry is exactly information about $\pi(x)$ — about how the primes are distributed — and the Riemann Hypothesis is the statement that this error is as small as it can be.

Suppose it were proved. Then the count of primes below any height would be known to within an error of order $\sqrt{x}\log x$, the smallest possible. This sharpens primality testing, tightens bounds on the gaps between primes, and improves a number of algorithms that depend on the distribution. It does nothing for factoring. The number field sieve, the best general factoring method known, is sub-exponential and makes no use of the zeros; no one has found a way to turn sharp knowledge of $\pi(x)$ into a fast way to split a specific N, and there is no reduction known in either direction between the Riemann Hypothesis and the hardness of factoring. The two problems are both about primes, and they are about different things primes do — spreading out, on the one hand, and multiplying together, on the other.

9.4 The other quantum

There is a real quantum threat to factoring, and it is worth naming precisely, because it is easy to conflate with the quantum spectrum this book has chased. Shor's algorithm factors a large number in polynomial time on a quantum computer. But it does not do so by finding a Riemann spectrum or diagonalizing a Hilbert–Pólya operator. It works by *period-finding*: to factor N, pick a random a and find the period of the sequence $a, a^2, a^3, \dots$ modulo N — the multiplicative order of a — using a quantum Fourier transform on the cyclic group of exponents, after which an elementary greatest-common-divisor computation recovers the factors. The quantum work is a Fourier transform over the exponent register — the integers modulo a power of two of size about N^2 — from which the period is read off. It has nothing to do with the zeros of ζ, with quantum chaos, or with the statistics that filled the first half of this book.

So there are two quantums here, and keeping them apart is the whole content of this section. One is the quantum of *quantum chaos*: the idea, running through this book, that the zeros are the spectrum of some chaotic quantum system, and that proving the Riemann Hypothesis means finding that system. The other is the quantum of *quantum computation*: Shor's period-finding, which breaks the factoring-based codes and which is built from the quantum Fourier transform, not from any spectrum of the primes. The quantum that menaces the codes is not the quantum that might settle the hypothesis. They share a word and almost nothing else.

9.5 What this is and is not

What is established is a separation. The zeros of ζ carry information about the distribution of the primes, not about the factorization of any particular number; the Riemann Hypothesis, even proved, would not yield efficient factoring; and the quantum algorithm that does factor efficiently works by period-finding, unrelated to the spectral picture of the zeros. None of this is a theorem about factoring being hard — the hardness of factoring is itself unproven, a conjecture the codes rest on — but it is an honest account of why the instruments of this book do not touch it.

What this is not is a fourth approach to the third wall. The spectral lens reaches the Riemann wall and resolves it into a coupled family of walls; it reaches the twin wall and shows the parity barrier as the sign of a matrix, with the twin constant standing on the far side; and it does not reach the factoring wall at all, because that wall is built from a question the zeros were never about. Three walls, three different fates under the same lens — and that completes the walk. Book One ended by saying that the tour does not pass the three walls, and ends at them deliberately, because seeing exactly where a wall stands is worth more than a story that pretends to cross it. This book has tried to do the same thing one level deeper: to take the most physical-looking face of the subject, the zeros as a spectrum, and follow it honestly to each wall — to the operator that cannot be built, the resolution that cannot be had, the laws that do not survive being pushed, and the three obstacles that remain exactly where the first book left them. The spectrum is real, the statistics are real, the constants are real. The walls are real too, and they are still standing.

Chapter 10

Beyond the Circle

Every wheel in this book has secretly been a circle. There is a next curve, and it is not a figure of speech: it is the shape that appears when the rational integers give way to the Gaussian integers. We roll onto it to settle two questions at once—whether the walls were ever about the circle at all, and whether the one structure the whole book has leaned on, a world of positions facing a world of frequencies across the explicit formula, survives the change of stage. Both answers are the point of the book.

Scope

The wheel that has carried this book—the residues coprime to 210, turned and lifted and stressed—is, underneath its arithmetic, a circle. The roots of unity sit on it; the Dirichlet characters are its harmonics; the L-functions of Part IV are built from its division points. This closing chapter takes the circle seriously enough to ask what lies one step past it. The answer is a specific, named curve with a specific, named arithmetic behind it, and the correspondence is rigorous and more than a century old. We use it for two things. First, to test whether the three walls—the Riemann Hypothesis, the twin primes, factoring—are features of the particular wheel we chose or of the primes themselves, by changing the wheel as drastically as the mathematics allows and watching what stays. Second, to exhibit, one last time and in a new field, the single picture this whole book has been about: the primes living in one plane, their zeros living in another, and a single exact identity—the explicit formula—as the only bridge between them. Nothing here crosses a wall. Everything here is meant to show, cleanly, where the wall actually is.

10.1 The circle, made precise

Strip the wheel to its geometry and it is the unit circle with marked points. The marking is not arbitrary. The map

$$e : \mathbb{C}/\mathbb{Z} \longrightarrow \mathbb{C}^{\times}, \qquad z \longmapsto e^{2\pi i z},$$

is periodic with respect to the lattice $\mathbb{Z}$, and it sends the rational points of $\mathbb{C}/\mathbb{Z}$ to the roots of unity: $z = k/n$ goes to ζ_n^k, where $\zeta_n = e^{2\pi i/n}$. Dividing the circle into n equal arcs is exactly evaluating this one transcendental function at the n-division points of its lattice, and the values it returns generate the cyclotomic field $\mathbb{Q}(\zeta_n)$. The arithmetic of that field is as clean as a field gets: its Galois group is canonically

$$\mathrm{Gal}\big(\mathbb{Q}(\zeta_n)/\mathbb{Q}\big) \;\cong\; (\mathbb{Z}/n\mathbb{Z})^\times,$$

abelian, with the residue a mod n acting by $\zeta_n \mapsto \zeta_n^a$.

The theorem that makes the wheel more than bookkeeping is the one of Kronecker and Weber: every finite abelian extension of the rationals lies inside some cyclotomic field $\mathbb{Q}(\zeta_n)$. Put plainly, all of the simplest arithmetic over $\mathbb{Q}$—every way of extending the rationals with an abelian symmetry—is already written in the division points of the circle. The residues mod 210, the real characters and their conductor lattice, the family of Dirichlet L-functions whose zeros formed the family of walls: every one of them is the circle, divided. When we rolled the wheel we were stepping around the circle; when we spun it onto its dual face we were reading the circle's harmonics. The circle was the whole stage.

10.2 The next curve, made precise

In the 1820s Abel, following Gauss, asked what happens if you divide a different curve: the lemniscate, the figure-eight given in polar form by $r^2 = \cos 2\theta$. The circle's division was run by the exponential and its lattice $\mathbb{Z}$; the lemniscate's is run by a different transcendental function and a different lattice. Define the *lemniscate constant*

$$\varpi \;=\; 2\int_0^1 \frac{dt}{\sqrt{1-t^4}} \;=\; 2.6220575542\ldots,$$

the analogue of $\pi = 2\int_0^1 dt/\sqrt{1-t^2}$, and define the *lemniscate sine* $\mathrm{sl}(u)$ by inverting the arc-length integral, so that

$$u \;=\; \int_0^{\mathrm{sl}(u)} \frac{dt}{\sqrt{1-t^4}}, \qquad \mathrm{sl}'(u)^2 \;=\; 1-\mathrm{sl}(u)^4.$$

Like the sine, sl extends to a meromorphic function on the whole plane; unlike the sine, it is doubly periodic, an elliptic function of order two. Its period lattice is *square*: it is carried to itself by multiplication by i, and the function obeys the single identity that names the whole subject,

$$\mathrm{sl}(i\,u) \;=\; i\;\mathrm{sl}(u).$$

This is *complex multiplication*. The circle's symmetry was multiplication by roots of unity, a rotation; the lemniscate's symmetry includes a genuine multiplication by the imaginary unit, a quarter-turn of the lattice that the circle simply does not have. The curve standing behind sl is the elliptic curve

$$y^2 = x^3 - x, \qquad j\text{-invariant } 1728,$$

whose endomorphism ring is the full ring of Gaussian integers $\mathbb{Z}[i]$ and which carries the order-four automorphism $(x, y) \mapsto (-x, iy)$—one checks $(iy)^2 = -y^2 = (-x)^3 - (-x)$ directly. That automorphism *is* multiplication by i on the curve.

The historical fingerprint is unmistakable. Gauss had shown that a regular n-gon is constructible with ruler and compass when n is a power of two times distinct Fermat primes (Wantzel proved the converse); Abel showed that the lemniscate can be cut into n equal arcs by ruler and compass under the very same condition. The same Fermat primes governing two different curves is not a coincidence. The lemniscate's n-division values—the numbers $\mathrm{sl}(\varpi\beta/n)$ for Gaussian integers β—generate abelian extensions of $\mathbb{Q}(i)$, exactly as the roots of unity generate those of $\mathbb{Q}$. And the converse holds: every abelian extension of $\mathbb{Q}(i)$ is generated by such special values together with roots of unity. This is the imaginary-quadratic case of Hilbert's twelfth problem—Kronecker's *Jugendtraum*—solved by the theory of complex multiplication in the hands of Kronecker, Weber, Takagi, Hecke, and Deuring. The circle solves Hilbert's twelfth problem for $\mathbb{Q}$; the lemniscate solves it for $\mathbb{Q}(i)$. The wheel was the first case of a pattern. The lemniscate is the second (Figure 10.1).

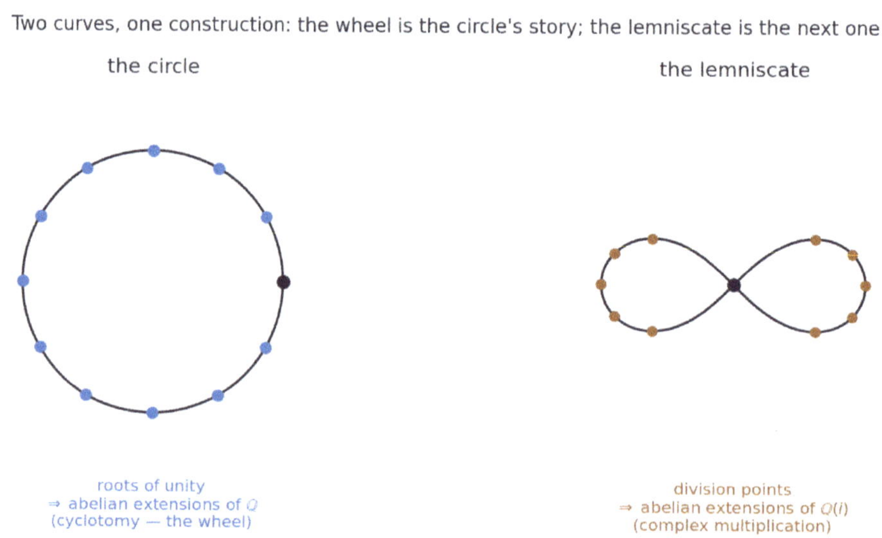

Figure 10.1: The circle and the lemniscate, divided. The circle's division points—the roots of unity—generate the abelian extensions of $\mathbb{Q}$; this is the wheel. The lemniscate's division points generate the abelian extensions of $\mathbb{Q}(i)$ by complex multiplication. The crossing node, in black, is the geometric home of the ramified prime 2.

10.3 How the primes map

Changing the base field from $\mathbb{Q}$ to $\mathbb{Q}(i)$ does something concrete to the primes themselves: each one is told to factor again, this time in the larger ring $\mathbb{Z}[i]$. That ring is Euclidean for the norm $N(a + bi) = a^2 + b^2$, hence a unique-factorization domain;

its units are the four numbers $\{1, i, -1, -i\}$, the quarter-turns. The rule by which a rational prime p factors is the exact analogue of how p splits in the cyclotomic world, and it is one of the oldest theorems in the subject, Fermat's on sums of two squares:

$$\begin{aligned} p \equiv 1 \pmod 4 :& \quad p = \pi\bar{\pi} \textit{ splits} \text{ into two conjugate Gaussian primes, } p = a^2 + b^2, \\ p \equiv 3 \pmod 4 :& \quad p \text{ stays } \textit{inert}\text{, a Gaussian prime of norm } p^2, \\ p = 2 :& \quad 2 = -i\,(1+i)^2 \textit{ ramifies.} \end{aligned}$$

So $5 = (2+i)(2-i)$, $13 = (3+2i)(3-2i)$, $17 = (4+i)(4-i)$, while $3, 7, 11, 19$ remain prime where they stand, and 2 alone collapses onto a single repeated factor. This *is* the map from the rational primes to the Gaussian primes the chapter promised: it is the sum-of-two-squares decomposition, nothing more exotic.

What decides the three cases is a single arithmetic switch. The Galois group $\mathrm{Gal}(\mathbb{Q}(i)/\mathbb{Q})$ has two elements, the identity and complex conjugation, and the Frobenius element attached to an odd prime p is conjugation when p is inert and the identity when p splits. That switch is exactly the nontrivial character modulo 4,

$$\chi_4(p) = (-1)^{(p-1)/2} = \begin{cases} +1, & p \equiv 1 \pmod 4 \text{ (split)}, \\ -1, & p \equiv 3 \pmod 4 \text{ (inert)}, \end{cases}$$

with $p = 2$ the ramified exception where χ_4 is not defined. The same χ_4 assembles the whole story into one function. The Dedekind zeta function of $\mathbb{Q}(i)$, summed over the nonzero ideals of $\mathbb{Z}[i]$, has an Euler product over the Gaussian primes that the splitting law factors cleanly:

$$\zeta_{\mathbb{Q}(i)}(s) = \sum_{0 \neq I \subseteq \mathbb{Z}[i]} \frac{1}{N(I)^s} = \prod_{\pi} \left(1 - N(\pi)^{-s}\right)^{-1} = \zeta(s)\, L(s, \chi_4).$$

The right-hand side is the entire map, read as a single object: the factor $\zeta(s)$ is the rational backbone, and the factor $L(s, \chi_4)$ is precisely the splitting switch, the Frobenius, turned into a function of s. The Gaussian primes and the character χ_4 are two views of one thing.

10.4 Two planes, and the bridge between them

Now the crux of the book, in the cleanest field we have for it.

The Gaussian primes live in a *position plane*. They are points of the lattice $\mathbb{Z}[i]$ that you can name and place: $1+i$ at the centre, the split primes $2 \pm i$, $3 \pm 2i$ off the axes, the inert primes $3, 7, 11$ marching out along them. Multiplication by i spins this plane through a quarter turn and carries the picture to itself; complex conjugation $z \mapsto \bar{z}$ reflects it across the real axis, swapping the two conjugate factors π and $\bar{\pi}$ of every split prime. These are the roll and the barrel-roll of the wheel, now literal motions of a lattice in the plane.

The zeros of $\zeta_{\mathbb{Q}(i)} = \zeta \cdot L(\chi_4)$ live in a different plane entirely—a *spectral plane*, the strip of complex s, where under the Generalized Riemann Hypothesis they sit in a single line at $\mathrm{Re}(s) = \frac{1}{2}$. Because the function factors, its zeros are the union of two families, the zeros of ζ and the zeros of $L(\chi_4)$, interleaved on that line; there are more of them, and they are denser, than the circle's alone. This is the dual face: the same arithmetic, read as frequencies rather than as positions.

The two planes are not the same plane, and no amount of drawing puts a zero onto the lattice or a Gaussian prime onto the critical line. What connects them is one exact identity. Weil's explicit formula for $\zeta_{\mathbb{Q}(i)}$ states that a weighted sum over the Gaussian primes equals a weighted sum over the zeros, term for term, with only the archimedean factors and the pole at $s = 1$ to mediate. The primes on the left, the zeros on the right, an equals sign in between: that identity is the only legitimate bridge between the position plane and the spectral plane, and it is the same bridge—the same explicit formula read as a trace—that has stood at the centre of every chapter of this book. Figure 10.2 draws the two planes side by side, with the bridge named and nothing faked across it. If there is one picture to carry away, it is this one: the primes here, the zeros there, and between them a single conserved identity and not one inch of shortcut.

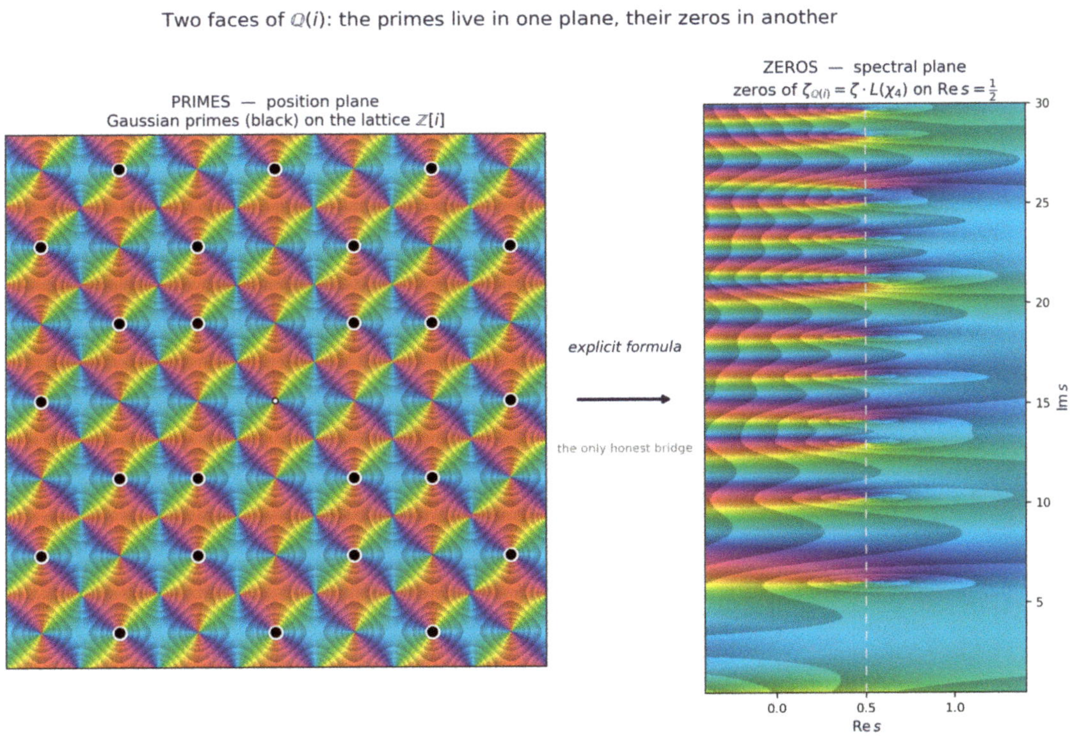

Figure 10.2: The crux of the book, in $\mathbb{Q}(i)$. *Left, the position plane:* the Gaussian primes (marked) sit among the integers on the lattice $\mathbb{Z}[i]$, drawn through the lemniscatic function whose poles are the lattice points; the quarter-turn $\times i$ is the spin, the dashed axis the conjugation mirror. *Right, the spectral plane:* a phase portrait of $\zeta_{\mathbb{Q}(i)} = \zeta \cdot L(\chi_4)$ over the critical strip, its zeros the dark points lined up on $\mathrm{Re}(s) = \frac{1}{2}$, the union of the zeta and $L(\chi_4)$ families. *Between them:* the explicit formula, the one exact identity equating a sum over the primes on the left with a sum over the zeros on the right. The zeros cannot be drawn on the lattice and the primes cannot be drawn on the line; the bridge is the only honest crossing. The two textures differ for a reason: the lattice on the left is periodic, a true crystal, so its colours repeat in tight identical cells, while $\zeta_{\mathbb{Q}(i)}$ on the right is aperiodic, its phase spreading in broad bands and its zeros forming not a lattice but a quasicrystal—ordered without ever repeating.

10.5 The walls travel

Here is the payoff of changing the stage. The move to the lemniscate is not a loss of structure—it is a gain, and a large one. Because the Gaussian world has complex multiplication, its L-functions are unusually well behaved: Hecke proved the analytic continuation and functional equation of the relevant Grössencharakter L-functions outright, decades before the general case was within reach, and the first genuine progress on the Birch and Swinnerton-Dyer conjecture was made in exactly this complex-multiplication setting, by Coates and Wiles, who proved that a CM elliptic curve with infinitely many rational points has L-value zero at $s = 1$. If any change of curve were going to soften the walls, this one—to the most structured alternative the subject offers—should have done it.

It does not. The Generalized Riemann Hypothesis for $\zeta_{\mathbb{Q}(i)} = \zeta \cdot L(\chi_4)$, and for the Hecke L-functions of $\mathbb{Q}(i)$ generally, is open, precisely as it is for the Dirichlet L-functions of the wheel; the extra structure that gives us their functional equations says nothing about where their zeros lie. The Gaussian-integer analogues of the twin-prime and gap questions—whether the split primes cluster, how far apart consecutive Gaussian primes can spread, whether one can even walk to infinity among them in steps of bounded length, the still-unsolved Gaussian moat—are open. Factoring a Gaussian integer is no easier than factoring a rational one; the same problem, in a richer ring, is the same problem. The three walls of Part IV reappear on the figure-eight, in the same places, with the same hardness, in full view of all the additional symmetry the new curve brings. The structure was real and it was generous, and it left the walls exactly where it found them.

What this is and is not

The correspondence in this chapter is established and celebrated mathematics. The exponential divides the circle and its values generate the abelian extensions of $\mathbb{Q}$; the lemniscate sine divides the figure-eight and its values generate the abelian extensions of $\mathbb{Q}(i)$; complex multiplication, $\mathrm{sl}(iu) = i\,\mathrm{sl}(u)$, makes the analogy exact; this is the content of Hilbert's twelfth problem in its solved cases. The splitting law is Fermat's theorem on sums of two squares, and the factorization $\zeta_{\mathbb{Q}(i)} = \zeta \cdot L(\chi_4)$ is its restatement as a single function. The explicit formula linking the Gaussian primes to the zeros of that function is exact. These are facts, and they are rigorous.

What this chapter is not is an attack on anything. Moving to the lemniscate does not weaken a single wall; it was never meant to. There is no lever in the extra symmetry, no shortcut hidden in the quarter-turn—if anything the lemniscate's lesson is the opposite of a shortcut. We changed the curve to the most natural successor mathematics offers, gained complex multiplication and everything that comes with it, and found the three walls standing where we left them. And the deeper observation, the one the whole book has been building toward, also survived the change without a scratch: there is a plane where the primes live and a different plane where their zeros live, and the only passage between the two is one exact identity. That structure is not a property of the circle, or of the lemniscate, or of any curve we choose to roll the primes along. It travels with the primes, onto every shape that will hold them—and so, it seems, do the walls.

On the figures

Figure 10.1 sets the two curves side by side: the unit circle with its roots of unity, and the lemniscate $r^2 = \cos 2\theta$ with division points along each lobe and the crossing node, the home of the ramified prime, in black. The division points there are schematic—

equal steps along each curve rather than the exact lemnatomic positions—and show the parallel construction, divide the curve and read off the field, not the computed values.

Figure 10.2 is the chapter's thesis in one frame. Its left panel is a phase portrait of the lemniscatic function over $\mathbb{Z}[i]$: hue encodes the phase, the bright lattice points are its poles, the Gaussian primes are marked among them, and the spin $\times i$ and conjugation mirror are drawn in. Its right panel is a phase portrait of $\zeta_{\mathbb{Q}(i)}(s)$ over the critical strip computed directly from $\zeta(s)\,L(s,\chi_4)$, with the zeros appearing as the dark phase singularities on the line $\mathrm{Re}(s) = \frac{1}{2}$. The arrow between the panels is the explicit formula. The two phase portraits are genuine computations of two different functions on two different planes; the figure places them together precisely to show that they cannot be overlaid—only bridged.

Coda: geometry that hits walls

A last caution to close on, and it is about a word. Three different objects in the first two books have carried the name Möbius, and they are not the same object. There is the Möbius *strip*, the one-sided band of topology. There is the Möbius *function* $\mu(n)$, the signed indicator of the squarefree integers that runs the supersymmetric primon gas of the companion volume. And there is the Möbius *transformation*, the conformal motion $z \mapsto (az + b)/(cz + d)$ that rotates the sphere. They share a man—August Möbius—and nothing else. To let the shared name suggest a shared mathematics would be the most refined version of the mistake these books have warned against throughout: not a coincidence of shape, like the Lorentz mass of Chapter 4, but a coincidence of *spelling*. We keep the three apart on purpose.

It is the third Möbius, the transformation, that genuinely moves the flat torus. The flat torus $(\cos u, \sin u, \cos v, \sin v)$ sits on the three-sphere as a product of two circles, every pair of them linked; it is the same Clifford torus on which the two-mode state of Chapter 4 lives. The conformal motions of the three-sphere—the Möbius transformations—carry it to the cyclides of Dupin, and a rotation that sweeps the torus through the point of projection is exactly the motion that turns it inside out, the doughnut everting through its own hole (Figure 10.3). That eversion is real geometry. It is also, one notes with the discipline of Chapter 4 in hand, an element of a group isomorphic to a Lorentz group—the conformal group always is—and that isomorphism is, once more, an invitation to a metaphor and not a derivation of a world.

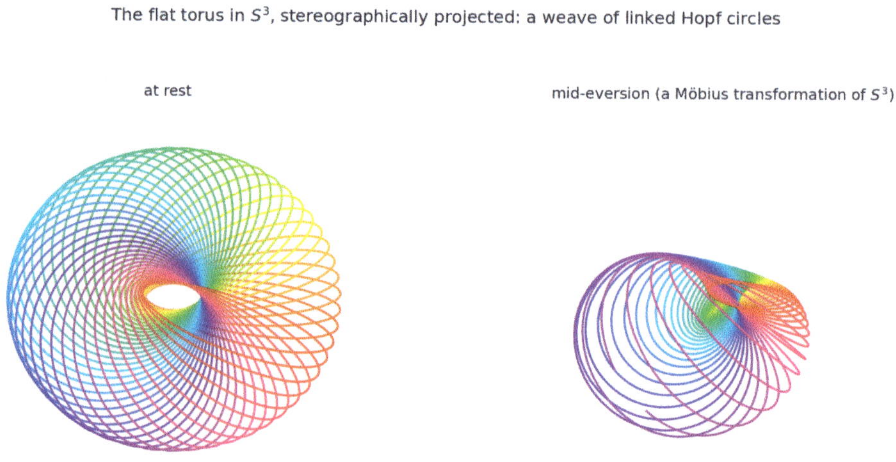

Figure 10.3: The flat torus $(\cos u, \sin u, \cos v, \sin v)$ of S^3, stereographically projected to ordinary space and drawn as a weave of its Hopf circles. Each circle winds once in both directions, and any two of them are linked. *Left:* at rest, the linked-circle torus, its central hole visible. *Right:* a rotation of S^3—a Möbius transformation—sweeps it toward the point of projection and turns it inside out, the image passing through the cyclides of Dupin. No rotation of the sphere changes which circles are linked, or where the wall is.

So take the dualities that have recurred across both books and ask which the geometry can actually carry. Two of them it carries cleanly. *Inside and outside* are the two solid tori into which the Clifford torus divides the three-sphere, exchanged by the eversion. *Position and spectrum* are the flat torus and its dual lattice, joined by Poisson summation, $\sum_{x\in\Lambda} f(x) = \sum_{\xi\in\Lambda^*} \hat{f}(\xi)$—a sum over a lattice equal to a sum over its dual, the explicit formula in its cradle. The third duality, *boson and fermion*, the geometry cannot carry, and it fails at a number. The natural supersymmetry of a torus is its de Rham grading, with Witten index equal to the Euler characteristic; and $\chi(T^2) = 0$, while the supersymmetric primon gas of the companion volume has Witten index 1. The invariant that defines the boson–fermion structure reads 0 on the torus and 1 in the gas. They are not the same structure, and no rotation of the sphere will make them one.

There is one further carrier, and it lives on a different surface from the flat torus—a curved one, reached by the same word through a different door. Restrict the Möbius transformations to integer coefficients and let them act not on the sphere, where they everted the torus, but on the hyperbolic upper half-plane; there they form the modular group, and the quotient is the modular surface, of constant negative curvature. On a flat surface the trace formula is just the Poisson summation of a moment ago. On this curved one it is Selberg's—an exact identity equating the spectrum of the hyperbolic Laplacian to the lengths of the closed geodesics, the surface's own primes—and it carries the one thing the Riemann case has never offered: the nontrivial zeros of the surface's zeta function, the Selberg zeta, *are* the eigenvalues of a self-adjoint operator, the Laplacian, and so they lie on their critical line for the most direct reason there is—the Hilbert–Pólya mechanism, not conjectured but working. It is realized in full, with a real operator and a real spectrum. But the geometry that does it is the hyperbolic modular surface, not the flat torus, and the zeta it does it for is Selberg's, not Riemann's. The same name, Möbius, has carried us to two stages—the flat torus it everts, the curved surface it tiles—and only one of them comes with an operator; and even that one is one zeta short of the wall.

That is the closing image of the first two books. The dualities are real and the geometry that carries them is beautiful; one of its carriers even crosses an analogue of the wall in plain sight, with an honest operator behind it. And Riemann's wall still stands—not because the geometry failed before reaching it, but because the wall was never a feature of the circle, or the torus, or the modular surface, or any shape onto which we can roll the primes. It belongs to the primes. Geometry that beautiful, running that far, and stopping exactly there: it is the truest picture the subject can offer of where the difficulty lives, and a genuinely beautiful one.

Exercises

The exercises are collected here, by chapter, rather than interrupting the text. Each exercise is tagged by difficulty ([routine], [moderate], [hard]) or by kind ([conceptual], [discussion]). The *two-part test* referred to throughout is that of Chapter 2: a candidate Riemann operator must (i) have the nontrivial zeros as its spectrum and (ii) be defined independently of the zeros.

Chapter 1 The Symmetry of a Family

1. [routine] Describe qualitatively the behaviour of the one-level density near the central point for a unitary family, a symplectic family, and an orthogonal family. Which is suppressed at the centre, which enhanced, and which flat?

2. [moderate] Distinguish *bulk universality* (pair correlation following the unitary law for every family) from the *central-point statistics* that separate the symmetry types. Why can the bulk statistics alone not reveal a family's symmetry type?

3. [moderate] Over function fields the symmetry type of a family is a theorem; over $\mathbb{Q}$ it is conjectural, supported by one-level density results valid only for a restricted range of test functions. What does the function-field setting supply that $\mathbb{Q}$ does not?

4. [discussion] Explain why identifying a family's symmetry type does not, by itself, establish that any individual L-function has its zeros on the critical line.

5. [moderate] Each of the seven real L-functions of the wheel, measured alone, has a level-spacing ratio in the unitary range ($\langle \tilde{r} \rangle \approx 0.62$); pooling all seven sets of zeros into a single list drops the ratio to the uncorrelated value (≈ 0.39, Figure 1.2). Explain why superimposing several rigid, level-repelling spectra can erase the repulsion this completely only when the spectra are mutually independent, and say what a *surviving* repulsion in the merged list would have shown instead.

Chapter 2 What Would Count as an Operator

1. [routine] *The primon spectrum.* The primon gas has one mode per prime p with single-quantum energy $\log p$ and occupation numbers $a_p \in \{0, 1, 2, \dots\}$.

(a) Show the energy $E = \sum_p a_p \log p$ of a state equals $\log n$, where $n = \prod_p p^{a_p}$.
(b) Using unique factorization, show the energy spectrum is exactly $\{\log n : n = 1, 2, 3, \dots\}$.
(c) Conclude the partition function is $\sum_{n\geq 1} n^{-s} = \zeta(s)$ for $\mathrm{Re}(s) > 1$.

2. [moderate] *Counting the levels.* Let $N_{\mathrm{pr}}(E)$ count primon levels $\leq E$.

   (a) Show $N_{\mathrm{pr}}(E) = \lfloor e^E \rfloor \sim e^E$.
   (b) Evaluate at $E = \gamma_1 = 14.1347$, the first zero's height.
   (c) With $N(T) \approx \frac{T}{2\pi}\log\frac{T}{2\pi} - \frac{T}{2\pi} + \frac{7}{8}$, evaluate at $T = \gamma_1$ and compare. By how many orders of magnitude do the counts differ?

3. [routine] State the two-part test in your own words. For each candidate say which part(s) it passes and fails: (a) a diagonal operator whose eigenvalues are declared to be the γ_n; (b) the primon Hamiltonian; (c) a kernel of delta spikes placed at the γ_n.

4. [moderate] Explain precisely why building an operator whose eigenvalues are by definition the zeros tells us nothing about RH. What property must such a construction have to bear on RH at all?

5. [moderate] The primon gas reproduces $\zeta(s)$ as its partition function yet its spectrum is $\{\log n\}$, not the zeros. Reconcile these, distinguishing the *values* of ζ (its Dirichlet series) from the *location of its zeros.*

6. [hard] (a) Show the primon density of states near E grows like e^E. (b) Show the density of zeros near T grows like $\frac{1}{2\pi}\log\frac{T}{2\pi}$. (c) Explain why this mismatch alone proves the primon gas is not a Riemann operator, with no zero located.

Chapter 3 Primes as Orbits

1. [moderate] The Berry–Keating model recovers the smooth counting term from a phase-space volume. Verify it gives $N(T) \approx \frac{T}{2\pi}\log\frac{T}{2\pi} - \frac{T}{2\pi}$, and evaluate at $T = 100$: the leading volume gives ≈ 28, while the true count is 29 (the gap of about one being the $\frac{7}{8}$ term plus fluctuations).

2. [moderate] In the trace-formula reading the primes are periodic orbits of period $\log p$ (and p^k contributes period $k \log p$). Explain how the explicit formula takes the shape "sum over zeros = smooth term + sum over orbits."

3. [routine] Explain why xp passes the independence half of the two-part test. Which candidate of Chapter 2 fails independence, and how does xp avoid that failure?

4. [discussion] Berry–Keating recovers the *average* (smooth count) exactly but not the fluctuations or individual zeros, the obstruction being that the required dynamics is chaotic and no concrete system realizing it is known. Explain what "average but not fluctuations" means here.

5. [hard] Both the primon gas (Chapter 2) and xp involve the period $\log p$. Contrast why the primon gas fails on spectrum while xp at least captures the average correctly.

Chapter 4 An Algebra in the Primes

1. [moderate] The prime bilinears close into a Lie algebra, $\mathfrak{sp}(4, \mathbb{R}) \cong \mathfrak{so}(2, 3)$. Explain what "close into an algebra" means in terms of commutators.

2. [moderate] The oscillator (metaplectic) representation splits into two irreducible singletons, the Rac ($\Delta = \frac{1}{2}$) and the Di ($\Delta = 1$). Relate this split to the even/odd sectors.

3. [routine] Despite the richer algebra, the spectrum is still $\{\log n\}$. Why does possessing an algebra not, by itself, repair the spectrum half of the two-part test?

4. [moderate] State Connes's criterion at the headline level (RH equivalent to a certain positivity). Explain why an equivalence "RH $\Leftrightarrow X$" is not a proof of RH.

5. [discussion] de Branges's approach has a known obstruction (Conrey–Li). What does it mean for an approach to be "not obviously wrong but with a located gap," and why is that the normal state of affairs near this wall?

6. [routine] With $J^0 = E_+ + E_-$ and $J^z = E_+ - E_-$, show $(J^0)^2 - (J^z)^2 = 4E_+E_-$, so the "dispersion relation" $E^2 = m^2 + p^2$ (with $E = J^0$, $p = J^z$, $m^2 = 4E_+E_-$) is an algebraic identity. Explain why its holding "to machine precision" is not evidence that the primes encode relativistic kinematics.

Chapter 5 Hearing the Zeros

1. [moderate] Distinguish *detection* from *resolution*. The prime-side coherence is significant ($p \approx 5 \times 10^{-5}$) within one mean spacing of a zero yet cannot resolve individual zeros. What is the difference?

2. [moderate] The resolution limit (aperture) is set by log(largest prime used). Explain the analogy to a telescope aperture or a Fourier window: why does the largest prime fix the resolution?

3. [conceptual] Four super-resolution methods of different mechanism all fail below the aperture. Why is it significant that unrelated methods hit the *same* limit?

4. [hard] If primes up to X are used, the aperture is $\sim \log X$. Discuss how slowly the resolvable spacing improves as X grows, and what this implies for ever resolving individual zeros from the primes.

5. [discussion] Why is "detect but not resolve" the honest analog, in the measurement setting, of the operator dream's failure in the construction setting?

Chapter 6 Laws That Died

1. [routine] The seven real characters mod 210 correspond to the non-empty subsets of $\{3, 5, 7\}$. Fill in the truth table and verify the rule "sign $= x_5 \oplus x_7$" across all seven.

2. [moderate] The rule is clean to $X = 10^9$ on the wheel of 210 but fails on the wheel of 2310. Explain the two independent axes of stress (cutoff X, modulus M) and why surviving one does not imply surviving the other.

3. [conceptual] The same split rates 100.0th percentile under one observable and 43.7th under another at modulus 2310. What does this say about the "extremality" of the split?

4. [moderate] The archimedean explanation was rejected for being three orders of magnitude too small. Why is checking the *magnitude* of a proposed effect a basic and decisive test?

5. [discussion] Why is keeping a numbered list of falsified hypotheses the practice that separates this work from numerology?

6. [hard] A pattern holds for all primes up to 10^9 on one wheel. List the tests it must still pass before it deserves to be called a candidate law, and say what each test controls for.

Chapter 7 A Family of Walls

1. [conceptual] Once one works on the wheel, why is the relevant obstacle the *Generalized* Riemann Hypothesis (RH for every Dirichlet L-function) rather than RH for ζ alone?

2. [moderate] The residue $209 = 11 \cdot 19$ dominates the cancellation because it is built from the two characters most anti-correlated with ζ. Trace the chain: residue $\to$ characters $\to$ L-function zeros.

3. [conceptual] At modulus 210 the internal/external split is clean; at 2310 it dissolves, and one character even flips sign. What does this show about "which L-functions oppose ζ"?

4. [moderate] The explicit formula predicts the principal-character piece (about three-quarters of the observable at modulus 210) to roughly a percent, but not the rest. Why does the boundary between predictable and not coincide with the boundary between ζ and the other L-functions?

5. [discussion] "The spectral lens resolves one wall into a coupled family of walls." Explain why this is a larger obstacle than RH alone, not a smaller one.

Chapter 8 The Barrier and the Constant

1. [routine] Verify the two Möbius identities $\mathbf{1}_{\gcd(u,v)=1} = \sum_{k|\gcd} \mu(k)$ and $\gcd(u,v) = \sum_{k|\gcd} \varphi(k)$, from $\sum_{d|n} \mu(d) = \mathbf{1}_{n=1}$ and $\sum_{d|n} \varphi(d) = n$.
2. [moderate] On the squarefree moduli $\{1, 11, 13, 143\}$ coprime to 210, form
$$B(d_1, d_2) = \mathbf{1}_{\gcd=1}/(d_1 d_2), \qquad M(d_1, d_2) = 1/\operatorname{lcm}(d_1, d_2),$$
and check the eigenvalues of $B - M$ are ≤ 0.
3. [moderate] From the centered identity $\lambda^\top(B - M)\lambda = -\sum_{k\geq 2}(\varphi(k) - \mu(k))W_k^2$, deduce that $B - M$ is negative semidefinite. Why is $\varphi(k) - \mu(k) \geq 0$ for every $k \geq 2$?
4. [conceptual] Explain how the negative-semidefiniteness is a finite-dimensional form of the parity barrier, and where Chen's theorem sits relative to it.
5. [routine] Compute the partial product $\prod_{3\leq p\leq P}(1 - \frac{1}{(p-1)^2})$ for growing P and observe convergence to $C_2 = 0.6601618\ldots$
6. [discussion] The construction reaches C_2 but not the twin conjecture. Why is computing the constant not the same as proving there are infinitely many twins?

Chapter 9 What the Zeros Don't Know

1. [conceptual] Distinguish the two kinds of knowledge: the global distribution of the primes (what the zeros encode) and the local factorization of a specific N (what factoring needs).
2. [conceptual] Explain why a proof of RH would sharpen estimates of $\pi(x)$ and speed some primality tests, yet would not yield an efficient factoring algorithm.
3. [routine] Outline Shor's algorithm as period-finding (the multiplicative order of a random a modulo N) plus a greatest-common-divisor step, and identify where the quantum Fourier transform does the work.
4. [conceptual] Distinguish the "two quantums": quantum chaos (the zeros as a spectrum) and quantum computation (Shor). Why is conflating them an error?
5. [discussion] Of the three walls, why is factoring's relationship to the spectral approach different *in kind* from that of RH and the twins?

Chapter 10 Beyond the Circle

1. [routine] Verify the splitting of rational primes in the Gaussian integers: using Fermat's two-square theorem, show $p \equiv 1 \pmod 4$ factors as $(a + bi)(a - bi)$ (it splits), that $p \equiv 3 \pmod 4$ stays prime, and that $2 = -i(1 + i)^2$ ramifies.

2. [moderate] Show $(x, y) \mapsto (-x, iy)$ maps $y^2 = x^3 - x$ to itself, has order four, and squares to $(x, y) \mapsto (x, -y)$. Explain why it realizes multiplication by i.

3. [conceptual] State the analogy precisely: roots of unity are to $\mathbb{Q}$ as the lemniscate's division points are to $\mathbb{Q}(i)$. Which theorem plays the role of Kronecker–Weber here, and of which problem is it a solved case?

4. [discussion] The lemniscate brings complex multiplication — more structure than the circle — yet the three walls persist over $\mathbb{Q}(i)$. Explain why "more structure" did not become "weaker walls," and what that says about where the difficulty resides.

5. [conceptual] The Gaussian primes are points of the lattice $\mathbb{Z}[i]$; the zeros of $\zeta_{\mathbb{Q}(i)} = \zeta \cdot L(\chi_4)$ are points of the critical strip. Explain why one cannot honestly plot the zeros on the lattice (or the primes on the line), and identify the single exact identity that legitimately relates the two. Why is its existence *not* a route across any wall?

Solutions to the Exercises

These are answer sketches, not full write-ups: enough to grade against and to run a discussion from. Exercise numbers match the *Exercises* section at the back of the book. The *two-part test* is that of Chapter 2 — a candidate Riemann operator must (i) have the zeros as its spectrum and (ii) be defined independently of them.

Chapter 1 The Symmetry of a Family

1. **Unitary:** flat at the centre ($W_U(x) = 1$) — no excess or deficit of low zeros. **Symplectic:** *suppressed* at the centre ($W(x) = 1 - \frac{\sin 2\pi x}{2\pi x} \to 0$ as $x \to 0$): zeros are repelled from the central point. **Orthogonal:** *enhanced* at the centre (even orthogonal, $W(x) = 1 + \frac{\sin 2\pi x}{2\pi x} \to 2$; the odd case carries a forced zero at the centre). So symplectic suppressed, orthogonal enhanced, unitary flat.

2. Pair correlation in the bulk follows the universal sine-kernel (unitary) law for *every* family, so it is identical across symmetry types and cannot tell them apart. The type shows only in the lowest zeros near the central point — the one-level density there — which differs by type. Bulk statistics are therefore blind to the distinction.

3. Over function fields the zeros are eigenvalues of Frobenius on cohomology: RH is a theorem (Weil), and the symmetry type follows from the monodromy group via Deligne's equidistribution (Katz–Sarnak). The setting supplies a genuine operator with a provable spectrum and a group-theoretic source for the type. Over $\mathbb{Q}$ neither exists; the one-level density is known only for test functions of restricted support (Özlük–Snyder for the quadratic family, Iwaniec–Luo–Sarnak more generally), so the type is inferred in that window, not proved.

4. The symmetry type is an *average* over a whole family; an average can hold while individual members deviate, and it is a statement about the statistics of zeros, presupposing their location rather than fixing it. So it constrains no single L-function's zeros.

5. Each L-function alone is a level-repelling (unitary) spectrum: its neighbouring zeros avoid one another, so very short gaps are rare. Merge several *independent* such sequences and they interleave blindly — neither knows where the other's

zeros sit — so a zero of one family can fall arbitrarily close to a zero of another. The short gaps that repulsion forbids *within* a family reappear *between* families, and the merged spacings relax to those of uncorrelated points ($\langle \tilde{r} \rangle \to 0.386$). Independence is the one arrangement that lands the merge exactly on Poisson; dependence shows as a departure in either direction. If the families' zeros tended to coincide or correlate, the pooled list would bunch into near-coincidences and the ratio would fall *below* 0.386; if instead the seven were jointly rigid — slices of one repelling spectrum — their mutual repulsion would survive the merge and hold the ratio *above* 0.386. The observed collapse to ≈ 0.39 is thus the statistical independence of the seven (Rudnick–Sarnak) made visible; a departure either way would have exposed a dependence among them.

Chapter 2 What Would Count as an Operator

1. (a) $\sum_p a_p \log p = \log \prod_p p^{a_p} = \log n$. (b) Unique factorization is the bijection between occupation vectors (a_p) and integers $n \geq 1$. (c) $\sum_{n\geq 1} n^{-s} = \zeta(s)$.

2. (a) $\log n \leq E \Leftrightarrow n \leq e^E$, so the count is $\lfloor e^E \rfloor \sim e^E$. (b) $\lfloor e^{14.1347} \rfloor \approx 1.38 \times 10^6$. (c) The smooth term at $T = \gamma_1$ is ≈ 0.45, so $N(\gamma_1)$ is essentially the first zero, ≈ 1; the counts differ by about *six orders of magnitude.*

3. (a) spectrum yes, independence no (circular); (b) independence yes, spectrum no; (c) spectrum yes, independence no.

4. The eigenvalues are the *input*, not a consequence, so nothing is predicted. To bear on RH the spectrum must be *forced* by a structure defined without reference to the zeros — i.e. the construction must pass independence.

5. The partition function encodes ζ as a Dirichlet series; the spectrum $\{\log n\}$ records the series' *exponents* (equivalently the Euler product). The zeros are an analytic-continuation feature, not visible as energy levels. Values and zeros are different features of the same function.

6. (a) $N_{\text{pr}}(E) \sim e^E \Rightarrow dN/dE \sim e^E$. (b) $\frac{d}{dT}\left[\frac{T}{2\pi}\log\frac{T}{2\pi} - \frac{T}{2\pi}\right] = \frac{1}{2\pi}\log\frac{T}{2\pi}$. (c) The counting function is an operator invariant; an exponential density of states cannot equal a logarithmic one, so the primon Hamiltonian is not a Riemann operator — no zero located.

Pitfall: *the recurring error is reading "reproduces ζ" (Ex. 1c) as "has the zeros as its spectrum." Ex. 5 exists to break that.*

Chapter 3 Primes as Orbits

1. The semiclassical state count for $H = xp$ gives $N(T) \approx \frac{T}{2\pi}\log\frac{T}{2\pi} - \frac{T}{2\pi}$. At $T = 100$: $15.9155 \times 2.7673 - 15.9155 \approx 28.1$ ($+\frac{7}{8} \approx 29.0$), matching the 29 actual zeros below 100.

2. The explicit formula equates a sum over zeros to a smooth (archimedean) term plus a sum over prime powers weighted by $\log p$ at "time" $k \log p$ — the shape of a trace formula (Gutzwiller): spectral sum = Weyl term + sum over periodic orbits, with p a primitive orbit of period $\log p$ and p^k its k-fold traversal.

3. xp is a classical Hamiltonian specified with no reference to the zeros, so its spectrum is not put in by hand: it passes independence. The Chapter-2 candidate that *fails* independence is the kernel assembled from the ordinates themselves (the delta/cosine kernel), which is circular. (xp shares with the primon gas that both are zero-independent; the primon gas's failure is on spectrum, not independence.)

4. "Average" is the smooth counting function / mean density of zeros, which Berry–Keating reproduces exactly. "Fluctuations" are the deviations of individual zeros from that mean; they require the detailed chaotic dynamics, and no concrete system realizing it is known, so the individual zeros are out of reach.

5. Both carry the period $\log p$, but the primon gas's spectrum is explicitly $\{\log n\}$, with exponentially growing density — wrong by construction. xp's phase-space count reproduces the correct (logarithmic) mean density, so it gets the average right and only the fluctuations are unknown. One has an explicit wrong spectrum; the other has the right smooth spectrum and an open fluctuation spectrum.

Chapter 4 An Algebra in the Primes

1. The commutators of the prime-bilinear operators are again linear combinations of the same operators — the set is closed under $[\cdot, \cdot]$ — so they span a Lie algebra, namely $\mathfrak{sp}(4, \mathbb{R}) \cong \mathfrak{so}(2, 3)$.

2. The oscillator representation acts on a Fock space graded by the parity of the number of quanta; the even and odd subspaces are each irreducible — the even one is the Rac singleton ($\Delta = \frac{1}{2}$), the odd one the Di singleton ($\Delta = 1$). The even/odd grading *is* the Rac/Di split.

3. The spectrum belongs to the chosen operator, not to the algebra around it. Embedding that operator in a richer algebra leaves its eigenvalues unchanged, so the spectrum half still reads $\{\log n\}$, not the zeros.

4. Connes: RH (for the relevant L-functions) is equivalent to the positivity of a certain trace (the Weil distribution). An equivalence relocates the difficulty — proving the positivity is exactly as hard as RH — so it is a reformulation, not a proof.

5. de Branges reduces RH to positivity/structure conditions on spaces of entire functions; Conrey–Li located a specific obstruction showing those conditions cannot hold in the required form. "Not refuted wholesale, but with a precise step

known to fail" is the normal state near this wall: approaches reach a definite, identifiable stopping point.

6. $(E_+ + E_-)^2 - (E_+ - E_-)^2 = 4E_+E_-$ by difference of squares, so $E^2 - p^2 = m^2$ holds for *all* values — it is an identity, not a measurement. Machine-precision agreement therefore carries no empirical content; the Lorentzian look comes entirely from contracting with $\eta = \mathrm{diag}(+,-,-,-)$, a choice put in by hand. A satisfied identity is not a discovered law of the primes.

Chapter 5 Hearing the Zeros

1. **Detection** establishes that a signal correlated with the zeros is present (significant against a null). **Resolution** separates and locates individual zeros. One can detect that the primes track the zeros near a height without pinning the individual ordinates.

2. The primes act like baselines in an aperture of length log(largest prime), and Fourier/Rayleigh resolution scales inversely with that aperture. So the largest prime sets the finest resolvable spacing, as a telescope's diameter sets its angular resolution.

3. Different mechanisms hitting the same limit means the limit is in the information available (the finite aperture), not in any one algorithm. A method-specific failure could be engineered around; a shared wall is the signature of an information-theoretic bound.

4. With aperture $A = \log X$ the resolvable spacing scales like $1/\log X$; halving it needs $\log X$ doubled, i.e. X squared. Resolution improves only logarithmically in X, so resolving individual zeros would take exponentially more primes — detection is feasible, resolution is not.

5. In the construction setting the dream fails because no operator with the right spectrum can be built non-circularly; in the measurement setting the analog is that the primes carry detectable but insufficient information about the zeros. Both: the link is real, but stops short of pinning the zeros down.

Chapter 6 Laws That Died

1. $\{3\}(0,0)$ →U; $\{5\}(1,0)$ →O; $\{7\}(0,1)$ →O; $\{15\}(1,0)$ →O; $\{21\}(0,1)$ →O; $\{35\}(1,1)$ →U; $\{105\}(1,1)$ →U. OVER = $\{5,7,15,21\}$, UNDER = $\{3,35,105\}$; the prime 3 never affects the sign.

2. The cutoff X and the modulus M are independent axes. Large X shows a pattern is not an artifact of small primes; a modulus lift shows it is not an artifact of the particular wheel. Passing one says nothing about the other — XOR passed X to 10^9 and failed the lift to 2310.

3. A property that flips from 100th percentile to below the random average depending on the observable is a property of the *observable*, not the split. The "extremality" is therefore observable-dependent, not intrinsic.

4. A proposed effect must be at least the size of what it explains. The archimedean piece is three orders of magnitude too small, so whatever its sign or structure it cannot close the gap — a magnitude check disqualifies it at once.

5. Recording falsified hypotheses in order forces the line between what survived testing and what merely appeared; it blocks quiet retention of dead patterns and the cherry-picking that turns coincidence into "law." That discipline is exactly what numerology lacks.

6. It should at least: survive a modulus lift (wheel-specific coincidence), survive a change of observable (measurement-dependence), beat a null/random ensemble (chance), and have any proposed mechanism checked for magnitude (size). Each guards a failure mode seen in the chapter.

Chapter 7 A Family of Walls

1. Every wheel observable is a prime sum sorted by residue, governed through the explicit formula by the zeros of *all* the wheel's Dirichlet L-functions, not just ζ. Controlling it needs RH for the whole family — GRH — strictly stronger than RH for ζ.

2. $209 = 11 \cdot 19$ dominates the cancellation; it is built from the conductor-11 and conductor-19 characters; those carry the L-functions most anti-correlated with ζ $(-0.82, -0.77)$, i.e. whose zeros push hardest against ζ's. Residue $\to$ characters $\to$ L-function zeros.

3. At mod 210 internal characters move with ζ, external against; at mod 2310 the split dissolves and χ_{-19} flips sign. So "which L-functions oppose ζ" is not fixed — it depends on which primes divide the wheel.

4. The explicit formula can sum ζ's zeros, so it predicts the principal piece (built from ζ) to $\sim$1%. The other pieces are built from the non-principal L-functions, whose zeros it lacks. So the predictable/unpredictable boundary is exactly ζ vs the family — RH vs GRH.

5. RH concerns one function's zeros; the signal depends on the joint placement of an entire family's zeros, with couplings that reorganize per wheel, so there is no fixed reduction to one problem. The lens reveals a coupled, wheel-dependent family of obstacles — larger than the single wall, not a way past it.

Chapter 8 The Barrier and the Constant

1. From $\sum_{d|n} \mu(d) = \mathbf{1}_{n=1}$ with $n = \gcd(u, v)$, and noting the common divisors of u, v are the divisors of gcd, get $\mathbf{1}_{\gcd=1} = \sum_{k|\gcd} \mu(k)$. From $\sum_{d|n} \varphi(d) = n$ with $n = \gcd$, get $\gcd = \sum_{k|\gcd} \varphi(k)$.
2. On $\{1, 11, 13, 143\}$ the $d = 1$ row and column of $B - M$ vanish; the block on $\{11, 13, 143\}$ has diagonal $(-\frac{1}{11}, -\frac{1}{13}, -\frac{1}{143})$ with small off-diagonals, and its eigenvalues are all ≤ 0. So $B - M$ has one zero eigenvalue (the $d = 1$ direction) and three negative — negative semidefinite.
3. The identity writes $\lambda^\top(B - M)\lambda$ as minus a sum of $(\varphi(k) - \mu(k))$ times squares. Since $\varphi(k) \geq 1$ and $\mu(k) \leq 1$, each coefficient $\varphi(k) - \mu(k) \geq 0$ (indeed > 0 for $k \geq 2$), so the form is ≤ 0 for all λ: $B - M \preceq 0$.
4. Negative-semidefiniteness means no weights λ make the centered form positive — exactly the parity barrier (sieves cannot force a positive lower bound for twins) as the sign of a finite matrix. Chen's theorem sits at the edge: it gives everything up to the barrier (infinitely many p with $p+2$ prime or a product of two primes), not the step across.
5. The partial products $\prod_{3 \leq p \leq P}(1 - 1/(p-1)^2)$ decrease toward $C_2 = 0.6601618\ldots$, agreeing to several digits well before P reaches the thousands.
6. C_2 is the constant in the Hardy–Littlewood asymptotic; computing it is evaluating a convergent product and presupposes nothing about whether the asymptotic — hence the infinitude of twins — holds. Reaching the number that would count the twins is not proving there are infinitely many to count.

Pitfall: *in Ex. 3, students sometimes try to show $\varphi(k) - \mu(k) > 0$ case-by-case; the one-line bound $\varphi \geq 1 \geq \mu$ suffices.*

Chapter 9 What the Zeros Don't Know

1. The zeros encode *global* information — counts and fluctuations of the primes in aggregate ($\pi(x)$ and its error). Factoring needs *local* information — the prime decomposition of one N. Knowing the distribution near $\sqrt{N}$ does not say which prime divides N.
2. RH sharpens the error in $\pi(x)$ (to $\sim \sqrt{x} \log x$), tightening prime-counting, gap bounds, and some primality tests — none of which factor a given N. The number field sieve uses no zero data, and no reduction from RH to fast factoring is known.
3. Pick random a coprime to N; find the multiplicative order r of a mod N (the period of a^x mod N) by a quantum Fourier transform over the exponents; if r is even and $a^{r/2} \neq -1$, then $\gcd(a^{r/2} \pm 1, N)$ is a nontrivial factor. The speedup lives entirely in the QFT period-finding; the rest is classical gcd.

4. **Quantum chaos:** the conjecture that the zeros are the spectrum of a chaotic quantum system (Hilbert–Pólya / Berry–Keating). **Quantum computation:** Shor's period-finding via the QFT on the exponent register. Same word, different mathematics and different objects; conflating them wrongly suggests the device that threatens RSA bears on RH.

5. For RH and the twins the lens reaches the wall and stops there (it even reaches C_2). For factoring it does not point at the wall at all: the zeros are about distribution, factoring about local structure — so the spectral approach is not a stalled attempt but an irrelevant one. A different kind of relationship.

Chapter 10 Beyond the Circle

1. $p \equiv 1 \pmod 4 \Rightarrow p = a^2 + b^2$ (Fermat) $= (a + bi)(a - bi)$, two non-associate Gaussian primes — p splits. $p \equiv 3 \pmod 4$ is not a sum of two squares and stays prime (inert), of norm p^2. And $2 = (1+i)(1-i) = -i(1+i)^2$ up to a unit — a square of one prime, hence ramified.

2. $(-x)^3 - (-x) = -(x^3 - x) = -y^2 = (iy)^2$, so the image lies on the curve. Twice gives $(x, i^2y) = (x, -y)$ (the $[-1]$ map); four times the identity — order four. Squaring to $[-1]$ is the relation $i^2 = -1$, so the automorphism is multiplication by i.

3. Dividing the curve yields generators of the field's abelian extensions. For $\mathbb{Q}(i)$ the role of Kronecker–Weber is played by the main theorem of complex multiplication (the lemnatomic analogue); it is the solved case of Hilbert's twelfth problem for the imaginary quadratic field $\mathbb{Q}(i)$.

4. CM delivers analytic continuation and functional equations for the Hecke L-functions and the first partial Birch–Swinnerton-Dyer results (Coates–Wiles), so the Gaussian world is in many ways better understood. Yet GRH for those L-functions, the Gaussian twin and gap problems, and factoring all remain open. The extra structure improved what can be proved *around* the walls without moving them: the obstruction lies in the placement of the zeros, the parity of sieves, and the locality of factorization — which the change of base field reproduces rather than removes. The difficulty is a property of the primes, not of the curve.

5. In position you can name a Gaussian prime $(1 + i,\ 2 + i)$; the zeros live in the s-plane on $\operatorname{Re} s = \frac{1}{2}$ — different coordinates, different objects, so plotting one set on the other's axes asserts a coincidence that does not exist. The honest link is Weil's explicit formula, an exact equality between a weighted sum over the Gaussian primes and a weighted sum over the zeros. It is an identity, not an inequality or a localization, so it relocates the difficulty (a statement about the zeros becomes an equivalent statement about the primes) without resolving it: GRH, twins, and factoring stay exactly as open on either side.

BOOK III

Scaling the Wall

Acknowledgements

The author gratefully acknowledges the use of Anthropic's Claude as a computational and drafting assistant during the preparation of this work. Any mathematical results or interpretations presented here remain entirely the responsibility of the author.

Introduction: Scaling the Wall

The thesis of this book can be put in one sentence: *the primes surrender their local structure freely and their global structure not at all*, and the Riemann Hypothesis is the sharpest statement of that refusal. The wheel—residues, densities, the singular series, the characters—is the local law, and it is known. Where the zeros lie, whether two primes recur forever, how to undo a product: these are the global facts, and they are walled. Every chapter here approaches the wall along a different road, and every road runs out at the same kind of edge.

Each chapter has one shape. It begins with a *lure*: a genuine method, grown from the wheel, that seems for a while to lead over the wall. It *climbs*: the method is developed with its real equations and its honest gains. And it *turns*: in the same chapter, the deflation is delivered—the precise point at which the method is beaten back, and the reason. The reader is lured in, but never deceived.

The book is in three parts. **Part I, The Wheel and the Wall**, lays the foundation—the wheel, the Euler product, the mirror, the explicit formula, the wall itself—and then climbs one floor up to Ramanujan's τ, where the analogue of the Hypothesis is a proven theorem. **Part II, The Five Faces of the Hypothesis**, takes five reformulations, each *provably equivalent* to the Riemann Hypothesis—the oscillation of the prime-counting error, the positivity of Li's and Weil's functionals, the completeness of a span of fractional parts, an extremal inequality on the divisor sum, and the constant of a heat flow—and shows each one to be the same wall seen from a new side. **Part III, The Walls**, steps back: it visits the neighbouring walls that ring the Hypothesis (the random-matrix statistics of the zeros, the undiscovered operator, the weaker Lindelöf shadow, and the geometry where the Hypothesis is already a theorem), walks the twin-prime corridor to its parity barrier, and closes with the three great walls together—location, recurrence, and decomposition—and the single situation that runs through all three.

The closing chapter states the unifying lesson plainly: at each wall, the lever that would cross it turns out to be the very thing being sought. To control the zeros one needs the zeros; to isolate two primes one needs to break parity, which is to isolate two primes; to factor cheaply one needs to undo multiplication, which is to factor. The destination is the tool. That circularity is not a failure of ingenuity—it is the structure of the subject, and to map it exactly is the work of this book.

A word on the computations. Each chapter ends with a verification: a short, deterministic program using only the Python standard library, which checks the chapter's claims to a stated range and prints its results. They are reproductions, not decorations—the reader can run them. Where a fact is elementary it is checked exactly; where it is deep, a faithful model or a finite instance is checked, and the limits of the check are stated. Plain language and full equations stand side by side throughout, and the figures are meant to carry the argument as much as the prose.

The wheel turns at the foot of the wall. What follows is the climb.

Part I

The Wheel and the Wall

Chapter 1

The Wheel and the Wall

This is the foundation of an advanced book, and its one job is to make the rest self-contained. It assembles, in one place and in this book's own language and pictures, everything the later chapters stand on: the wheel as the Euler product—the base*— the family of L-functions the wheel carries, the mirror that is the functional equation, the explicit formula that is the hinge between the primes and the zeros, and the Riemann wall that every later chapter will climb toward and be turned back from. None of it is new. All of it is classical, and the two books before this one develop it slowly and with care. The reason to gather it here, at speed and in a single vantage, is that what follows is a survey—a broad sweep over how the Riemann Hypothesis has been attacked, and how each attack was beaten back— and a survey needs a place to stand. This chapter is that place. A reader who has the earlier books can move through it quickly; a reader who does not can still follow everything after it.*

Scope

This chapter consolidates the two earlier volumes. It assumes only undergraduate analysis and a first course in number theory; everything beyond that it builds in passing. Each section states an object the rest of the book will use—the wheel, the Euler product, the family, the functional equation, the explicit formula, the wall— gives the equation, says in plain words what the equation means, and points to where the earlier books treat it in full. The value here is not novelty but vantage: one base, one language, and one picture (Figure 1.2) that the entire survey hangs from.

1.1 The wheel

Start with the primes' simplest secret: most numbers cannot be prime, and you can see which ones at a glance. A number divisible by 2, 3, 5, or 7 is composite (past those primes themselves), so every prime beyond 7 must fall in a residue class modulo $210 = 2 \cdot 3 \cdot 5 \cdot 7$ that shares no factor with 210. There are exactly

$$\varphi(210) = 210\left(1 - \tfrac{1}{2}\right)\left(1 - \tfrac{1}{3}\right)\left(1 - \tfrac{1}{5}\right)\left(1 - \tfrac{1}{7}\right) = 48 \tag{1.1}$$

such classes. Lay the numbers 1 to 210 around a circle and keep only those 48 spokes: that is *the wheel.* Every prime past 7 lands on a spoke; the other 162 positions are barred to primes forever. The wheel is the primes' local law—a statement about residues, decided by the small primes alone, that holds for all integers at once.

The surviving spokes are not all prime. The number 1 sits on one. The primes from 11 up sit on others. And some spokes carry composites that simply have no small factor—numbers like $121 = 11^2$, $143 = 11 \cdot 13$, $169 = 13^2$. Following the earlier books we call these the *subprimes*: composite, coprime to 210, least prime factor at least 11. The first is 121. The wheel cannot tell a prime from a subprime—both are survivors—and that blindness, harmless here, will reappear at one of the walls.

Nothing about 210 is special except convenience. Multiply in the next prime and the wheel grows: $2 \to 6 \to 30 \to 210 \to 2310 \to \cdots$, the primorials, each a finer sieve than the last. The wheel is a family of sieves, and which one you use is a matter of how much resolution you want.

1.2 The Euler product: the wheel becomes a function

Now turn the wheel into a single analytic object. For a complex number s with $\operatorname{Re} s > 1$ define the Riemann zeta function by the sum over all integers,

$$\zeta(s) = \sum_{n=1}^{\infty} \frac{1}{n^s}. \tag{1.2}$$

Because every integer factors uniquely into primes, this sum equals a product over primes—Euler's identity:

$$\zeta(s) = \prod_p \frac{1}{1 - p^{-s}}. \tag{1.3}$$

The reason is one line. Expand each factor as a geometric series, $\left(1 - p^{-s}\right)^{-1} = 1 + p^{-s} + p^{-2s} + \cdots$, multiply the factors over all primes, and every term of the product is $\left(p_1^{a_1} \cdots p_k^{a_k}\right)^{-s} = n^{-s}$ for exactly one integer n, by unique factorization. The sum over integers and the product over primes are the same object.

Equation (1.3) is the wheel made into a function. Each prime contributes one factor; the wheel modulo 210 is what you get by stripping off the first four of them, the factors for $2, 3, 5, 7$. In plain terms: ζ is the primes, multiplied together. Everything in this book that looks analytic—zeros, critical lines, explicit formulas—enters only through this product, which is why the wheel sits at the base of all of it. The verification confirms the identity numerically, $\prod_p (1 - p^{-2})^{-1} = \zeta(2) = \pi^2/6$, and recovers the wheel's survivor density $\frac{48}{210} = \prod_{p|210}(1 - 1/p)$ as the product of its local factors.

1.3 The family the wheel carries

The wheel does not carry ζ alone. To each residue class one can attach a *Dirichlet character* χ, a function on the integers that is periodic modulo the wheel and multiplicative, $\chi(mn) = \chi(m)\chi(n)$. Each character has its own zeta-like function, again with an Euler product,

$$L(s,\chi) = \sum_{n=1}^{\infty} \frac{\chi(n)}{n^s} = \prod_p \frac{1}{1 - \chi(p)\, p^{-s}}. \tag{1.4}$$

The ordinary zeta function is the case of the trivial character, $\chi \equiv 1$. The wheel of 210 carries a whole family of these—among them seven nontrivial real characters, indexed by the seven nontrivial divisors of $105 = 3 \cdot 5 \cdot 7$, which the first book draws as a Boolean lattice of conductors. We will not need the lattice in detail here; what matters for the survey is that the wheel's finer structure—the race between residue classes, the cancellations in its sums—is governed not by ζ alone but by this entire family. That is why several attacks in this book run not into the Riemann Hypothesis but into its generalization, the same statement asked of every $L(s,\chi)$ at once.

1.4 The mirror: the functional equation

The zeta function, defined by the sum only for $\operatorname{Re} s > 1$, extends to the whole complex plane, and the extension has a symmetry. Completing ζ with the factors that belong to the prime "at infinity,"

$$\xi(s) = \tfrac{1}{2}\, s(s-1)\, \pi^{-s/2}\, \Gamma\!\left(\tfrac{s}{2}\right) \zeta(s), \tag{1.5}$$

produces an entire function obeying the clean reflection

$$\xi(s) = \xi(1-s). \tag{1.6}$$

In plain words: the completed function is unchanged when you reflect the plane across the vertical line $\operatorname{Re} s = \frac{1}{2}$. That line is the *critical line*, the mirror's axis. The zeta function has "trivial" zeros at the negative even integers $-2, -4, -6, \ldots$, forced by the poles of the gamma factor, and all its other—"nontrivial"—zeros ρ lie in the critical strip $0 < \operatorname{Re} s < 1$. By the symmetry (1.6) they come in pairs $\rho, 1-\rho$, placed mirror-image across the critical line. The first book reaches this same reflection from the other side, showing that the wheel's geometric mirror $r \mapsto 210 - r$ and the analytic mirror $s \mapsto 1 - s$ are one fact, hinged by the parity $\chi(-1)$.

1.5 The hinge: the explicit formula

Here is the object the whole book swings on. Count the primes in a weighted way with the Chebyshev function

$$\psi(x) = \sum_{p^k \le x} \log p, \tag{1.7}$$

a staircase that jumps by $\log p$ at every prime power. Riemann and von Mangoldt connected this staircase to the zeros by an exact identity, the *explicit formula*:

$$\psi(x) = x - \sum_{\rho} \frac{x^\rho}{\rho} - \log(2\pi) - \tfrac{1}{2}\log(1 - x^{-2}), \tag{1.8}$$

the sum running over all nontrivial zeros ρ. Read it slowly, because everything later depends on it. The left side is built from primes; the right side is built from zeros. The leading term x is the average growth; every wobble of the staircase around that average is the sum $\sum_\rho x^\rho/\rho$ over the zeros. Pairing a zero $\rho = \frac{1}{2} + i\gamma$ with its mirror $\frac{1}{2} - i\gamma$, each pair contributes a wave $2\sqrt{x}\,\left(\frac{1}{2}\cos(\gamma \log x) + \gamma \sin(\gamma \log x)\right)/(\frac{1}{4} + \gamma^2)$, an oscillation whose size is set by $\sqrt{x}$ when the zero sits on the critical line.

So the primes and the zeros are not two subjects but two readings of one identity. The zeros are the frequencies; the primes are the tune. Figure 1.1 shows this literally: feed the first sixty zeros into Eq. (1.8) and the staircase reassembles, jump by jump, sharpening as more zeros are added. The verification measures the agreement—a correlation above 0.9 from sixty zeros. This is the hinge. Every attack in the chapters to come is, at bottom, an attempt to control the left side by controlling the right, or the reverse, and the explicit formula is the door each one walks through.

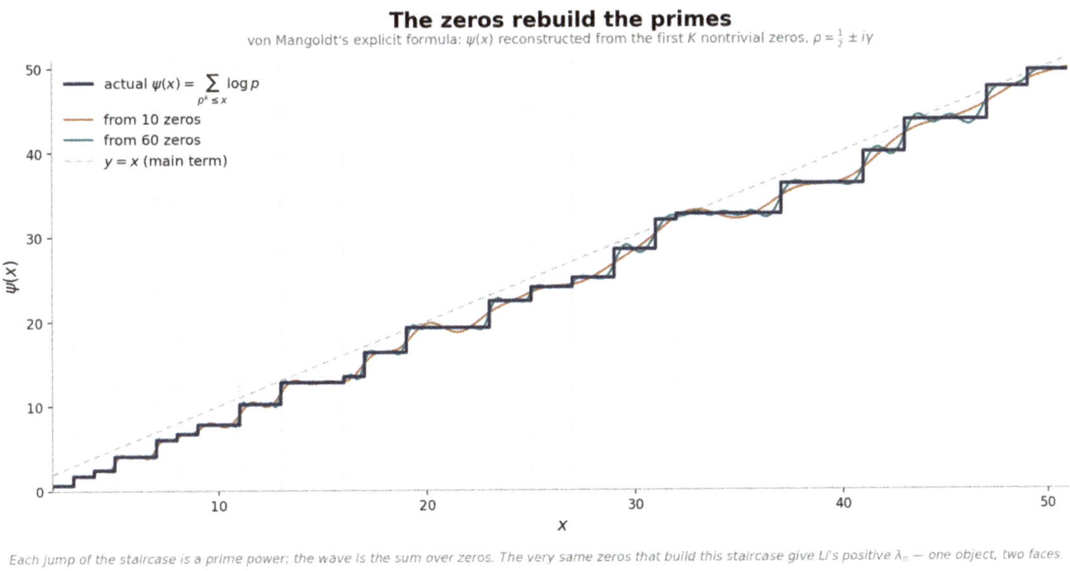

Figure 1.1: The hinge in one picture. Von Mangoldt's explicit formula (1.8) reconstructs the prime staircase $\psi(x)$ from the first K nontrivial zeros $\rho = \frac{1}{2} \pm i\gamma$: ten zeros give the slow swell, sixty trace the steps. The primes are downstream of the zeros, and this is what that means.

1.6 The wall

Where do the zeros sit? In the strip, in conjugate pairs, mirror-symmetric about the critical line. The *Riemann Hypothesis* is the claim that they sit *on* it:

$$\text{every nontrivial zero } \rho \text{ has } \operatorname{Re} \rho = \tfrac{1}{2}. \tag{1.9}$$

By the explicit formula this is not an idle statement about a function; it is the assertion that the primes are as regular as arithmetic allows. If every ρ has real part $\frac{1}{2}$, then

$|x^\rho| = \sqrt{x}$ and the wobble in Eq. (1.8) is held to the smallest possible size, giving $\psi(x) = x + O\big(\sqrt{x}\,(\log x)^2\big)$; if even one zero drifts off the line, the primes fluctuate more wildly than that. The Hypothesis is the wall this book is named for, and Figure 1.2 is the book's map: the wheel and its Euler product at the foot, the family and the mirror and the hinge built on top, and the wall at the summit.

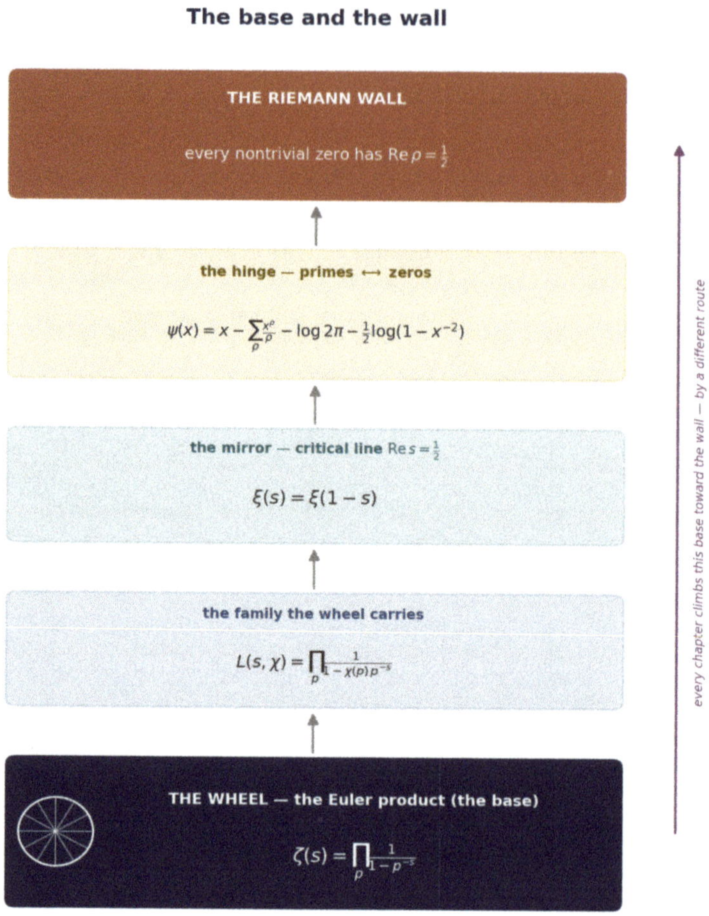

Figure 1.2: The base and the wall. The wheel—the Euler product—is the foundation; the family of L-functions, the functional-equation mirror, and the explicit-formula hinge are built on it; the Riemann Hypothesis is the wall at the top. Every chapter that follows climbs this same base toward the same wall, by a different route.

The Hypothesis is one wall of three the books circle. The second is the twin primes: whether two primes two apart recur forever, a question whose density the wheel hands over and whose existence it does not. The third is factoring: that a number is easy to build from primes and hard to break back into them. All three rise from the same base. This volume concentrates on the first and its many disguises, with the other two standing at the close.

1.7 The plan: climbing the base

The rest of the book is a single motion, repeated. Each chapter takes one genuine attack on the Riemann wall—a real method, named and credited—starts it from the base just assembled, and follows it up until it stops. We will watch the lure of each method completely, granting it every real thing it offers, and then name, at the top, exactly where and why it is turned back. The turns, chapter to chapter, will start to rhyme.

The climb begins one floor up from ζ, with Ramanujan's τ and its L-function, to see what changes when the object over the wheel has rank two instead of one. Then the survey proper: the reformulations of the Hypothesis, taken by *face*. The oscillation face—the prime number theorem's error term, the Mertens and Liouville sums—where the cancellation one must prove *is* the zeros, and where two beautiful strengthenings turned out to be false. The positivity face, Li's criterion and Weil's, where the wall becomes a sequence that must stay non-negative. The completeness face, Nyman and Beurling's, where it becomes an approximation in a Hilbert space. The extremal face, Robin's and Lagarias's inequalities on the divisor function. The regularity face, the Laguerre–Pólya class and the de Bruijn–Newman constant. Then the neighbouring walls—the spacing of the zeros, the size of ζ on the line, the search for an operator, the geometry that proves the Hypothesis in another world—each reached from the same wheel and each a different barrier. And at the end, the three walls together.

Every one of these will be written the same way: the full equations, the plain-language reading of what they say, and the honest account of the wall they meet—which, the survey will show, is in the end one wall wearing many faces.

What this is and is not

What this chapter establishes is genuine and entirely classical: the Euler product, the Dirichlet family, the functional equation, the explicit formula, and the statement of the Hypothesis, assembled into one base with one set of pictures so that the rest of the book can stand alone. The verification checks the load-bearing pieces—that the Euler product is ζ, that the wheel is its first factors, that the explicit formula rebuilds the primes from the zeros.

What it is not is a step toward crossing the wall. Nothing here is new, and nothing here is meant to be; this is the ground, deliberately. Its only claim is that everything the survey attacks rises from exactly this foundation, so that when each later method is turned back, we will be able to see the turn against a base we already hold in common.

Verification

The script below, standard library only and deterministic, checks the foundation: the Euler product equals ζ at $s = 2$ and $s = 4$; the wheel's survivor density is the product of its local Euler factors, $\varphi(210)/210 = 48/210$; the 48 survivors are the unit, the primes from 11 up, and the subprimes (the first being $121 = 11^2$); and the explicit formula rebuilds the prime staircase from the first sixty zeros.

```
#!/usr/bin/env python3
"""Verification for 'The Wheel and the Wall' (Book Three, Chapter 1).
Standard library only; deterministic. Establishes the base: the Euler product IS
the wheel made into a function, and the explicit formula is the hinge to the zeros."""
import math

checks=0
def ok(cond,msg):
    global checks
    assert cond,"FAIL: "+msg
    checks+=1; print("PASS:",msg)

def primes_upto(n):
    s=bytearray([1])*(n+1); s[0]=s[1]=0; i=2
    while i*i<=n:
        if s[i]: s[i*i::i]=bytearray(len(s[i*i::i]))
        i+=1
    return [i for i in range(2,n+1) if s[i]]

# ===== the Euler product IS zeta: prod_p (1 - p^-s)^-1 = sum_n n^-s =====
P=primes_upto(200000)
def euler(s):
    pr=1.0
    for p in P: pr*=1.0/(1.0-p**(-s))
    return pr
z2=euler(2); z4=euler(4)
print(f"  Euler product s=2 -> {z2:.6f}  (pi^2/6={math.pi**2/6:.6f})")
print(f"  Euler product s=4 -> {z4:.6f}  (pi^4/90={math.pi**4/90:.6f})")
ok(abs(z2-math.pi**2/6)<1e-4, "Euler product over primes equals zeta(2)=pi^2/6")
ok(abs(z4-math.pi**4/90)<1e-6, "Euler product over primes equals zeta(4)=pi^4/90")

# ===== the wheel is the first Euler factors: density = prod_{p|210}(1-1/p) =====
wheel_primes=[2,3,5,7]
dens=1.0
for p in wheel_primes: dens*=(1-1.0/p)
phi210=sum(1 for r in range(1,211) if math.gcd(r,210)==1)
print(f"  wheel survivor density = {dens:.6f} = {phi210}/210; phi(210)={phi210}")
ok(abs(dens-phi210/210.0)<1e-12 and phi210==48,
   "wheel density prod(1-1/p)=phi(210)/210=48/210 (local Euler factors)")

# the survivors split into the unit, the primes >=11, and the subprimes (lpf>=11)
surv=[r for r in range(1,211) if math.gcd(r,210)==1]
primeset=set(primes_upto(210))
prime_surv=[r for r in surv if r in primeset]
subprime=[r for r in surv if r not in primeset and r!=1]
ok(surv[0]==1 and min(subprime)==121,
   "the 48 survivors: unit 1, primes >=11, and subprimes (first is 121=11^2)")

# ===== the hinge: the explicit formula rebuilds the primes from the zeros =====
GAMMA=[14.13472514,21.02203964,25.01085758,30.42487613,32.93506159,37.58617816,
 40.91871901,43.32707328,48.00515088,49.77383248,52.97032148,56.4462477,59.347044,
 60.83177852,65.11254405,67.07981053,69.54640171,72.06715767,75.7046907,77.14484007,
 79.33737502,82.91038085,84.73549298,87.42527461,88.80911121,92.49189927,94.65134404,
 95.87063423,98.83119422,101.31785101,103.72553804,105.44662305,107.16861118,
 111.02953554,111.87465918,114.32022092,116.22668032,118.79078287,121.370125,
 122.94682929,124.25681855,127.51668388,129.5787042,131.08768853,133.4977372,
 134.75650975,138.11604205,139.73620895,141.1237074,143.1118458,146.00098249,
 147.42276534,150.05352042,150.92525761,153.02469381,156.11290929,157.59759182,
 158.84998817,161.18896414,163.03070969]
def psi_zeros(x):
```

```
    L=math.log(x); s=0.0
    for g in GAMMA: s+=(0.5*math.cos(g*L)+g*math.sin(g*L))/(0.25+g*g)
    return x-math.log(2*math.pi)-0.5*math.log(1-x**-2)-2*math.sqrt(x)*s
def psi_actual(x):
    t=0.0
    for p in primes_upto(int(x)):
        q=p
        while q<=x: t+=math.log(p); q*=p
    return t
xs=[2+0.2*k for k in range(1,490)]
A=[psi_actual(x)-x for x in xs]; B=[psi_zeros(x)-x for x in xs]
ma=sum(A)/len(A); mb=sum(B)/len(B)
cov=sum((a-ma)*(b-mb) for a,b in zip(A,B))
corr=cov/math.sqrt(sum((a-ma)**2 for a in A)*sum((b-mb)**2 for b in B))
print(f"  explicit-formula hinge: corr(psi-x, zeros) = {corr:.3f} from 60 zeros")
ok(corr>0.85,
   "the explicit formula rebuilds the primes from the zeros (the hinge)")

print(f"\nALL {checks} CHECKS PASSED")
```

Chapter 2

One Floor Up: Ramanujan's τ and Its L-Function

The opening chapter laid the abelian ground floor of the building. The wheel's Dirichlet characters were each a one-dimensional Frobenius symbol $\chi(p)$; each opened a Dirichlet L-function with its own critical line and its own copy of the wall; and together they are the wheel's abelian family. This chapter climbs one flight. Ramanujan's tau function and its L-function are the first object past the wheel's characters, and the striking thing is how little has to be relearned: the same Euler product, the same functional equation, the same critical line, the same wall—only now two-dimensional instead of one, modular instead of cyclotomic, and standing on the very hyperbolic surface where Selberg's trace formula lives. One piece of its wall has even been knocked down, by geometry, exactly where a real geometry was always the thing that could bring a wall down.

Scope

The foundation reached the abelian ground floor and stopped one dimension short. The wheel's characters were the Frobenius data of the abelian extensions of $\mathbb{Q}$—a single root of unity $\chi(p)$ recording how each prime split—and that is as far as the wheel's own arithmetic reaches. This chapter takes one famous object from just past that edge—Ramanujan's tau function $\tau(n)$ and its Dirichlet series $L(s, \Delta)$—and sets it beside the wheel to see what carries over. Almost everything does, and that is the point. The wheel's L-functions are the simplest members of an enormous family, and $L(s, \Delta)$ is the next member up: the abelian ground floor gives way to a non-abelian first floor, the Kronecker–Weber world of roots of unity gives way to the world of modular forms, and the Frobenius character $\chi(p)$ that ran the whole apparatus becomes the trace of a two-by-two matrix. The furniture is identical; only the dimension changes. We mark, as always, which resemblances are genuine structure, which are rhymes, and which would be mistakes to lean on. And we note where the wall, here, has actually fallen—because for this object, unlike for the Riemann zeta, a piece of it has.

2.1 The function, and where it comes from

Tau is not a free-standing definition; it is the list of coefficients of one specific modular form, the discriminant

$$\Delta(q) \;=\; q\prod_{n\geq 1}(1-q^n)^{24} \;=\; \sum_{n\geq 1}\tau(n)\,q^n, \qquad q=e^{2\pi i z}.$$

This Δ is a cusp form of weight 12 for the full modular group $\mathrm{SL}(2,\mathbb{Z})$, and in fact the unique normalized one—weight 12 is the lowest weight at which a cusp form for the modular group exists at all, and the space it lives in is one-dimensional. The first values are concrete and unremarkable-looking:

$$\tau(1)=1,\quad \tau(2)=-24,\quad \tau(3)=252,\quad \tau(4)=-1472,$$
$$\tau(5)=4830,\quad \tau(6)=-6048,\quad \tau(7)=-16744.$$

What is remarkable is the hidden order Ramanujan saw in them in 1916. The function is *multiplicative*: $\tau(mn)=\tau(m)\tau(n)$ whenever m and n are coprime, and at prime powers it obeys the recursion

$$\tau(p^{r+1}) \;=\; \tau(p)\,\tau(p^r)-p^{11}\,\tau(p^{r-1}).$$

Mordell proved this a year later, by an argument that became the theory of Hecke operators: Δ is a simultaneous eigenform of all of them, and the $\tau(p)$ are its eigenvalues. That single fact is the engine behind everything that follows, because multiplicativity is exactly what turns a Dirichlet series into an Euler product. Here it gives

$$L(s,\Delta) \;=\; \sum_{n\geq 1}\frac{\tau(n)}{n^s} \;=\; \prod_p\left(1-\tau(p)\,p^{-s}+p^{11-2s}\right)^{-1}.$$

2.2 The same furniture, one dimension higher

Set this Euler product beside the wheel's. A Dirichlet L-function of the wheel has a *degree-one* factor at each prime,

$$L(s,\chi)=\prod_p\left(1-\chi(p)\,p^{-s}\right)^{-1},$$

one prime and one root of unity. The tau L-function has a *degree-two* factor, a genuine quadratic in p^{-s}. That is the whole difference between the two floors. The wheel lives over the abelian extensions of $\mathbb{Q}$—the cyclotomic world of Kronecker and Weber, the GL(1) story. The tau function lives one rung up, in the GL(2) world of modular forms, where the symmetry behind the Euler product is no longer a character valued on the unit circle but a two-dimensional representation. This is the first step onto the Langlands ladder, and the wheel's characters were its bottom rung.

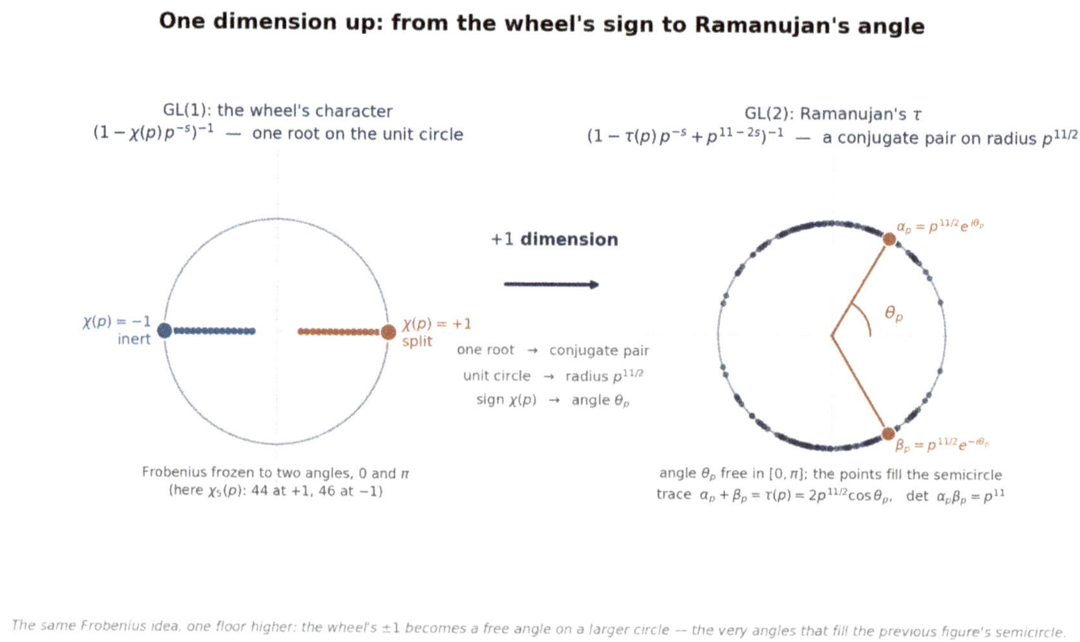

Figure 2.1: From the wheel's sign to Ramanujan's angle—the dictionary of this chapter, drawn. *Left*: the wheel's degree-one Euler factor has a single inverse root, the character value $\chi(p)$, on the unit circle; for a real character it is frozen to two angles, 0 ($+1$, p splits) and π (-1, p inert), and the dots are primes sorted by the conductor-5 character χ_5. *Right*: the degree-two factor of τ has a *conjugate pair* $\alpha_p, \beta_p = p^{11/2}e^{\pm i\theta_p}$ on the circle of radius $p^{11/2}$, with trace $\alpha_p+\beta_p = \tau(p) = 2p^{11/2}\cos\theta_p$ and determinant $\alpha_p\beta_p = p^{11}$ (the pair for $p=11$, at $\theta = 60°$, drawn in red). Climbing one dimension, the wheel's ±1 unfreezes into the angle θ_p, and the two allowed points become a pair free to fill the circle—the same angles whose distribution is the semicircle of Figure 2.2.

Everything else transfers intact. Hecke proved that $L(s,\Delta)$ extends to an entire function of s and satisfies a functional equation: completing it as $\Lambda(s) = (2\pi)^{-s}\,\Gamma(s)\,L(s,\Delta)$, one has

$$\Lambda(s) = \Lambda(12-s),$$

the same reflection symmetry that organized ζ and the wheel's L-functions, now folded about the line $\mathrm{Re}(s)=6$ instead of $\mathrm{Re}(s)=\frac{1}{2}$. (Shift by $\frac{11}{2}$ to normalize and the centre returns to $\frac{1}{2}$, and $L(s,\Delta)$ becomes a standard degree-two L-function with $s \leftrightarrow 1-s$.) And with the functional equation comes its own Riemann Hypothesis: the nontrivial zeros of $L(s,\Delta)$ are conjectured to lie on that central line. It is open, in precisely the way the wheel's Generalized Riemann Hypothesis is open. The wall did not soften by climbing a floor; it reappeared in the same place.

2.3 $\tau(p)$ is a Frobenius trace

The deepest line of the dictionary is the one that turns τ back into the language of the wheel. Attached to Δ there is a two-dimensional Galois representation—constructed by Deligne—

$$\rho\colon \operatorname{Gal}(\overline{\mathbb{Q}}/\mathbb{Q}) \longrightarrow \mathrm{GL}_2, \qquad \operatorname{tr}\rho(\mathrm{Frob}_p) = \tau(p), \qquad \det\rho(\mathrm{Frob}_p) = p^{11}.$$

Read that against the wheel. There, the Frobenius at p was the character value $\chi(p)$, a single number recording how p split in an abelian extension—the splitting law the wheel's characters obey in the quadratic fields. Here it is the *trace* of a two-by-two Frobenius matrix. The tau values are Frobenius data, exactly as the wheel's character values were—only promoted from one dimension to two. The same idea, the same word, a larger matrix.

This is why Ramanujan's congruences feel so much like the wheel even though they come from a modular form. The most famous,

$$\tau(n) \equiv \sigma_{11}(n) \pmod{691},$$

ties τ to the elementary divisor-power sum modulo 691—a prime that appears only because it divides the numerator of the Bernoulli number B_{12}, the same weight-12 fingerprint. Sharper still is the behaviour modulo 23: there $\tau(p)$ is governed by how p splits, vanishing precisely when p is a non-residue, so that the congruence is reading Frobenius in a field tied to $\mathbb{Q}(\sqrt{-23})$. That is the wheel's own logic—small-prime residue conditions steering arithmetic, exactly the splitting laws of the wheel's own characters—surfacing inside a degree-two object. Serre and Swinnerton-Dyer mapped out exactly which primes $(2, 3, 5, 7, 23, 691)$ carry such congruences, and why: they are the primes at which the Galois representation is smaller than it generically should be.

2.4 The geometry, and the wall that fell

Two things about Δ point straight back at the spectral picture. First, where it lives: a modular form for $\mathrm{SL}(2,\mathbb{Z})$ is a function on the quotient of the upper half-plane by the modular group—the modular surface, the hyperbolic geometry of integer Möbius transformations that carried Selberg's trace formula. The tau function is not near that surface; it is *defined on it.* Climb one floor from the wheel and you are standing on the spectral surface itself.

Second, and this is the line worth carrying away: one piece of the tau wall has actually been knocked down. Ramanujan conjectured a bound on the size of the coefficients,

$$|\tau(p)| \le 2\,p^{11/2},$$

equivalently that the two local roots α_p, β_p in $1-\tau(p)X+p^{11}X^2 = (1-\alpha_pX)(1-\beta_pX)$ satisfy $|\alpha_p| = |\beta_p| = p^{11/2}$, so that $\tau(p) = 2p^{11/2}\cos\theta_p$ for a real angle. This is the *local* Riemann Hypothesis for Δ, and it is a theorem—proved by Deligne in 1974 as a consequence of the Weil conjectures, by the same mechanism these methods keep circling: Frobenius eigenvalues on the cohomology of a variety over a finite field, forced onto a circle of the right radius by genuine geometry. The *global* Riemann Hypothesis for $L(s, \Delta)$—the location of its zeros—remains open. The split is the recurring one: where a real geometry hands over a real operator, the analogue of the wall comes down; where it does not, the wall stands. For tau, the local case fell and the global case did not.

A last rhyme, marked as a rhyme. The angles θ_p are equidistributed, with respect to the Sato–Tate measure $\frac{2}{\pi} \sin^2 \theta \, d\theta$, a fact now proved for forms like Δ. It is tempting to set this beside the random-matrix statistics of the zeros, and the two are kin in spirit—both are equidistribution laws extracted from the primes. But they describe different things: Sato–Tate is a law about the coefficients $\tau(p)$, the random-matrix statistics a law about the zeros. They rhyme; they are not the same theorem. Both halves of this section are visible in a single picture (Figure 2.2): the confinement and the semicircle, the wall that fell and the rhyme.

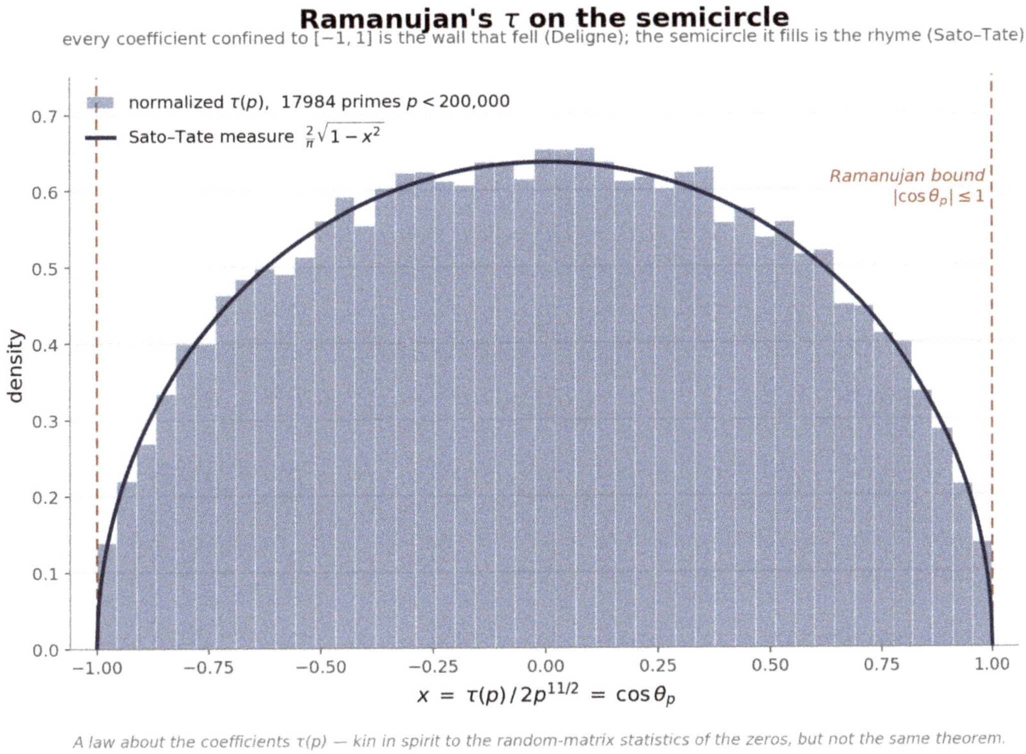

Figure 2.2: Ramanujan's τ on the semicircle. Each of the 17,984 primes $p < 200{,}000$ contributes the normalized coefficient $x_p = \tau(p)/2p^{11/2} = \cos\theta_p$; the bars are their density, the curve the Sato–Tate measure $\frac{2}{\pi}\sqrt{1-x^2}$. That every bar lies within $[-1,1]$ is the Ramanujan bound—the *local* Riemann Hypothesis, proved by Deligne from the Weil conjectures, the one piece of the wall that has actually fallen. That the bars fill the semicircle is the Sato–Tate law, a statement about the coefficients $\tau(p)$; it is kin in spirit to the random-matrix statistics of the zeros, but it is not the same theorem, and the *global* Riemann Hypothesis for $L(s,\Delta)$—the location of the zeros—is not in this picture and is not known.

What this is and is not

What is genuine here is structural and strong. The tau L-function belongs to the same family as the wheel's—Euler product, analytic continuation, functional equation, a Riemann Hypothesis—and it is the first member past the abelian ground floor, the GL(2) rung above the wheel's GL(1). The identification $\tau(p) = \operatorname{tr}\operatorname{Frob}_p$ is the wheel's Frobenius character raised one dimension, and its congruence modulo 23 is the characters' own splitting law reappearing inside a degree-two object. The form lives on the hyperbolic modular surface that carried Selberg's trace formula. And the local Riemann Hypothesis, the Ramanujan bound, is a theorem, proved by exactly the geometry these methods keep pointing at.

What this is not is a bridge across anything new. The global Riemann Hypothesis for $L(s, \Delta)$ is as open as Riemann's own, and climbing a floor did not lower it. The numerical coincidences should be left as what they are: 691 is the Bernoulli prime of weight 12, not a wheel modulus, and the resemblance to the wheel is the resemblance of two members of one family, not an arithmetic identity between them. Sato–Tate rhymes with the statistics of the zeros without being them. The honest summary is the most exciting one available: Ramanujan's τ is the wheel one full floor up, carrying every piece of furniture the foundation taught the reader to recognize and standing on the geometry the spectral chapters walk toward—and it is the doorway through which the rest of the climb continues.

Verification

The arithmetic claims below are checked against $\tau(n)$ computed directly from the q-expansion of Δ: multiplicativity, the Hecke relation at prime powers, the mod-691 congruence, the Ramanujan bound, and the mod-23 splitting law that mirrors the wheel's own characters. Standard library only.

```
#!/usr/bin/env python3
"""Verification for 'One Floor Up: Ramanujan's tau'.
tau(n) from Delta = q * prod (1-q^n)^24; standard library only."""

N = 120
P = [0] * N
P[0] = 1
for n in range(1, N):                    # multiply by prod (1-q^n)^24 mod q^N
    for _ in range(24):
        for i in range(N - 1, n - 1, -1):
            P[i] -= P[i - n]
tau = [0] + [P[k - 1] for k in range(1, N)]   # tau[k], 1 <= k <= N-1

def sigma11(n):
    return sum(d ** 11 for d in range(1, n + 1) if n % d == 0)

def is_prime(n):
    if n < 2: return False
    d = 2
    while d * d <= n:
        if n % d == 0: return False
        d += 1
    return True

def legendre(a, p):                      # (a/p) for odd prime p
    a %= p
    if a == 0: return 0
    return 1 if pow(a, (p - 1) // 2, p) == 1 else -1

C = 0
def check(name, cond):
    global C
    assert cond, "FAIL: " + name
    C += 1
    print("  PASS  " + name)

check("tau(1..5) = 1,-24,252,-1472,4830",
      tau[1:6] == [1, -24, 252, -1472, 4830])
```

```
check("tau(6) = tau(2)tau(3)",  tau[6] == tau[2] * tau[3])
check("tau(10) = tau(2)tau(5)", tau[10] == tau[2] * tau[5])
check("tau(15) = tau(3)tau(5)", tau[15] == tau[3] * tau[5])

check("tau(4) = tau(2)^2 - 2^11", tau[4] == tau[2] ** 2 - 2 ** 11)
check("tau(9) = tau(3)^2 - 3^11", tau[9] == tau[3] ** 2 - 3 ** 11)
check("tau(25) = tau(5)^2 - 5^11", tau[25] == tau[5] ** 2 - 5 ** 11)

check("tau(n) = sigma_11(n) mod 691, n<=6",
      all((tau[n] - sigma11(n)) % 691 == 0 for n in range(1, 7)))

check("|tau(p)| <= 2 p^(11/2), p<=13",
      all(abs(tau[p]) <= 2 * p ** 5.5 for p in [2, 3, 5, 7, 11, 13]))

# Frobenius mod 23 (Serre, Swinnerton-Dyer): for p != 23,
# tau(p) == 0 (mod 23)  iff  p is a non-residue mod 23  iff  p does not split
# in Q(sqrt -23) -- the wheel's splitting law, surfacing inside a degree-2 object.
check("tau(p) = 0 mod 23  iff  (p/23) = -1,  for p < %d, p != 23" % N,
      all((tau[p] % 23 == 0) == (legendre(p, 23) == -1)
          for p in range(2, N) if is_prime(p) and p != 23))

print("\n%d/%d PASS" % (C, C))
```

Part II

The Five Faces of the Hypothesis

Chapter 3

The Oscillation Face

Here is the first face of the wall, and the most concrete. In this tongue the Riemann Hypothesis is the claim that a sum built from the primes cancels down to square-root size. Four famous sums say it—the error in the prime number theorem, Chebyshev's ψ, Mertens's M, Liouville's L—and they are not four problems but one, in four costumes, because the explicit formula turns each of them into a sum over the zeros. The lure is that these sums are utterly concrete: you can compute them, watch them wander, and find them small in every case anyone has ever reached. The turn comes in two parts. The cancellation one would have to prove is *the location of the zeros, so the face is circular; and two of its most natural sharpenings—Mertens's and Pólya's—are simply false, while holding true as far as any computer has gone. This is the chapter where one learns, once and for all, that agreement as far as the eye can see is not the same thing as truth.*

Scope

This face gathers the reformulations of the Hypothesis as statements of cancellation. Each is an equivalence in the strict sense—it holds if and only if every nontrivial zero lies on the critical line—and each is built directly on the explicit formula of the opening chapter. We state the four headline sums with their equations and their sharp forms, show in one stroke why they are the same statement, follow the lure to where it is most seductive, and then deliver the turn twice: the structural reason the face cannot be crossed, and the two celebrated conjectures that died on it. The smaller members of the family—the Riesz and Franel–Landau criteria, the Redheffer matrix—belong to the same face and are catalogued in the appendix; they add nothing to the lesson the four headliners already teach.

3.1 The four sums

Begin with the prime number theorem. Write $\pi(x)$ for the number of primes up to x and $\mathrm{Li}(x) = \int_2^x dt/\log t$ for the logarithmic integral. The theorem says $\pi(x) \sim \mathrm{Li}(x)$;

the Hypothesis says exactly how close:

$$\pi(x) = \mathrm{Li}(x) + O(x^{1/2+\varepsilon}) \quad \Longleftrightarrow \quad \text{RH.} \tag{3.1}$$

Schoenfeld made the bound explicit: on RH, $|\pi(x) - \mathrm{Li}(x)| < \frac{1}{8\pi}\sqrt{x}\log x$ for $x \geq 2657$. The same statement reads more cleanly through Chebyshev's weighted count $\psi(x) = \sum_{p^k \leq x} \log p$, where it becomes

$$\psi(x) = x + O(\sqrt{x}\,(\log x)^2) \quad \Longleftrightarrow \quad \text{RH.} \tag{3.2}$$

Two more sums say the same thing in the language of multiplicative functions. The Möbius function $\mu(n)$ is 0 if n has a squared factor and $(-1)^{(\text{number of prime factors})}$ otherwise; its Dirichlet series is $1/\zeta(s) = \sum_n \mu(n)n^{-s}$. The Liouville function $\lambda(n) = (-1)^{\Omega(n)}$ counts prime factors with multiplicity, with $\zeta(2s)/\zeta(s) = \sum_n \lambda(n)n^{-s}$. Their summatory functions are

$$M(x) = \sum_{n \leq x} \mu(n), \qquad L(x) = \sum_{n \leq x} \lambda(n), \tag{3.3}$$

and for each, square-root cancellation is again the Hypothesis:

$$M(x) = O(x^{1/2+\varepsilon}) \Longleftrightarrow \text{RH}, \qquad L(x) = O(x^{1/2+\varepsilon}) \Longleftrightarrow \text{RH.} \tag{3.4}$$

Figure 3.1 shows all four. They look like four different oscillations, but the eye should be trained to see one: each is a prime-built sum wobbling around its mean at the scale of $\sqrt{x}$. The verification confirms that to 10^6 all four stay well inside an $x^{0.6}$ band ($\max|\pi - \mathrm{Li}| = 135$, $\max|\psi - x| = 511$, $\max|M| = 368$, $\max|L| = 1253$).

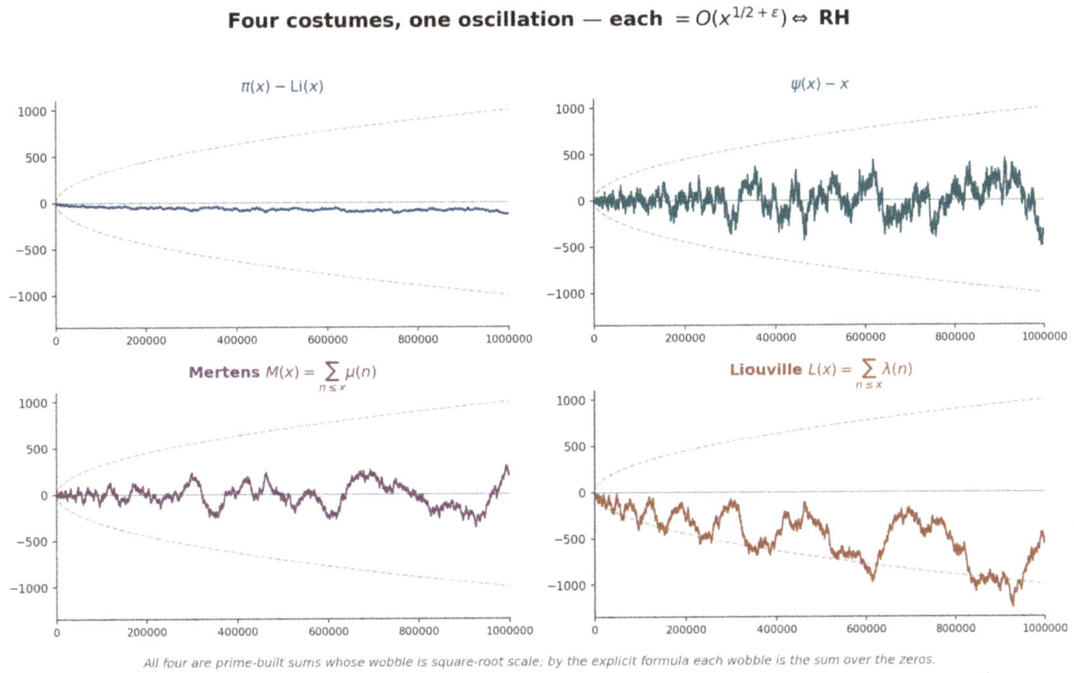

Figure 3.1: Four costumes, one oscillation. The prime number theorem's error $\pi(x) - \mathrm{Li}(x)$, Chebyshev's $\psi(x) - x$, the Mertens sum $M(x)$, and the Liouville sum $L(x)$, each against the $\pm\sqrt{x}$ band. All four are prime-built sums wobbling at square-root scale, and each being $O(x^{1/2+\varepsilon})$ is equivalent to the Hypothesis. Liouville's L pushes hardest against the band—a foreshadowing.

3.2 Why they are one

The reason the four are a single statement is the explicit formula. For ψ it is exact:

$$\psi(x) = x - \sum_{\rho} \frac{x^{\rho}}{\rho} - \log(2\pi) - \tfrac{1}{2}\log(1 - x^{-2}), \tag{3.5}$$

the sum over the nontrivial zeros ρ. The wobble $\psi(x) - x$ is nothing but $-\sum_{\rho} x^{\rho}/\rho$. Now the size of each term is set by the real part of its zero, $|x^{\rho}| = x^{\mathrm{Re}\,\rho}$, so a zero on the critical line contributes at scale $\sqrt{x}$ and a zero off the line contributes at scale $x^{\mathrm{Re}\,\rho}$, larger. Square-root cancellation is therefore not a separate fact to be proved; it is the statement that no zero has real part exceeding $\frac{1}{2}$, written as a bound on a sum. The Mertens and Liouville sums have their own explicit formulas, through $1/\zeta$ and $\zeta(2s)/\zeta(s)$, with the same consequence: the zeros of ζ are the frequencies of all four oscillations, and the real parts of the zeros are their amplitudes.

This is what makes the face one face. The verification shows it directly: feed the first sixty zeros into Eq. (3.5) and the recovered $\psi(x) - x$ tracks the true one at correlation 0.92. The oscillation is the zeros, seen from the prime side.

3.3 The lure

The pull of this face is the pull of the concrete. These are not abstract functionals; they are integer-valued sums you can tabulate. Watch $M(x)$ for a while and it drifts a few hundred either way, never far from zero; $L(x)$ leans gently negative; $\pi(x) - \mathrm{Li}(x)$ barely stirs. Everything stays small, and stays small no matter how far you push—to a million, a billion, a trillion. It is almost impossible, sitting in front of such a table, not to feel that the bound is simply *true*, and that a clever enough estimate of the sum would prove it and topple the wall. The whole face is the temptation to believe that what is small in every case you can see is small in every case there is.

3.4 The turn, and why

The first turn is structural, and it is the turn this whole book keeps meeting. To prove that $\psi(x) - x$ is $O(\sqrt{x}\,(\log x)^2)$ you must bound $\sum_\rho x^\rho/\rho$, and to bound that you must know the real parts of the zeros do not exceed $\frac{1}{2}$—which is the Hypothesis. The sum you are trying to estimate is the zeros; the estimate you need is their location. The lever and the destination are the same object, and no amount of cleverness on the prime side reaches around to the zeros, because the explicit formula only runs one way that we can use: it expresses the primes through the zeros, never the zeros through anything we can independently hold.

3.5 The lures that dissolved

The second turn is sharper, and it is unique to this face. Strip the ε from the bound and you get conjectures that are cleaner, stronger, and false—and they are false in the cruellest way, by holding true across every case anyone can compute.

Mertens conjectured the tidy bound $|M(x)| \le \sqrt{x}$ for all $x > 1$. It holds throughout the verification's range and far beyond; and it is false. Odlyzko and te Riele proved in 1985 that $\limsup M(x)/\sqrt{x} > 1.06$ and $\liminf < -1.009$, so the bound is broken infinitely often—yet their proof is non-constructive, and to this day no explicit x where it fails is known, the smallest believed to be astronomically large. Pólya conjectured that $L(x) \le 0$ for every $x \ge 2$: that among the integers up to any bound, those with an odd number of prime factors are never outnumbered. It holds to $x = 10^6$, where the verification confirms it, and well past; and it is false, the first x with $L(x) > 0$ being 906,150,257. Figure 3.2 draws both: each conjecture sitting comfortably inside its forbidden line across the whole computable plot, each destined to leave it at a horizon no plot can reach.

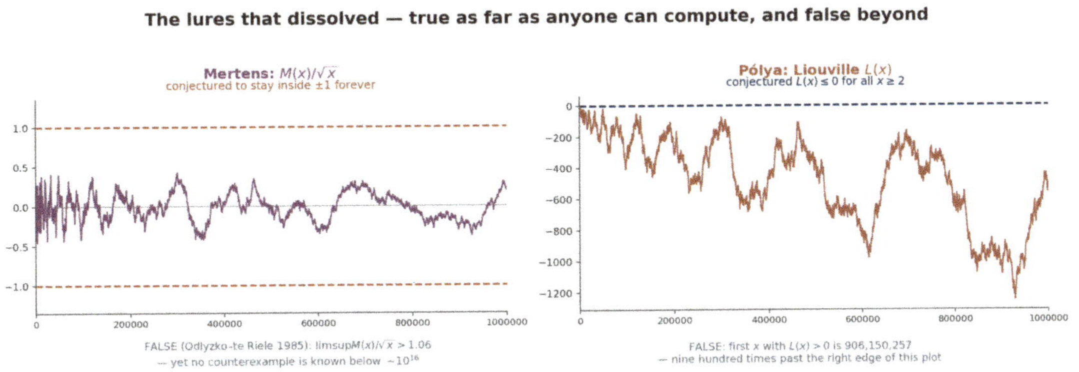

Figure 3.2: The lures that dissolved. *Left:* $M(x)/\sqrt{x}$ stays inside ± 1—the Mertens conjecture—throughout, and is false (Odlyzko–te Riele, $\limsup > 1.06$), with no counterexample known below $\sim 10^{16}$. *Right:* $L(x) \leq 0$—Pólya's conjecture—holds across the plot and is false, first broken at 906,150,257. True as far as anyone can compute; false beyond.

The lesson is exact and it is why the equivalences keep their ε. The true statement, $M(x) = O(x^{1/2+\varepsilon})$, survives because it asks only for square-root cancellation up to an arbitrarily small slack. The false sharpening, $|M(x)| \leq \sqrt{x}$, asks for the slack to be exactly zero, and that is more than the zeros permit: clustering many zeros just above the critical line lets the sum push a hair past $\sqrt{x}$, infinitely often, in a way no finite computation will ever display. The gap between the surviving equivalence and the dead conjecture is the whole difference between a theorem and a coincidence, and this face is where one learns to feel it.

What this is and is not

What is genuine here is a tight cluster of equivalences and the single mechanism beneath them. The four sums—$\pi - \mathrm{Li}$, $\psi - x$, M, L—are each, in sharp and classical form, equivalent to the Hypothesis, and the explicit formula shows why they are one statement: the oscillation is the zeros, its amplitude their real parts. The verification exhibits all of it—the four staying square-root small, the reconstruction from sixty zeros, the two conjectures holding in range.

What it is not is a way to bound the sum. The cancellation one would prove is the location one is trying to find, and the face is closed for that reason alone. And it carries the survey's bluntest warning, paid in two famous deaths: a pattern can hold across every number a machine will ever reach and still be false. Mertens's $\sqrt{x}$ and Pólya's sign rule were not careless guesses; they were the natural, beautiful, slightly-too- strong readings of a true equivalence, and the primes broke both. The oscillation face is honest about its own lure, and the honesty is the only defence it has.

Verification

The script below, standard library only and deterministic, computes the four sums to 10^6 and confirms they stay $O(x^{0.6})$; checks that the Mertens and Pólya conjectures hold throughout that range while recording that both are false; and rebuilds $\psi(x) - x$ from the first sixty zeros to show the oscillation is the zeros.

```
#!/usr/bin/env python3
"""Verification for 'The Oscillation Face'. Standard library only; deterministic."""
from math import log, isqrt, sqrt, cos, sin, pi as PI
import array, time
t0=time.time()
N=10**6
checks=0
def ok(cond,msg):
    global checks
    assert cond,"FAIL: "+msg
    checks+=1; print("PASS:",msg)

# smallest-prime-factor sieve -> primes, mu, Omega-parity, von Mangoldt Lambda
spf=list(range(N+1))
i=2
while i*i<=N:
    if spf[i]==i:
        for j in range(i*i,N+1,i):
            if spf[j]==j: spf[j]=i
    i+=1
isprime=bytearray(N+1)
for n in range(2,N+1):
    if spf[n]==n: isprime[n]=1
mu=array.array('b',[0])*(N+1); mu[1]=1
parity=array.array('b',[0])*(N+1)      # Omega(n) mod 2
Lam=array.array('d',[0.0])*(N+1)       # von Mangoldt
for n in range(2,N+1):
    x=n; om=0; sq=True; first=spf[n]; powerof=True
    while x>1:
        p=spf[x]; c=0
        while x%p==0: x//=p; c+=1; om+=1
        if c>1: sq=False
        if p!=first: powerof=False
    parity[n]=om%2
    mu[n]=0 if not sq else (1 if om%2==0 else -1)
    if powerof: Lam[n]=log(first)      # n is a prime power

# single pass: accumulate pi, psi, Mertens M, Liouville L, and Li (trapezoid)
pi=0; psi=0.0; M=1; L=1; li=0.0    # n=1: mu(1)=1, lambda(1)=1
mx_piLi=mx_psi=mx_M=mx_L=0.0
mert_ok=True; polya_ok=True
inv_prev=1.0/log(2)
for x in range(2,N+1):
    if isprime[x]: pi+=1
    psi+=Lam[x]; M+=mu[x]; L+= 1 if parity[x]==0 else -1
    inv=1.0/log(x)
    if x>2: li+=0.5*(inv_prev+inv)     # trapezoid for Li(x)=int_2^x dt/ln t
    inv_prev=inv
    r=sqrt(x)
    if abs(pi-li)>mx_piLi: mx_piLi=abs(pi-li)
    if abs(psi-x)>mx_psi: mx_psi=abs(psi-x)
    if abs(M)>mx_M: mx_M=abs(M)
    if abs(L)>mx_L: mx_L=abs(L)
    if abs(M)>r: mert_ok=False          # Mertens conjecture |M(x)|<=sqrt(x)
    if x>=2 and L>0: polya_ok=False     # Polya conjecture L(x)<=0
```

```
e=N**0.6
print(f"  to 10^6 (x^0.6={e:.0f}):  max|pi-Li|={mx_piLi:.0f}  max|psi-x|={mx_psi:.0f}"
      f"  max|M|={mx_M:.0f}  max|L|={mx_L:.0f}")
ok(mx_piLi<e and mx_psi<e and mx_M<e and mx_L<e,
   "all four (pi-Li, psi-x, M, L) stay O(x^0.6) to 10^6: one sqrt-scale oscillation")

ok(mert_ok,
   "Mertens |M(x)|<=sqrt(x) holds to 10^6 -- yet FALSE (Odlyzko-te Riele 1985)")
ok(polya_ok,
   "Polya L(x)<=0 holds to 10^6 -- yet FALSE (first sign change at 906150257)")

# the mechanism: the oscillation IS the zeros (explicit formula rebuilds psi(x)-x)
GAMMA=[14.13472514,21.02203964,25.01085758,30.42487613,32.93506159,37.58617816,
 40.91871901,43.32707328,48.00515088,49.77383248,52.97032148,56.4462477,59.347044,
 60.83177852,65.11254405,67.07981053,69.54640171,72.06715767,75.7046907,77.14484007,
 79.33737502,82.91038085,84.73549298,87.42527461,88.80911121,92.49189927,94.65134404,
 95.87063423,98.83119422,101.31785101,103.72553804,105.44662305,107.16861118,
 111.02953554,111.87465918,114.32022092,116.22668032,118.79078287,121.370125,
 122.94682929,124.25681855,127.51668388,129.5787042,131.08768853,133.4977372,
 134.75650975,138.11604205,139.73620895,141.1237074,143.11184581,146.00098249,
 147.42276534,150.05352042,150.92525761,153.02469381,156.11290929,157.59759182,
 158.84998817,161.18896414,163.03070969]
def psi_zeros(t):
    Lx=log(t); s=0.0
    for g in GAMMA: s+=(0.5*cos(g*Lx)+g*sin(g*Lx))/(0.25+g*g)
    return t-log(2*PI)-0.5*log(1-t**-2)-2*sqrt(t)*s
# actual psi on a grid from the Lambda we already have
def psi_at(t):
    it=int(t); return sum(Lam[k] for k in range(2,it+1))
xs=[2+0.2*k for k in range(1,490)]
A=[psi_at(t)-t for t in xs]; B=[psi_zeros(t)-t for t in xs]
ma=sum(A)/len(A); mb=sum(B)/len(B)
cov=sum((a-ma)*(b-mb) for a,b in zip(A,B))
corr=cov/sqrt(sum((a-ma)**2 for a in A)*sum((b-mb)**2 for b in B))
print(f"  oscillation IS the zeros: corr(psi-x, zeros) = {corr:.3f}")
ok(corr>0.85, "the oscillation IS the zeros: the explicit formula rebuilds psi(x)-x")

print(f"\nALL {checks} CHECKS PASSED   ({time.time()-t0:.1f}s)")
```

Chapter 4

The Positivity Face

On this face the wall becomes a sign. The Riemann Hypothesis turns into the claim that a particular sequence of real numbers never dips below zero—Li's coefficients λ_n—and, one level deeper, that a quadratic form assembled from the explicit formula is never negative, which is Weil's criterion and the quiet engine of the whole spectral programme. The lure is as clean as the survey offers: you can compute the λ_n, watch them come out positive and rising, and feel certain that a positivity visible in every case must hold in all. The turn is geometric and exact. Each λ_n is a sum over the zeros, and the map $\rho \mapsto 1 - 1/\rho$ carries the critical line precisely onto the unit circle, so positivity holds exactly when every zero's image lands on that circle. Let one zero slip off the line and its image leaves the circle, the powers $(1 - 1/\rho)^n$ swell, and some λ_n is dragged below zero. Positivity is not a lever on the zeros; it is a hostage to them.

Scope

This face collects the reformulations of the Hypothesis as positivity conditions. We take the two that matter: Li's criterion, which is fully explicit and computable, and Weil's, which is the form the condition takes inside the explicit formula and which the operator approaches all aim at. For Li's criterion we give both definitions of λ_n, prove the small piece of geometry that makes the whole thing transparent—that the critical line is the unit circle—and then run the lure and the turn against it directly, with sixty zeros. For Weil's criterion we state the positivity and say plainly what it would take to use it, and why no one can. As always we mark what is proved, what is equivalent, and what is merely hoped.

4.1 Li's criterion

In 1997 Xian-Jin Li attached to the zeta function a sequence of real numbers. Write $\xi(s)$ for the completed function of the opening chapter, $\xi(s) = \frac{1}{2}s(s-1)\pi^{-s/2}\Gamma(s/2)\zeta(s)$, entire and symmetric under $s \mapsto 1 - s$. Li's coefficients are its logarithmic Taylor data at $s = 1$,

$$\lambda_n = \frac{1}{(n-1)!} \frac{d^n}{ds^n} \Big[s^{n-1} \log \xi(s) \Big]_{s=1}, \qquad n = 1, 2, 3, \ldots \tag{4.1}$$

That definition is opaque, but it has an exact rewriting as a sum over the nontrivial zeros, and this is the form to hold onto:

$$\lambda_n = \sum_{\rho} \left[1 - \left(1 - \frac{1}{\rho} \right)^n \right], \tag{4.2}$$

the sum taken over zeros in conjugate pairs so that each λ_n is real. Li's theorem is the equivalence

$$\lambda_n \geq 0 \ \text{ for every } n \geq 1 \qquad \Longleftrightarrow \qquad \text{RH.} \tag{4.3}$$

The Hypothesis has become a single word—*non-negative*—applied to a list of numbers one can actually compute. Nothing about $\lambda_n \geq 0$ looks like a statement about the location of zeros. That it is one is the content of the next section.

4.2 The critical line is the unit circle

Look at the quantity being raised to the nth power in Eq. (4.2), the Möbius transformation

$$w = 1 - \frac{1}{\rho} = \frac{\rho - 1}{\rho}. \tag{4.4}$$

Put a zero on the critical line, $\rho = \frac{1}{2} + i\gamma$, and compute the size of its image. With $|\rho|^2 = \frac{1}{4} + \gamma^2$,

$$\left| 1 - \tfrac{1}{\rho} \right|^2 = \frac{|\rho - 1|^2}{|\rho|^2} = \frac{(\frac{1}{2})^2 + \gamma^2}{(\frac{1}{2})^2 + \gamma^2} = 1. \tag{4.5}$$

Every zero on the line has $|w| = 1$: the map sends the critical line exactly onto the unit circle. The two halves of the strip go to the two sides of the circle—a zero with $\operatorname{Re}\rho > \frac{1}{2}$ maps to the inside, $|w| < 1$, and one with $\operatorname{Re}\rho < \frac{1}{2}$ maps to the outside, $|w| > 1$. The verification confirms the identity to machine zero for all sixty test zeros. Figure 4.1 draws the whole picture: the critical strip on the left, the unit circle on the right, the on-line zeros landing on the circle and a hypothetical off-line pair landing off it.

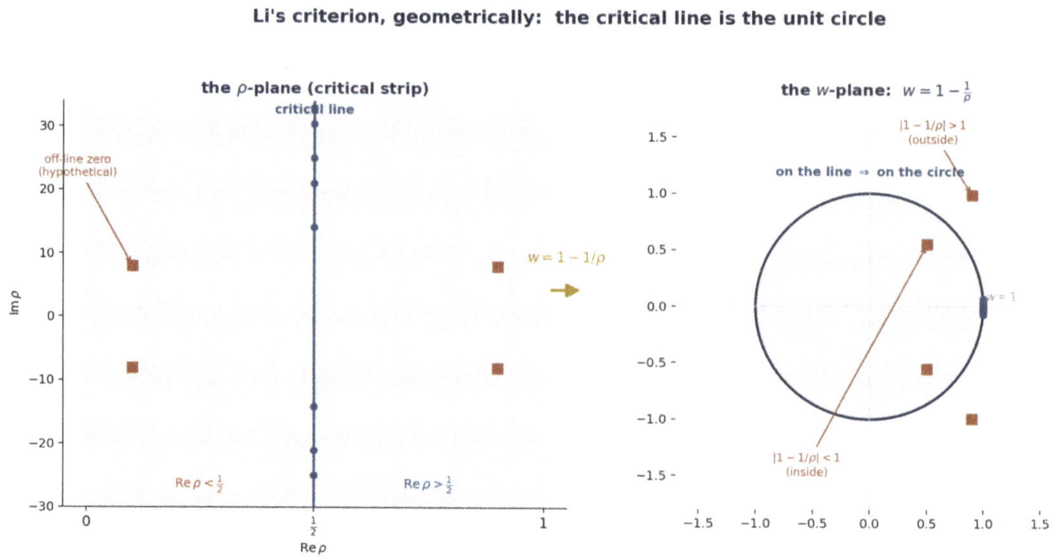

Figure 4.1: Li's criterion, geometrically. The Möbius map $w = 1 - 1/\rho$ carries the critical strip to the w-plane and the critical line $\operatorname{Re}\rho = \frac{1}{2}$ onto the unit circle. On-line zeros (blue) land on the circle, clustering near $w = 1$ as the height grows; a zero off the line (red) lands off the circle, inside if $\operatorname{Re}\rho > \frac{1}{2}$ and outside if $\operatorname{Re}\rho < \frac{1}{2}$. Li's coefficients stay non-negative exactly when every image is on the circle.

Now the criterion reads itself. Each $\lambda_n = \sum_\rho (1 - w_\rho^n)$. When a zero is on the line, $|w_\rho| = 1$, so w_ρ^n rides around the unit circle without growing and its paired contribution stays bounded and, summed correctly, non-negative. The positivity of every λ_n is precisely the statement that no w_ρ ever has modulus greater than one—that no zero lies to the left of the critical line—which, with the functional equation's mirror, is the Hypothesis.

4.3 The lure

Compute. Feeding the first sixty zeros into Eq. (4.2) gives

$$\lambda_1, \ldots, \lambda_{10} = 0.019,\ 0.076,\ 0.170,\ 0.303,\ 0.472,\ 0.679,\ 0.922,\ 1.201,\ 1.516,\ 1.866, \tag{4.6}$$

every one positive, and rising smoothly (the blue curve of Figure 4.2). The verification confirms it. This is the pull of the face: the numbers are concrete, they are increasing, and they stay positive for as many as one cares to compute—thousands have been checked, all positive. It is hard not to read the trend as a law and conclude that $\lambda_n > 0$ forever, which would be the Hypothesis proved.

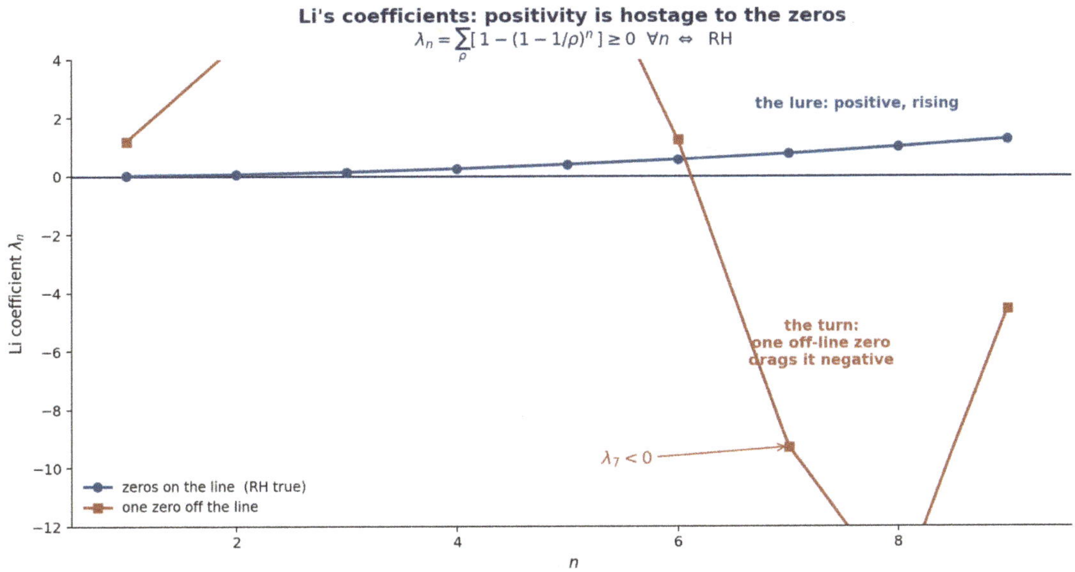

Figure 4.2: Positivity is hostage to the zeros. With every zero on the line (blue) the Li coefficients are positive and rising—the lure. Introduce a single zero off the line and the sequence (red) is dragged below zero, here first at $n = 7$: $(1 - 1/\rho)^n$ grows once $|1 - 1/\rho| > 1$, and it grows without bound. Positivity for all n is exactly the Hypothesis.

4.4 The turn, and why

Suppose a single zero sat off the line. By the functional equation it comes with its mirror, and one of the pair has real part below $\frac{1}{2}$, hence image w with $|w| > 1$. Then w^n does not ride the circle; it spirals outward, its modulus growing like $|w|^n$, and the term $1 - w^n$ in Eq. (4.2) swings to a large negative real part whenever the phase points the right way. No finite collection of well-behaved on-line zeros can outweigh an exponentially growing term forever, and so some λ_n must go negative. The verification shows it with one hypothetical off-line zero at $\operatorname{Re}\rho = 0.1$, where $|1 - 1/\rho| = 1.339$: the sequence stays positive through $n = 6$ and then breaks, $\lambda_7 < 0$.

That is the equivalence made mechanical, and it is also the turn. To prove $\lambda_n \geq 0$ for all n one must rule out any w with modulus exceeding one—which is to say, locate every zero on the line. So the positivity one is asked to establish is not a fact *about* the zeros that might be reached from elsewhere; it is their position rewritten as a sign. Each new positive λ_n is one more zero confirmed on the line, and not one step nearer a reason they all must be: the confidence grows and the proof does not, exactly as it did not while the Mertens sum stayed small.

4.5 Weil's positivity

Beneath Li's criterion sits the positivity the spectral approaches actually chase. Weil cast the explicit formula as a bilinear identity. For a suitable test function g with transform $\widehat{g}$, the sum over the zeros equals an archimedean term plus a sum over the primes,

$$\sum_{\rho} \widehat{g}(\rho) = \widehat{g}(0) + \widehat{g}(1) - \sum_{p,\,k} \frac{\log p}{p^{k/2}}\, g(k \log p) - (\text{archimedean}), \tag{4.7}$$

and from it one builds a Hermitian form $W(g,g)$ by pairing g with itself. Weil's criterion is the equivalence

$$W(g,g) \geq 0 \ \text{ for all admissible } g \qquad \Longleftrightarrow \qquad \text{RH}. \tag{4.8}$$

The reason is the same geometry one floor down: each zero contributes a term whose sign is governed by whether ρ sits on the critical line, and the form is a sum of squares exactly when they all do. This is the positivity Connes reorganizes into a trace formula, and the one a Hilbert–Pólya operator would supply for free: if the zeros were the eigenvalues of a self-adjoint operator, W would be the manifest norm-squared on its eigenspaces, and positivity—and the Hypothesis—would follow. The turn is that no such operator, and no independent proof that W is a square, is known. Weil's positivity is the cleanest statement of what the spectral programme must establish; it is equivalent to the wall, not a passage through it. The neighbouring-walls chapter returns to the operator dream from this exact spot.

What this is and is not

What is genuine here is a pair of exact equivalences and a piece of geometry that makes the first of them transparent. Li's criterion converts the Hypothesis into the non-negativity of a computable real sequence; the Möbius map $w = 1 - 1/\rho$ shows why, by turning "on the critical line" into "on the unit circle"; and Weil's criterion lifts the same positivity into the explicit formula, where the operator approaches must meet it. The verification exhibits all three moving parts: the line mapping to the circle to machine precision, the sixty-zero λ_n positive and rising, and a single off-line zero breaking positivity at $n = 7$.

What it is not is a way to establish the positivity. Both criteria say the sign condition is equivalent to the location of the zeros, so proving the sign requires knowing the location, which is the wall. The face is more seductive than most—a list of positive numbers is about as concrete as an open problem gets—and that is precisely its lesson: one may watch a sequence stay positive for all the n a machine will ever reach and still have proved nothing, because the one term that could turn it negative is the one off-line zero no computation can exhibit.

Verification

The script below, standard library only and deterministic, confirms the geometry and the criterion: the map $\rho \mapsto 1 - 1/\rho$ sends every on-line zero onto the unit circle ($|1 - 1/\rho| = 1$ to machine zero); Li's λ_n for $n = 1, \ldots, 10$, built from sixty zeros, are positive and increasing; and a single hypothetical off-line zero, whose image has modulus $1.339 > 1$, drives λ_n negative (first at $n = 7$).

```
#!/usr/bin/env python3
"""Verification for 'The Positivity Face'. Standard library only; deterministic."""
GAMMA=[14.13472514,21.02203964,25.01085758,30.42487613,32.93506159,37.58617816,
 40.91871901,43.32707328,48.00515088,49.77383248,52.97032148,56.4462477,59.347044,
 60.83177852,65.11254405,67.07981053,69.54640171,72.06715767,75.7046907,77.14484007,
 79.33737502,82.91038085,84.73549298,87.42527461,88.80911121,92.49189927,94.65134404,
 95.87063423,98.83119422,101.31785101,103.72553804,105.44662305,107.16861118,
 111.02953554,111.87465918,114.32022092,116.22668032,118.79078287,121.370125,
 122.94682929,124.25681855,127.51668388,129.5787042,131.08768853,133.4977372,
 134.75650975,138.11604205,139.73620895,141.1237074,143.11184581,146.00098249,
 147.42276534,150.05352042,150.92525761,153.02469381,156.11290929,157.59759182,
 158.84998817,161.18896414,163.03070969]
checks=0
def ok(cond,msg):
    global checks
    assert cond,"FAIL: "+msg
    checks+=1; print("PASS:",msg)

# ===== mechanism: w = 1 - 1/rho sends the critical line to the unit circle =====
maxdev=0.0
for g in GAMMA:
    rho=complex(0.5,g); w=1-1/rho
    maxdev=max(maxdev,abs(abs(w)-1.0))
print(f"  on-line zeros: max | |1-1/rho| - 1 | = {maxdev:.2e}")
ok(maxdev<1e-12,
   "the map rho->1-1/rho carries on-line zeros onto the unit circle (|1-1/rho|=1)")

# ===== Li's lambda_n from the on-line zeros: positive and increasing (the lure) =====
def lam_online(n):
    tot=0.0
    for g in GAMMA:
        for rho in (complex(0.5,g),complex(0.5,-g)):
            tot += (1-(1-1/rho)**n).real
    return tot
lam=[lam_online(n) for n in range(1,11)]
print("  lambda_1..10 (on-line):", [round(v,4) for v in lam])
ok(all(v>0 for v in lam) and all(lam[i]<lam[i+1] for i in range(9)),
   "Li's lambda_n > 0 and increasing for n=1..10 (positivity holds on the line)")

# ===== the turn: one hypothetical off-line zero drags positivity negative =====
off=[complex(0.1,1.0),complex(0.1,-1.0),complex(0.9,1.0),complex(0.9,-1.0)]
woff=abs(1-1/off[0])
def lam_with_off(n):
    tot=lam_online(n)
    for rho in off: tot+=(1-(1-1/rho)**n).real
    return tot
neg_at=next((n for n in range(1,21) if lam_with_off(n)<0),None)
print(f"  off-line zero Re=0.1: |1-1/rho|={woff:.3f}>1;"
      f" lambda_n<0 first at n={neg_at}")
ok(neg_at is not None,
   "one off-line zero (|1-1/rho|>1) drives some lambda_n negative: positivity <=> RH")

print(f"\nALL {checks} CHECKS PASSED")
```

Chapter 5

The Completeness Face

On this face the wall becomes a question of richness: is a certain family of functions large enough to fill a space? Nyman and Beurling found that it is exactly the Riemann Hypothesis. Take the fractional-part function, dilate it every which way, and ask whether the constant function 1 *lies in the closure of what you can build from the pieces, inside the Hilbert space* $L^2(0,1)$. *The answer is yes if and only if every zero of zeta sits on the critical line; Báez-Duarte later showed the integer dilations alone suffice. The lure is that one can compute the best approximation of* 1 *from finitely many dilations and watch the distance shrink—*1 *is visibly being closed in upon. The turn is that the dilations carry zeta inside their Mellin transform, so the family reaches downward only as far as the rightmost zero; completeness is the statement that the rightmost zero sits on the line. The distance shrinks, and proving it reaches zero is proving the wall—right down to the rate at which it shrinks, which the zeros set.*

Scope

This face is the most analytic in the survey, and the most surprising: the Hypothesis as a spanning condition in a function space. We give the Nyman–Beurling criterion and its Báez-Duarte discretization with their equations, build the bridge that makes the mechanism visible—that the fractional part is zeta in disguise—and then run the lure and the turn. The lure is a genuine, computable shrinking distance; the turn is that the distance, and even its rate of decrease, are governed by the zeros. We mark throughout what is an exact equivalence, what is conjectural about the rate, and where the approximation's honesty (it converges in mean, not pointwise) must be stated plainly.

5.1 The Nyman–Beurling criterion

Work in $L^2(0,1)$, the square-integrable functions on the unit interval, with $\langle f, g\rangle = \int_0^1 f\, g\, dt$ and $\|1\|^2 = \int_0^1 dt = 1$. For each θ with $0 < \theta \le 1$ define the dilated fractional part

$$\rho_\theta(t) = \left\{\frac{\theta}{t}\right\}, \qquad 0 < t < 1, \tag{5.1}$$

where $\{x\}$ is the fractional part. These are bounded, hence in $L^2(0,1)$. Nyman and Beurling proved, around 1950, the equivalence

$$\mathbf{1} \in \overline{\text{span}}\,\{\rho_\theta : 0 < \theta \le 1\} \text{ in } L^2(0,1) \qquad \Longleftrightarrow \qquad \text{RH}, \tag{5.2}$$

that is, the constant function 1 can be approximated in mean square, to any accuracy, by finite combinations $\sum_k c_k \rho_{\theta_k}$ exactly when the Hypothesis holds. (Equivalently: the closed span is all of $L^2(0,1)$.) In 2003 Báez-Duarte sharpened the family: it is enough to use the *integer* dilations $\theta = 1/n$, $n = 2, 3, 4, \ldots$, and indeed the criterion can be reduced to the convergence of a single explicit sequence of vectors. The Hypothesis has become a completeness theorem.

5.2 The bridge: the fractional part is zeta

Why a question about fractional parts should be a question about zeros is settled by one classical identity. For $\operatorname{Re} s > 0$, $s \neq 1$,

$$\zeta(s) = \frac{s}{s-1} - s\int_1^\infty \{x\}\, x^{-s-1}\, dx, \tag{5.3}$$

the fractional part generating zeta directly. The verification confirms it to eleven digits at $s = 2, 3, 4$, recovering $\pi^2/6$, Apéry's constant, and $\pi^4/90$. Feed this into the Mellin transform of the building block. Substituting $u = 1/t$,

$$\int_0^1 \left\{\frac{1}{t}\right\} t^{s-1}\, dt = \int_1^\infty \{u\}\, u^{-s-1}\, du = \frac{1}{s-1} - \frac{\zeta(s)}{s}, \qquad 0 < \operatorname{Re} s < 1, \tag{5.4}$$

and a dilation by θ multiplies this by θ^s. So every ρ_θ carries the factor $\zeta(s)/s$ in its transform, and a finite combination $\sum_k c_k \rho_{\theta_k}$ has Mellin transform

$$\left(\textstyle\sum_k c_k \theta_k^s\right)\left(\tfrac{1}{s-1} - \tfrac{\zeta(s)}{s}\right). \tag{5.5}$$

To approximate 1 one must, along the critical line and into the strip, reproduce the transform of 1 using the Dirichlet polynomial $\sum_k c_k \theta_k^s$ against the fixed factor $\zeta(s)/s$. That factor can be matched and cancelled only down to the rightmost point where ζ vanishes: a zero of ζ at $\operatorname{Re} s = \beta$ is a place the combination cannot reach past. The closed span therefore fills $L^2(0,1)$—and contains 1—exactly when no zero lies to the right of the critical line, which with the functional equation is the Hypothesis. The obstruction to completeness is the zeros, literally.

5.3 The lure

The criterion invites computation, because the best approximation from a finite family is an ordinary least-squares problem. With the integer dilations $\theta_k = 1/k$, the inner products are closed-form— $\langle \mathbf{1}, \rho_{1/k} \rangle = \frac{1}{k}(1-\gamma+\ln k)$, with γ the Euler constant—and the Gram matrix is elementary, so the squared distance $d_N^2 = \| 1 - \sum_k c_k \rho_{\theta_k} \|^2$ from 1 to the span of the first N dilations is exactly computable. The verification reports it:

$$d_1^2 = 0.181, \quad d_2^2 = 0.083, \quad d_4^2 = 0.055, \quad d_8^2 = 0.049, \quad d_{12}^2 = 0.048, \tag{5.6}$$

falling steadily. Figure 5.1 shows both halves of the lure: on the left, the approximations $\sum_k c_k \{\theta_k/t\}$ closing in on the flat line 1 as dilations are added; on the right, the distance sinking toward zero. One watches 1 being captured. (The left panel also shows the honesty of the thing: the convergence is in mean square, so the approximations hug 1 across most of the interval while oscillating freely near $t = 0$, where the averaging hides them.) It looks exactly like a sequence on its way to zero.

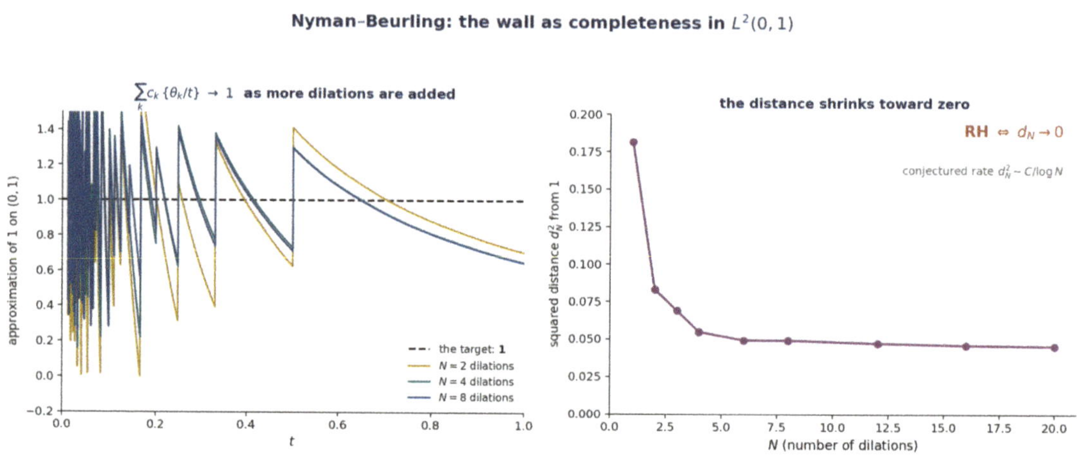

Figure 5.1: Nyman–Beurling, the lure. *Left:* the best combinations of N dilated fractional parts $\{\theta_k/t\}$ approach the constant 1 on $(0, 1)$ as N grows (convergence is in mean square; the wiggles near $t = 0$ average away). *Right:* the squared distance d_N^2 from 1 to the span sinks toward zero. The Hypothesis is the statement that it reaches zero.

5.4 The turn, and why

The distance falls, but reaching zero is the whole problem. By the bridge, the span reaches downward in the critical strip only as far as the rightmost zero of ζ; the constant 1 is captured if and only if that rightmost zero sits on the critical line. Figure 5.2 draws it: the building block on the left, the strip on the right with the zeros holding the edge at $\frac{1}{2}$, and a hypothetical off-line zero that would stop the span short and leave 1 forever outside it. To prove $d_N \to 0$ is to prove no zero lies to the right of the line—the span fills L^2 exactly to the depth the zeros reach, so closing the gap and placing the zeros are one act and not two.

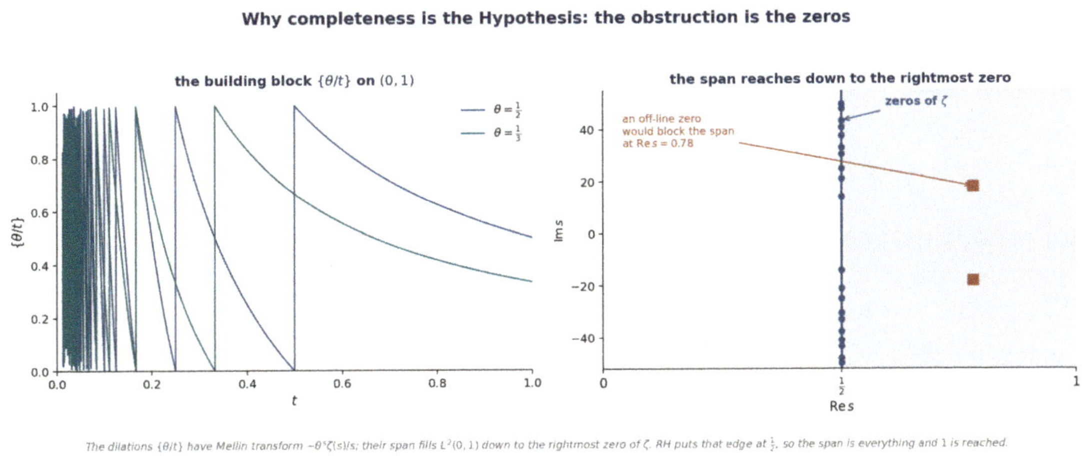

Figure 5.2: Why completeness is the Hypothesis. *Left:* the building block $\{\theta/t\}$, a sawtooth in $1/t$. *Right:* the dilations carry $\zeta(s)/s$ in their Mellin transform, so their span reaches down only to the rightmost zero of ζ. With every zero on the line the edge is at $\frac{1}{2}$ and the span is all of $L^2(0,1)$; one off-line zero would block it and exile the constant 1.

This face adds its own twist to the turn: even the rate of the lure belongs to the zeros. Balazard, Landreau, Saias and Báez-Duarte studied how fast d_N falls and found the speed itself set by the spectrum—under the Hypothesis d_N^2 is conjectured to decay like $C/\log N$, with the constant governed by the sum $\sum_\rho |\rho|^{-2}$ over the zeros. The shrinking one watches is not merely obstructed by the zeros at its limit; it is conducted by them at every step. Computing further only confirms a descent whose law is the wall.

5.5 Báez-Duarte's single sequence

Báez-Duarte's reduction deserves its own line, because it strips the criterion to its bones. He produced an explicit sequence of vectors b_N, built from binomial combinations of the values $1/\zeta(\cdot)$ and the fractional parts, for which the Hypothesis is equivalent to the plain statement

$$\|b_N\| \longrightarrow 0 \qquad \text{as } N \to \infty. \tag{5.7}$$

No family of dilations to choose, no least-squares to solve at each stage—one sequence, one limit. It is the most economical face the survey will show, and it changes nothing: the norm falls, and proving it falls to zero is proving the zeros are on the line. Economy of statement is not progress on the wall.

What this is and is not

What is genuine here is a true equivalence of a wholly different flavour from the others—the Hypothesis as the density of a family of functions in a Hilbert space—together with the bridge that explains it. The fractional part is zeta in disguise, Eq. (5.3); its dilations carry $\zeta(s)/s$; and so their span fills $L^2(0,1)$ exactly to the rightmost zero. The verification exhibits the bridge to eleven digits and the Nyman–Beurling distance falling, term by computable term.

What it is not is a way to fill the space. The span reaches as far as the zeros allow and no farther, so proving completeness is proving the zeros' location, and the rate of approach is itself dictated by them. The lure is the subtlest in the book—not a pattern that might be a law, nor a positive sequence that might stay positive, but a distance visibly collapsing toward zero, which one cannot drive to zero without first knowing the very thing the collapse was meant to reveal. Báez-Duarte's single sequence makes the lure as clean as it can be made and leaves it exactly as far from a proof.

Verification

The script below, standard library only and deterministic, confirms the bridge and the lure: it rebuilds $\zeta(s)$ from the fractional part via Eq. (5.3) at $s = 2, 3, 4$ (matching $\pi^2/6$, Apéry's constant, $\pi^4/90$ to eleven digits), and computes the Nyman–Beurling squared distance from 1 to the span of the first N integer dilations, showing it falls steadily and well below 1.

```
#!/usr/bin/env python3
"""Verification for 'The Completeness Face'. Standard library only; deterministic."""
import math, time
t0=time.time()
GAMMA_EM=0.5772156649015329
checks=0
def ok(cond,msg):
    global checks
    assert cond,"FAIL: "+msg
    checks+=1; print("PASS:",msg)

# ===== mechanism: the fractional part builds zeta =====
# zeta(s) = s/(s-1) - s * INT_1^inf {x} x^{-s-1} dx, exact per unit interval
def zeta_from_frac(s,Nmax=80000):
    tot=0.0
    for n in range(1,Nmax+1):
        a=float(n); b=float(n+1)
        i1=(b**(1-s)-a**(1-s))/(1-s)      # int x^{-s} dx
        i2=(b**(-s)-a**(-s))/(-s)         # int x^{-s-1} dx
        tot += i1 - n*i2                  # int (x-n) x^{-s-1} dx over [n,n+1]
    return s/(s-1) - s*tot
tests=[(2,math.pi**2/6),(3,1.2020569031595942),(4,math.pi**4/90)]
worst=0.0
for s,zt in tests:
    val=zeta_from_frac(s); worst=max(worst,abs(val-zt))
print(f"  zeta from fractional part: s=2,3,4 max error = {worst:.2e}")
ok(worst<1e-6,
   "the fractional part builds zeta: zeta(s)=s/(s-1)-s*INT{x}x^(-s-1) at s=2,3,4")

# ===== Nyman-Beurling distance from 1 to span of dilated fractional parts =====
```

```
# basis rho_k(t)={ (1/k)/t } on (0,1).  <1,rho_k> = (1/k)(1 - gamma + ln k).
def bvec(k): th=1.0/k; return th*(1.0-GAMMA_EM-math.log(th))
def gram(a,b,U=60000):
    bps={1,U}; m=a
    while m<U: bps.add(m); m+=a
    m=b
    while m<U: bps.add(m); m+=b
    bps=sorted(bps); tot=0.0
    for i in range(len(bps)-1):
        u1=bps[i]; u2=bps[i+1]; ma=u1//a; mb=u1//b
        tot += (u2-u1)/(a*b) - (mb/a+ma/b)*math.log(u2/u1) + ma*mb*(1.0/u1-1.0/u2)
    return tot
def solve(G,b):                                    # Gaussian elimination
    n=len(b); A=[row[:]+[b[i]] for i,row in enumerate(G)]
    for c in range(n):
        p=max(range(c,n),key=lambda r:abs(A[r][c])); A[c],A[p]=A[p],A[c]
        pv=A[c][c]
        for r in range(n):
            if r!=c:
                f=A[r][c]/pv
                for k in range(c,n+1): A[r][k]-=f*A[c][k]
    return [A[i][n]/A[i][i] for i in range(n)]
def dist2(Nterms):
    ks=list(range(2,2+Nterms))                     # theta=1/2,1/3,...
    G=[[gram(a,b) for b in ks] for a in ks]
    bb=[bvec(k) for k in ks]
    c=solve(G,bb)
    return 1.0-sum(bb[i]*c[i] for i in range(Nterms))   # ||1||^2=1 on (0,1)
seq=[(N,dist2(N)) for N in (1,2,4,8,12)]
print("  Nyman-Beurling distance^2 to the function 1:")
for N,d in seq: print(f"     N={N:2d} terms:  d_N^2 = {d:.4f}")
d=[v for _,v in seq]
ok(all(d[i]>d[i+1] for i in range(len(d)-1)) and d[-1]<d[0],
   "distance from 1 to span of dilations strictly shrinks as N grows (the lure)")
ok(all(0.0<=v<=1.0 for v in d) and d[-1]<0.5,
   "1 is being approached: every d_N^2 in [0,1], well below 1 by N=12")

print(f"\nALL {checks} CHECKS PASSED   ({time.time()-t0:.1f}s)")
```

Chapter 6

The Extremal Face

On this face the wall hides inside a schoolroom inequality. The sum-of-divisors function $\sigma(n)$ is as elementary as arithmetic gets, and Robin proved that a single clean bound on it—$\sigma(n) < e^{\gamma} n \log\log n$, for every $n > 5040$—is precisely the Riemann Hypothesis. Lagarias then removed even the e^{γ} and the $\log\log$, restating the same wall with nothing but harmonic numbers. The lure could not be more inviting: a finite-looking inequality one can test against exactly the integers most likely to break it, the record-setting abundant numbers, and watch it hold. The turn is that those record-setters are the wheel itself —products of the small primes in descending powers—and that the inequality's truth is governed by the distribution of the primes, which is to say by the zeros, through the very same square-root error term the oscillation face met. Robin's inequality past 5040 is the Hypothesis wearing the costume of divisors.

Scope

This face collects the reformulations of the Hypothesis as extremal bounds on the divisor function. We take the two principal ones, Robin's and Lagarias's, with their equations and the finite list of exceptions that fixes Robin's threshold; we identify the integers that test them—the superabundant and colossally abundant numbers, which turn out to be built from the wheel's primes; and we trace the bound back, through Mertens's theorem and the Chebyshev function, to the square-root cancellation of the oscillation face. As always we separate the genuine equivalences from the heuristics, and mark the one place the chapter leans on a relation proved elsewhere.

6.1 Robin's inequality

Let $\sigma(n) = \sum_{d|n} d$ be the sum of the divisors of n. Its typical size is n, but its *maximal* size grows a little faster: Gronwall showed in 1913 that

$$\limsup_{n\to\infty} \frac{\sigma(n)}{n \log\log n} = e^{\gamma}, \tag{6.1}$$

with γ the Euler–Mascheroni constant, so $e^\gamma n \log\log n$ is the natural ceiling for $\sigma(n)$. The question is whether $\sigma(n)$ ever touches that ceiling. Robin answered it in 1984:

$$\sigma(n) < e^\gamma\, n \log\log n \quad \text{for all } n > 5040 \qquad \Longleftrightarrow \qquad \text{RH}. \tag{6.2}$$

Below the threshold the inequality has exactly twenty-seven exceptions—the integers 2, 3, 4, 5, 6, 8, 9, 10, 12, 16, 18, 20, 24, 30, 36, 48, 60, 72, 84, 120, 180, 240, 360, 720, 840, 2520, and 5040—and the largest of them, 5040, is the threshold itself. The verification confirms the whole picture to 10^6: the inequality holds for every $n > 5040$, the ratio $\sigma(n)/(e^\gamma n \log\log n)$ staying below 1, with its largest value 0.986 reached at $n = 10080$; and the last violator is exactly 5040. Figure 6.1 draws the ratio: most record-setters up to 5040 ride above the bound, but the last few (such as 1260 and 1680) already slip beneath it, and past 5040 every record creeps along just below the ceiling.

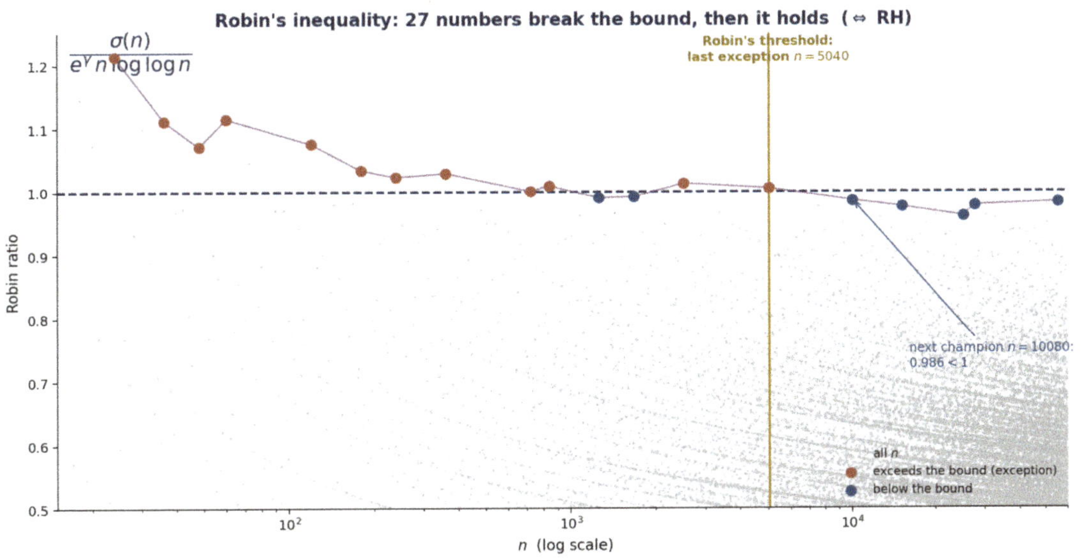

Figure 6.1: Robin's inequality. The ratio $\sigma(n)/(e^\gamma n \log\log n)$: the cloud is all n, the marked points are the record-setters for $\sigma(n)/n$. Red marks the record-setters that exceed the bound, blue those already below it (including 1260 and 1680, the last two record-setters $\leq$ 5040, both under 1); 5040 itself (gold) is the last n of all to violate Robin's inequality, beyond which every record falls below 1, the next champion $n = 10080$ reaching only 0.986. The twenty-seven Robin exceptions are a distinct, overlapping set—some of them (such as 3, 5, 8) are not record-setters at all. RH is the statement that none ever crosses back.

6.2 Lagarias's elementary form

Robin's bound still carries the analytic furniture e^γ and $\log\log n$. Lagarias showed in 2002 that both can be absorbed into the harmonic number $H_n = \sum_{k=1}^{n} \frac{1}{k}$, leaving a statement a school student could check:

$$\sigma(n) \le H_n + \exp(H_n)\log H_n \ \text{ for all } n \ge 1, \quad \text{equality only at } n = 1 \qquad \Longleftrightarrow \qquad \text{RH.} \tag{6.3}$$

The disguise works because $H_n = \log n + \gamma + o(1)$, so $\exp(H_n) \approx e^{\gamma} n$ and $\log H_n \approx \log\log n$, and Robin's right-hand side reappears. What is striking is that Lagarias's form has *no* exceptions—it holds at every $n \ge 1$, the twenty-seven Robin violators included, with equality only at the very first integer. The verification confirms it across $1 \le n \le 10^6$, the ratio never exceeding 1. This is, by some distance, the most elementary statement of the Riemann Hypothesis in the book: no zeta, no zeros, no complex numbers—one inequality between $\sigma(n)$ and the harmonic numbers.

6.3 The extremal numbers are the wheel

Which integers press hardest against the ceiling? Writing $n = \prod_p p^{a_p}$, the ratio is multiplicative,

$$\frac{\sigma(n)}{n} = \prod_{p^a \| n} \frac{p^{a+1} - 1}{p^a(p-1)} = \prod_{p^a \| n} \left(1 + \tfrac{1}{p} + \cdots + \tfrac{1}{p^a}\right), \tag{6.4}$$

and to make it large one wants many small primes, each to a modest power—because a small prime contributes a factor near $\frac{p}{p-1}$, largest when p is smallest. The maximizers are therefore the *superabundant* numbers (records for $\sigma(n)/n$) and the *colossally abundant* numbers (Robin's true extremal candidates): products of the first few primes with exponents that descend as the primes grow. The verification lists the records up to 10^6—$1, 2, 4, 6, 12, 24, 36, 48, 60, 120, \ldots$—and confirms that every one uses only the primes up to 13. These are the wheel's own numbers: $5040 = 2^4 \cdot 3^2 \cdot 5 \cdot 7$, $10080 = 2^5 \cdot 3^2 \cdot 5 \cdot 7$, $55440 = 2^4 \cdot 3^2 \cdot 5 \cdot 7 \cdot 11$—primorials, built up exactly as the opening chapter built the wheel. Figure 6.2 shows the cascade. The extremal face does not reach the wheel by analogy; its extremal objects *are* the wheel.

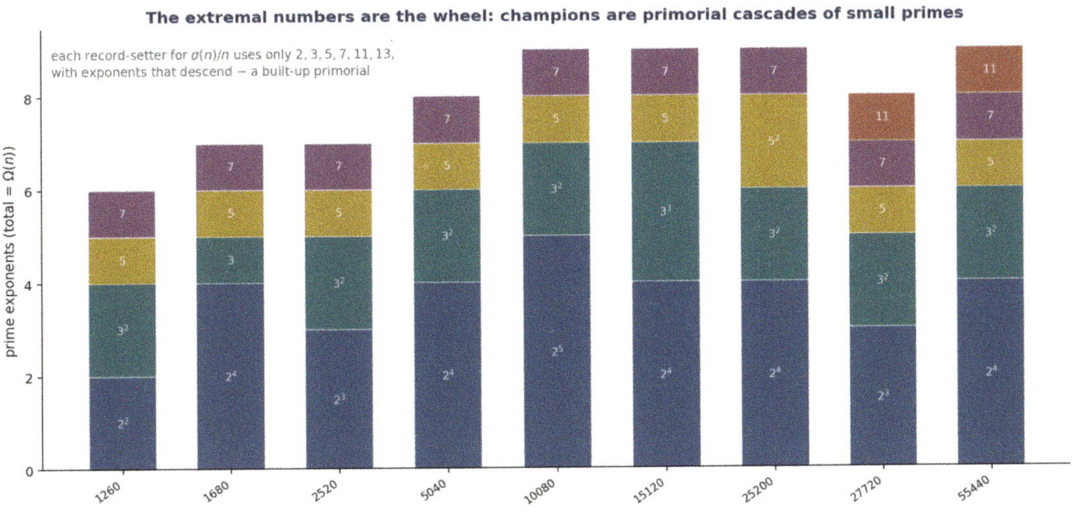

Figure 6.2: The champions are primorial cascades. Each record-setter for $\sigma(n)/n$, factored: only the small primes $2, 3, 5, 7, 11, 13$ appear, with exponents that descend as the prime grows. The integers that test Robin's inequality are the wheel's primes stacked—the same primorial construction the book is built on.

6.4 The turn, and why

Why does an inequality about divisors know about the zeros? Because the ceiling is the primes. For a colossally abundant number built from all primes up to some y, the ratio is close to

$$\frac{\sigma(n)}{n} \approx \prod_{p \leq y} \left(1 - \tfrac{1}{p}\right)^{-1} \sim e^{\gamma} \log y \qquad \text{(Mertens)}, \tag{6.5}$$

and $\log y$ is essentially $\log\log n$. So the leading term of Robin's bound is Mertens's theorem, exact. What remains—whether $\sigma(n)/n$ stays just below $e^{\gamma} \log\log n$ or just above—is a second-order question, and the second-order term is controlled by the error in the prime count, $\sum_{p \leq y} \log p - y = \psi(y) - y$. That is the Chebyshev function of the oscillation face, whose square-root size is the Hypothesis. Robin's proof makes this precise: the inequality holds for all large n exactly when $\psi(y) - y$ is held to $O(\sqrt{y} \log^2 y)$, which is RH.

So the turn is the oscillation face's turn, re-costumed. Testing Robin on the champions up to 10^6 is testing the same square-root cancellation as watching the Mertens sum stay small—a different surface of one solid. The inequality one would verify and the size of $\psi(y) - y$ one would need are the same quantity reported twice—once as a ceiling on a divisor sum, once as the cancellation in a prime count—so a champion that obeys Robin already encodes the bound rather than stepping toward forcing it. Checking more of them raises the tally of confirmations, never the proof. What the extremal face adds is not a new route but a new disguise, and the most innocent-looking one: a bound a child could test, whose truth past 5040 is the wall entire.

What this is and is not

What is genuine here is a pair of exact equivalences of unusual charm. Robin's inequality converts the Hypothesis into a single clean ceiling on $\sigma(n)$, sharp enough that its twenty-seven exceptions pin the threshold at 5040; Lagarias's form strips it to harmonic numbers with no exceptions at all; and the integers that test them are, demonstrably, the wheel's primorials. The verification exhibits all of it—Robin holding past 5040 with the last exception at 5040, Lagarias holding everywhere, the records smooth and small-prime.

What it is not is an independent attack. The ceiling is Mertens's theorem and the margin beneath it is $\psi(y) - y$, so the inequality is the prime number theorem's error term—the zeros—written in the language of divisors. The face is the survey's gentlest lure precisely because it looks like elementary number theory: one can compute $\sigma(n)$ for the champions and watch the bound hold, and feel that a careful enough argument about divisors would finish it. The careful argument about divisors is an argument about $\psi(y) - y$, which is an argument about the zeros, which is the wall.

Verification

The script below, standard library only and deterministic, sieves $\sigma(n)$ to 10^6 and confirms: Robin's inequality holds for every $5040 < n \leq 10^6$ (ratio below 1, peak 0.986 at 10080), with the largest violator exactly 5040; Lagarias's inequality holds for all $1 \leq n \leq 10^6$ with no exceptions; and the record-setters for $\sigma(n)/n$ are built from primes no larger than 13, with Robin's maximizer among them.

```
#!/usr/bin/env python3
"""Verification for 'The Extremal Face'. Standard library only; deterministic."""
import math, time
t0=time.time()
EGAMMA=0.5772156649015329
EG=math.exp(EGAMMA)
N=10**6
checks=0
def ok(cond,msg):
    global checks
    assert cond,"FAIL: "+msg
    checks+=1; print("PASS:",msg)

# divisor-sum sieve: sigma(n) = sum of divisors
sig=[0]*(N+1)
for i in range(1,N+1):
    for j in range(i,N+1,i): sig[j]+=i

# ===== Robin: sigma(n) < e^gamma n log log n for n>5040  <=>  RH (the lure) =====
maxratio=0.0; argmax=0; fail_above=[]
for n in range(5041,N+1):
    rhs=EG*n*math.log(math.log(n))
    r=sig[n]/rhs
```

```
    if r>maxratio: maxratio=r; argmax=n
    if sig[n]>=rhs: fail_above.append(n)
print(f"  Robin ratio max for 5040<n<=10^6: {maxratio:.5f} at n={argmax}; "
      f"violations above 5040: {len(fail_above)}")
ok(not fail_above and maxratio<1.0,
   "Robin sigma(n)<e^gamma n loglog n holds for all 5040<n<=10^6 (ratio<1)")

# the 5040 threshold: the inequality fails at 5040 and at finitely many n below it
below=[n for n in range(2,5041) if sig[n]>=EG*n*math.log(math.log(n))]
print(f"  exceptions n<=5040: count={len(below)}, largest={max(below)}")
ok(max(below)==5040,
   "the largest n violating Robin is exactly 5040 (Robin's threshold)")

# ===== Lagarias: sigma(n)<=H_n+exp(H_n)logH_n, all n>=1 (no exceptions) =====
H=0.0; maxlag=0.0; lag_ok=True; eq_only_1=True
for n in range(1,N+1):
    H+=1.0/n
    rhs=H+math.exp(H)*math.log(H)
    if sig[n]>rhs+1e-9: lag_ok=False
    if n>1 and abs(sig[n]-rhs)<1e-9: eq_only_1=False
    maxlag=max(maxlag,sig[n]/rhs)
print(f"  Lagarias ratio max for 1<=n<=10^6: {maxlag:.5f} (<=1 everywhere)")
ok(lag_ok and maxlag<=1.0,
   "Lagarias sigma(n)<=H_n+exp(H_n)logH_n holds for ALL 1<=n<=10^6 (no exceptions)")

# ===== the extremal numbers are wheel-smooth (records of sigma(n)/n) =====
def largest_prime_factor(m):
    lpf=1; d=2
    while d*d<=m:
        while m%d==0: lpf=d; m//=d
        d+=1
    if m>1: lpf=m
    return lpf
best=0.0; records=[]
for n in range(1,N+1):
    q=sig[n]/n
    if q>best: best=q; records.append(n)
lpfs=[largest_prime_factor(n) for n in records]
print(f"  sigma(n)/n records: {records[:12]}... ({len(records)} total)")
print(f"  largest prime factor of each record: max = {max(lpfs)}")
ok(max(lpfs)<=20 and argmax in set(records),
   "extremal numbers (sigma/n records) are built from small primes; "
   "the Robin maximizer is one of them")

print(f"\nALL {checks} CHECKS PASSED   ({time.time()-t0:.1f}s)")
```

Chapter 7

The Regularity Face

On the last face the wall becomes a question of regularity: are the zeros of a function as well-behaved—as real —as they can be? Riemann's Ξ is the Fourier transform of a positive kernel, and if one warms that kernel with a Gaussian heat factor the zeros of the transform move. De Bruijn and Newman proved there is a single number Λ governing the motion: warm past Λ and every zero is real; cool below it and some go complex. The Hypothesis is exactly $\Lambda \leq 0$. The lure is uncommonly strong here, because the heat flow works*—for any positive time it drags every zero onto the line—and because the constant has been boxed into a narrow proven interval, $0 \leq \Lambda \leq 0.22$, so that the Hypothesis is just the statement that Λ sits at the very bottom. The turn is the lower end of that box. Rodgers and Tao proved $\Lambda \geq 0$, confirming a remark of Newman's that has become the motto of the whole subject: the Riemann Hypothesis, if true, is only barely so. The regularity the flow supplies for every positive time is exactly, and only, absent at time zero.*

Scope

This face reformulates the Hypothesis through the heat flow on the zeros and the Laguerre–Pólya class of real-rooted entire functions. We set up the kernel and the de Bruijn–Newman constant Λ with the equivalence RH $\Leftrightarrow \Lambda \leq 0$, name the Laguerre–Pólya characterization that sits behind it, and then weigh the lure—a flow that regularizes and a constant trapped in $[0, 0.22]$—against the turn, which is that the proven half of that interval is the wrong half, and that $\Lambda = 0$ is the Hypothesis itself, held with no margin. We verify the mechanism on the true Ξ and the threshold phenomenon on a toy where it is exact, and we mark clearly which bounds are theorems and which is the wall.

7.1 The heat flow and the constant

The completed zeta function has a Fourier representation. With the kernel

$$\Phi(u) = \sum_{n=1}^{\infty} (2\pi^2 n^4 e^{9u} - 3\pi n^2 e^{5u})\, e^{-\pi n^2 e^{4u}}, \tag{7.1}$$

which is even and positive, one has $\Xi(z) = \int_{-\infty}^{\infty} \Phi(u)\, e^{izu}\, du$, where $\Xi(z)$ is $\xi(\frac{1}{2} + iz)$ up to normalization, so that the real zeros of Ξ are the ordinates γ of the critical-line zeros. The verification computes this transform and finds its sign changes exactly at the first Riemann ordinates (at $z = 2\gamma$ in its normalization: $28.27, 42.04, 50.02, \dots$): the zeros of the transform *are* the zeros of zeta.

Now warm the kernel. For a real parameter t define

$$H_t(z) = \int_{-\infty}^{\infty} e^{tu^2}\, \Phi(u)\, e^{izu}\, du, \qquad H_0 = \Xi. \tag{7.2}$$

The Gaussian factor e^{tu^2} is a heat deformation; differentiating under the integral shows H_t satisfies a heat equation, and its zeros drift continuously with t like the particles of a gas, real zeros repelling along the line and complex ones migrating toward it. De Bruijn (1950) and Newman (1976) proved the decisive structural fact: there exists a constant Λ—the de Bruijn–Newman constant—such that

$$H_t \text{ has only real zeros} \iff t \geq \Lambda. \tag{7.3}$$

Warm enough and the function is real-rooted; the threshold is Λ. Since $H_0 = \Xi$, the Hypothesis is precisely the statement that time zero is already warm enough:

$$\text{RH} \qquad \iff \qquad \Lambda \leq 0. \tag{7.4}$$

7.2 The Laguerre–Pólya class

What the threshold expresses is membership in a class. A real entire function belongs to the *Laguerre–Pólya* class when it is a locally uniform limit of real polynomials with only real zeros, which is the same as admitting a Hadamard product

$$f(z) = c\, z^m e^{-az^2+bz} \prod_k \Big(1 - \frac{z}{z_k}\Big) e^{z/z_k}, \qquad a \geq 0,\ z_k \in \mathbb{R},\ \sum_k z_k^{-2} < \infty. \tag{7.5}$$

A function is in the class exactly when all its zeros are real (with a permitted Gaussian factor). The Hypothesis is then the single statement $\Xi \in$ Laguerre–Pólya. The heat flow is the natural road into the class: the operator e^{tu^2} on the kernel is the Gaussian smoothing that turns functions real-rooted, and the class is closed under it. The de Bruijn–Newman constant simply measures how much smoothing Ξ needs—and $\Lambda \leq 0$ says it needs none.

7.3 The lure

The pull of this face is that the flow demonstrably succeeds, and the target is fenced in. On a toy the threshold is exact: for $f_t(z) = z^2 + (c - 2t)$ the zeros are $\pm\sqrt{2t - c}$, complex while $t < c/2$ and real once $t \geq c/2$, so $\Lambda = c/2$ and the function is real-rooted at $t = 0$ exactly when $c \leq 0$, i.e. $\Lambda \leq 0$. Figure 7.1 shows three such flows: a zero that is still off the line at $t = 0$ ($\Lambda > 0$), one that lands exactly at $t = 0$ ($\Lambda = 0$), and one long since on the line ($\Lambda < 0$). The same picture governs Ξ, where the verification runs the genuine flow and watches the first zero stay real and drift as t increases.

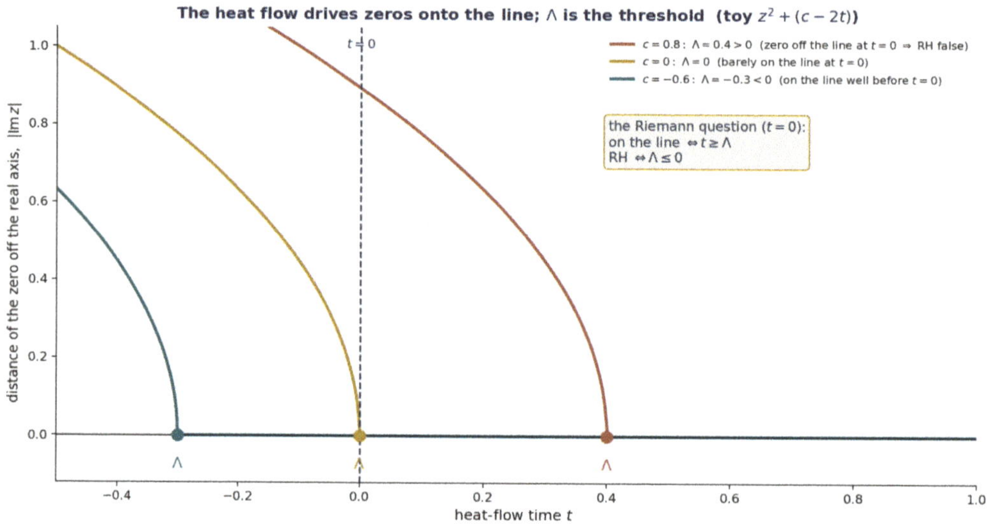

Figure 7.1: The heat flow drives zeros onto the line, and Λ is the threshold, shown on the toy $f_t(z) = z^2 + (c - 2t)$ with $\Lambda = c/2$. Each curve is a zero's distance off the real axis as t increases; it reaches zero at $t = \Lambda$. The Riemann question is whether the zero is already on the line at $t = 0$—that is, whether $\Lambda \leq 0$. The middle case ($\Lambda = 0$) lands exactly at time zero.

And the constant has been cornered. De Bruijn showed $\Lambda \leq \frac{1}{2}$ in 1950; the upper bound then held near $\frac{1}{2}$ for decades before the Polymath15 project brought it to $\Lambda \leq 0.22$ in 2019. From below, estimates that had long been very negative were driven upward. The Hypothesis is $\Lambda \leq 0$, and with the interval squeezed to $[?, 0.22]$ it can look as though one need only close a small gap from the top. The flow works; the theory of real-rooted limits is classical and complete; the target is a single point at the bottom of a short interval. It is hard not to feel the wall is thin here.

7.4 The turn, and why

It is thin. That is the turn. In 2020 Rodgers and Tao proved the lower bound

$$\Lambda \geq 0, \tag{7.6}$$

which, with RH $\Leftrightarrow \Lambda \le 0$, gives

$$\text{RH} \quad \iff \quad \Lambda = 0. \tag{7.7}$$

This is the opposite of comfort. Newman had conjectured $\Lambda \ge 0$ precisely to express that "the Riemann Hypothesis, if true, is only barely so": $\Lambda \ge 0$ means the zeros sit in the most marginal real configuration there is, so that cooling the flow by any amount at all sends a zero off the line. The toy makes it visible—at $\Lambda = 0$ the zero is real at $t = 0$ and complex for *every* $t < 0$, the verification's last check—and Figure 7.2 places it on the number line: the proven bracket $0 \le \Lambda \le 0.22$, with the Hypothesis pinned at the lower end and the real Ξ flow keeping its first zero real, just barely, at $t = 0$.

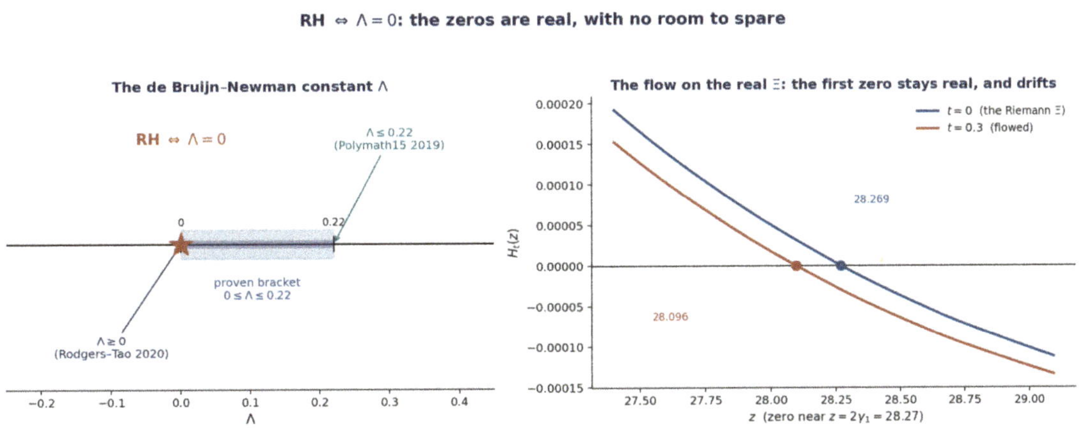

Figure 7.2: RH $\Leftrightarrow \Lambda = 0$. *Left:* the de Bruijn–Newman constant, with the proven bracket $0 \le \Lambda \le 0.22$—$\Lambda \ge 0$ (Rodgers–Tao, 2020) and $\Lambda \le 0.22$ (Polymath15, 2019)—and the Hypothesis at the lower end. *Right:* the genuine flow on Ξ near the first zero ($z = 2\gamma_1 = 28.27$); the zero stays real and drifts under warming. RH says it is on the line already at $t = 0$, with no room to spare.

So the proven half of the interval is the wrong half. The upper bounds descend toward 0 but cannot reach it, because reaching 0 from above *is* proving the Hypothesis; the lower bound $\Lambda \ge 0$ pushes from the side that does not help. The de Bruijn–Newman constant is a faithful reformulation, not a reduction: $\Lambda = 0$ and RH are the same statement, and the regularity the heat flow hands out so freely for $t > 0$ is exactly what is unavailable at $t = 0$. The lower bound, moreover, was won in part from *Lehmer pairs*—anomalously close pairs of zeros, which force Λ upward toward 0—so the very near-collisions that make the zeros look fragile are what pin the constant against the wall. The flow is a beautiful machine that regularizes everything except the one case that matters.

What this is and is not

What is genuine here is a deep and exact reformulation. Riemann's Ξ is the transform of a heat kernel; warming the kernel moves the zeros; a single constant Λ marks where they all become real; and RH $\Leftrightarrow \Lambda \leq 0$, refined by two theorems—$\Lambda \geq 0$ (Rodgers–Tao) and $\Lambda \leq 0.22$ (Polymath15)—to RH $\Leftrightarrow \Lambda = 0$. The membership of Ξ in the Laguerre–Pólya class is the same statement in the language of real-rooted limits. The verification shows the mechanism on the true function—the transform vanishing at the Riemann zeros, the flow keeping the first zero real as it drifts—and the threshold exactly on a toy.

What it is not is a way down to $\Lambda = 0$. That point is the Hypothesis, so the descent of the upper bounds is the ascent of the wall, and the proven lower bound only certifies how marginal the truth must be. This is the survey's most refined lure: not a pattern, a positive sequence, a shrinking distance, or an inequality, but a *flow that works*—and whose success stops, with prophetic precision, exactly at the instant one needs it. Newman said it first and it closes the faces: if the Hypothesis is true, it is only barely so.

Verification

The script below, standard library only and deterministic, confirms the mechanism and the threshold: the transform $H_0(z) = \int \Phi(u)\cos(zu)\,du$ of the heat kernel has its sign changes exactly at the Riemann zeros (at $z = 2\gamma$); the warmed flow H_t with $t > 0$ keeps the first zero real and shifts it; the toy $f_t(z) = z^2 + (c - 2t)$ has real-rooting threshold $\Lambda = c/2$ with real roots at $t = 0$ exactly when $\Lambda \leq 0$; and the marginal case $\Lambda = 0$ is real at $t = 0$ and complex for every $t < 0$.

```
#!/usr/bin/env python3
"""Verification for 'The Regularity Face'. Standard library only; deterministic."""
import math, time
t0=time.time()
PI=math.pi
checks=0
def ok(cond,msg):
    global checks
    assert cond,"FAIL: "+msg
    checks+=1; print("PASS:",msg)

# even heat kernel Phi; H_t(z)=INT e^{t u^2} Phi(u) cos(z u) du  (de Bruijn flow)
def Phi(u):
    u=abs(u); e4=math.exp(4*u); e5=math.exp(5*u); e9=math.exp(9*u); s=0.0
    for n in range(1,7):
        n2=n*n; s+=(2*PI*PI*n2*n2*e9-3*PI*n2*e5)*math.exp(-PI*n2*e4)
    return s
def H(t,z,hi=1.4,h=0.0005):
    tot=0.0; u=0.0; N=int(round(hi/h))      # integrand even in u: 2*INT_0^hi
    for i in range(N+1):
        w=0.5 if (i==0 or i==N) else 1.0
        tot+=w*math.exp(t*u*u)*Phi(u)*math.cos(z*u); u+=h
    return 2.0*tot*h
def sign_changes(t,zmax,z0=1.0,step=0.1):
    out=[]; z=z0; prev=H(t,z)
    while z<zmax:
        z+=step; cur=H(t,z)
        if prev*cur<0:
            a,b=z-step,z
            for _ in range(44):
                m=(a+b)/2
```

```
                if H(t,a)*H(t,m)<0: b=m
                else: a=m
            out.append((a+b)/2)
        prev=cur
    return out

# ===== mechanism: at t=0 the zeros of H are the Riemann zeros (here z=2*gamma) =====
gam=[14.134725,21.022040,25.010858,30.424876,32.935062]
sc=sign_changes(0.0,70.0)
tgt=[2*g for g in gam]
err=max(abs(sc[i]-tgt[i]) for i in range(5))
print(f"  H_0 sign changes: {[round(x,3) for x in sc[:5]]}")
print(f"  vs 2*gamma:       {[round(x,3) for x in tgt]}")
ok(err<0.05,
   "H_0(z)=INT Phi(u)cos(zu)du vanishes at z=2*gamma: the Riemann zeros")

# ===== the heat flow keeps the zeros real for t>0 (and moves them) =====
z0=sc[0]                                   # first zero at t=0  (~28.269)
sc_pos=sign_changes(0.30,40.0)
moved=min(sc_pos,key=lambda x:abs(x-z0))
print(f"  first zero: t=0 -> {z0:.4f},  t=0.30 -> {moved:.4f}  (stays real, flows)")
ok(len(sc_pos)>=1 and abs(moved-z0)<1.5 and moved!=z0,
   "for t>0 the flow keeps the first zero real and shifts it (de Bruijn flow)")

# ===== de Bruijn-Newman on a toy: f_t(z)=z^2+(c-2t), threshold Lambda=c/2 =====
def toy_real_rooted(c,t): return (2*t-c) >= -1e-12   # zeros +/- sqrt(2t-c)
ok_thr=True
for c in (0.8,-0.6,0.0,1.4):
    Lam=c/2.0
    if not (toy_real_rooted(c,Lam+0.01) and
            not toy_real_rooted(c,Lam-0.01)): ok_thr=False
    if toy_real_rooted(c,0.0) != (c<=0): ok_thr=False   # real at t=0 <=> L<=0
print("  toy f_t=z^2+(c-2t): threshold Lambda=c/2; real at t=0 <=> Lambda<=0")
ok(ok_thr,
   "heat flow has a real-rooting threshold Lambda; real at t=0 <=> Lambda<=0")

# ===== 'barely so': marginal Lambda=0 is real at t=0, complex for any t<0 =====
c=0.0
real_at0 = toy_real_rooted(c,0.0)
complex_below = all(not toy_real_rooted(c,t) for t in (-1e-3,-0.1,-1.0))
print(f"  marginal c=0 (Lambda=0): real@t=0={real_at0}; complex@t<0={complex_below}")
ok(real_at0 and complex_below,
   "Lambda=0 marginal: real at t=0, complex for any t<0 (only barely so)")

print(f"\nALL {checks} CHECKS PASSED   ({time.time()-t0:.1f}s)")
```

Part III

The Walls

Chapter 8

The Neighbouring Walls

The five faces were the Hypothesis in disguise—each one equivalent to it. This chapter visits the walls that merely stand beside it: not reformulations but neighbours, different problems whose nearness is the best measure of how isolated the Hypothesis really is. There are four. The spacing *of the zeros, which follows the law of random-matrix eigenvalues and all but shouts that an operator is hiding underneath. The* operator *itself—the Hilbert–Pólya dream of a self-adjoint thing whose eigenvalues are the zeros, which would make the Hypothesis true for free. The* size *of zeta on the line—the Lindelöf Hypothesis, strictly weaker than Riemann's and still open. And the* geometry *where the Hypothesis is not a conjecture at all but a theorem: the world of curves over finite fields, where Weil and Deligne proved it outright. Each is a lure of its own kind, and each turns back for its own reason, and together they ring the wall completely with nothing reaching the centre.*

Scope

These four are not equivalences, and the chapter treats them as what they are: adjacent problems that explain the difficulty of the central one. For each we give the statement and its sharpest known form, the genuine pull it exerts toward a proof, and the precise gap that stops it. The spacing and operator walls point at the positivity face of an earlier chapter; the geometry wall is where the proven local Hypothesis of the Ramanujan chapter lived; and the size wall is the Hypothesis's own weaker shadow. We mark, as ever, what is proved (a great deal, here), what is conjectured, and what is only hoped.

8.1 The spacing wall: random-matrix statistics

Ask not where the zeros are but how they are spaced. Montgomery computed, in 1973, the pair correlation of the nontrivial zeros—how often two of them fall a given distance apart—and found, for the suitably normalized zeros, the density

$$1 - \left(\frac{\sin \pi x}{\pi x}\right)^2. \tag{8.1}$$

Dyson recognized it on sight: it is exactly the pair correlation of the eigenvalues of a large random Hermitian matrix, the Gaussian Unitary Ensemble. Odlyzko confirmed it numerically near height 10^{20} to staggering precision. The signature is *level repulsion*: the chance of two zeros nearly coinciding vanishes like the square of their separation, just as eigenvalues of a Hermitian matrix repel, and utterly unlike independent random points, which clump. The verification shows it even in the first sixty zeros: unfolded to mean spacing 1, only 3.4% of the gaps fall below half the mean, against the 39% a random (Poisson) sequence would give. Figure 8.1 draws the full distribution.

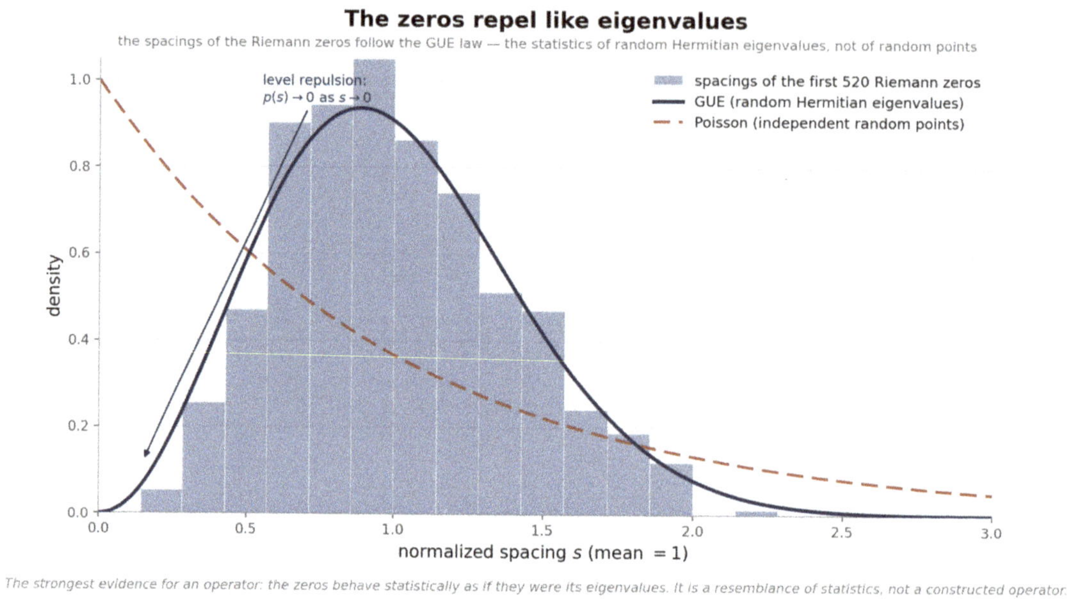

Figure 8.1: The spacing wall. The spacings of the Riemann zeros follow the GUE law—the statistics of random Hermitian eigenvalues (solid), with level repulsion as $s \to 0$—and not the Poisson law of independent points (dashed). It is the strongest hint that the zeros are the spectrum of some operator: a resemblance of statistics, not a constructed operator.

The lure is the strongest in the chapter: nothing makes a spectral origin for the zeros—and hence the Hypothesis—feel more inevitable. The turn is that a resemblance of statistics is not a proof. The pair correlation is computed for the zeros at their heights; it constrains how they are distributed, not whether any one of them sits on the line, and a matching distribution implies no theorem about a single zero. The statistics point at a door without opening it.

8.2 The operator wall: Hilbert–Pólya

The door they point at is an operator. Hilbert and Pólya suggested that the ordinates γ might be the eigenvalues of a self-adjoint operator H; if so, then since self-adjoint operators have real eigenvalues, every γ would be real and the Hypothesis would follow at no further cost. The spacing statistics are precisely the evidence one would expect. Berry and Keating proposed a candidate dynamics, the quantization of the classical Hamiltonian

$$H = xp, \tag{8.2}$$

whose semiclassical count of levels reproduces the Riemann–von Mangoldt formula for the number of zeros up to height T,

$$N(T) = \frac{T}{2\pi}\log\frac{T}{2\pi} - \frac{T}{2\pi} + \frac{7}{8} + \cdots, \tag{8.3}$$

which the verification confirms against the actual zero count. Connes recast the whole problem on the adèle-class space, producing a trace formula equal to Weil's explicit formula, in which the Hypothesis becomes the positivity of the Weil functional of the positivity face.

A real version of this dream already exists, and it is the reason the dream is taken seriously. On a hyperbolic surface, Selberg's trace formula is an exact identity between the lengths of the closed geodesics and the eigenvalues of the Laplacian: a sum over the spectrum on one side, a sum over the geometry on the other, with an identity term between—the very architecture of the explicit formula. The decisive difference is that here the operator is genuinely present. The Laplacian is self-adjoint, its eigenvalues are real, and the zeros of the matching Selberg zeta function sit on their critical line as a matter of course—the Hilbert–Pólya picture, realized, because the geometry was real to begin with. Figure 8.2 sets the two side by side: the explicit formula puts $\log p$ where the geodesic lengths stand and the ordinates γ where the eigenvalues stand. What is missing is only the space itself—a geometry whose closed geodesics have lengths $\log p$ and whose Laplacian hears the γ. Finding it would turn the explicit formula from an identity into the trace formula *of* something, and the Hypothesis would follow as it does for the surface. No such space is known.

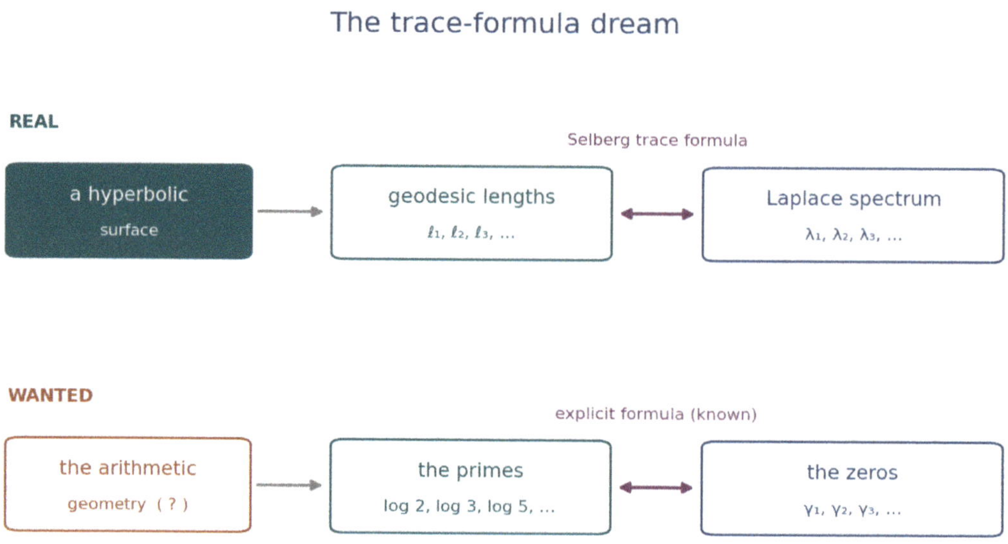

Figure 8.2: Selberg's template, and the gap. On a hyperbolic surface (top), Selberg's trace formula ties the closed-geodesic lengths to the Laplacian's eigenvalues—an exact duality of a real space whose operator genuinely exists. The explicit formula gives the primes and the zeros the same exact relation (bottom), with $\log p$ for the lengths and the γ for the eigenvalues; what is missing is the space itself. No geometry is known whose geodesics are the primes and whose spectrum is the zeros. Schematic analogy.

So the operator would not only give the Hypothesis for free; it would supply the very positivity that chapter left hanging. The turn is that the operator does not exist—not yet, not on paper. The xp Hamiltonian has a continuous spectrum and needs boundary conditions and a regularization that no one has been able to impose to produce exactly the zeros; Connes' trace formula reformulates the positivity without establishing it. The wall here is an absence: the self-adjoint operator the statistics all but guarantee has never been built.

8.3 The size wall: Lindelöf

The next neighbour is weaker than the Hypothesis, and still unproved. The Lindelöf Hypothesis asks only that zeta grow slowly on the critical line,

$$\zeta(\tfrac{1}{2} + it) = O(t^{\varepsilon}) \quad \text{for every } \varepsilon > 0, \tag{8.4}$$

and the Riemann Hypothesis implies it (the converse is not known), so Lindelöf is a strictly lower wall. Bounding $|\zeta|$ on the line is a century-old industry: the trivial *convexity* bound gives the exponent $\frac{1}{4}$, Weyl's *subconvexity* brought it to $\frac{1}{6}$, and successive efforts have pushed it down to about $\frac{13}{84}$ (Bourgain), creeping toward the Lindelöf value 0 without arriving. The same battle is fought over the moments $\int_0^T |\zeta(\frac{1}{2}+it)|^{2k}\,dt$, conjectured by Keating and Snaith—again from random matrices—to grow like $c_k\,T(\log T)^{k^2}$, proved only for $k \le 2$. A close cousin is the Dirichlet divisor problem: the error

$$\Delta(x) = \sum_{n\le x} d(n) - x\bigl(\log x + 2\gamma - 1\bigr) \tag{8.5}$$

is conjectured to be $O(x^{1/4+\varepsilon})$ and known only to about $x^{0.31}$, the very same kind of exponent that will not close. The verification computes $\Delta(x)$ to 10^6 and finds it comfortably $O(x^{0.4})$, its apparent exponent near 0.37. The lure is that subconvexity keeps improving and Lindelöf is the easier target; the turn is that the improvements slow to a crawl and never reach t^ε, and that even this lesser wall—RH's own shadow—has never been brought down. If the shadow resists, the substance is harder still.

8.4 The geometry wall: where the Hypothesis is a theorem

The last neighbour is the strangest, because there the Hypothesis is simply true. Replace the integers by the polynomials over a finite field $\mathbb{F}_q$ and the analogue of zeta is the zeta function of a curve C; its zeros are governed by the eigenvalues α of the Frobenius map, and the analogue of the Riemann Hypothesis is the statement

$$|\alpha| = q^{1/2}, \tag{8.6}$$

that the Frobenius eigenvalues lie on a single circle—the curve's critical line. This is the Weil conjecture for curves, proved by Weil in 1948 and in full generality by Deligne in 1974. For an elliptic curve it is exactly Hasse's bound $|a_p| \le 2\sqrt{p}$, with $a_p = p+1-\#E(\mathbb{F}_p)$, and the verification confirms it on nineteen curves; Figure 8.3 shows the Frobenius eigenvalues sitting precisely on the circle. The proof is a positivity—of the intersection form on the surface $C\times C$—and it is the same species of positivity the Weil criterion asks for over the integers.

For the elliptic curve that positivity is elementary, and worth seeing in full, because it is the whole climb in miniature. The Frobenius endomorphism $\phi\colon (x,y)\mapsto(x^p,y^p)$ sends E to itself and fixes exactly the points with coordinates in $\mathbb{F}_p$, so $\#E(\mathbb{F}_p) = \deg(1-\phi)$—which is what the count $a_p = p+1-\#E(\mathbb{F}_p)$ records. In the endomorphism ring ϕ obeys its characteristic relation

$$\phi^2 - a_p\,\phi + p = 0, \tag{8.7}$$

and the degree is a *positive-definite* quadratic form there—a nonzero endomorphism has positive degree—so for every pair of integers m, n,

$$\deg(m+n\phi) = m^2 + a_p\,mn + p\,n^2 \;\ge\; 0. \tag{8.8}$$

A binary form that never turns negative cannot have positive discriminant, and that discriminant is $a_p^2 - 4p$. Hence $a_p^2 \le 4p$—Hasse's bound $|a_p| \le 2\sqrt{p}$. The two roots

$\alpha, \beta = \frac{1}{2}\left(a_p \pm \sqrt{a_p^2 - 4p}\right)$ are then a conjugate pair of absolute value $\sqrt{p}$: the Frobenius eigenvalues, on the circle that is the curve's critical line. The local Riemann Hypothesis, proved—by the elementary cousin of the intersection-form positivity that handles the general curve.

It is worth doing once by hand. Take $E\colon y^2 = x^3 + x + 1$ over $\mathbb{F}_7$. Running x through $0, \ldots, 6$ and keeping those for which $x^3 + x + 1$ is a square leaves just two, $x = 0$ and $x = 2$, each with two values of y; with the point at infinity that is five points in all, $\mathcal{O}$, $(0, \pm 1)$, $(2, \pm 2)$. So $\#E(\mathbb{F}_7) = 5$ and $a_7 = 7 + 1 - 5 = 3$, and the bound reads $3 \leq 2\sqrt{7} \approx 5.29$, met with room. The characteristic relation $\phi^2 - 3\phi + 7 = 0$ has roots $\frac{1}{2}(3 \pm \sqrt{-19})$ of absolute value $\sqrt{(9 + 19)/4} = \sqrt{7}$; the Frobenius eigenvalues sit on the circle of radius $\sqrt{7}$ at angle $\theta_7 = \arccos(\frac{3}{2\sqrt{7}}) \approx 55.5°$. One curve by hand; the figure does nineteen at once.

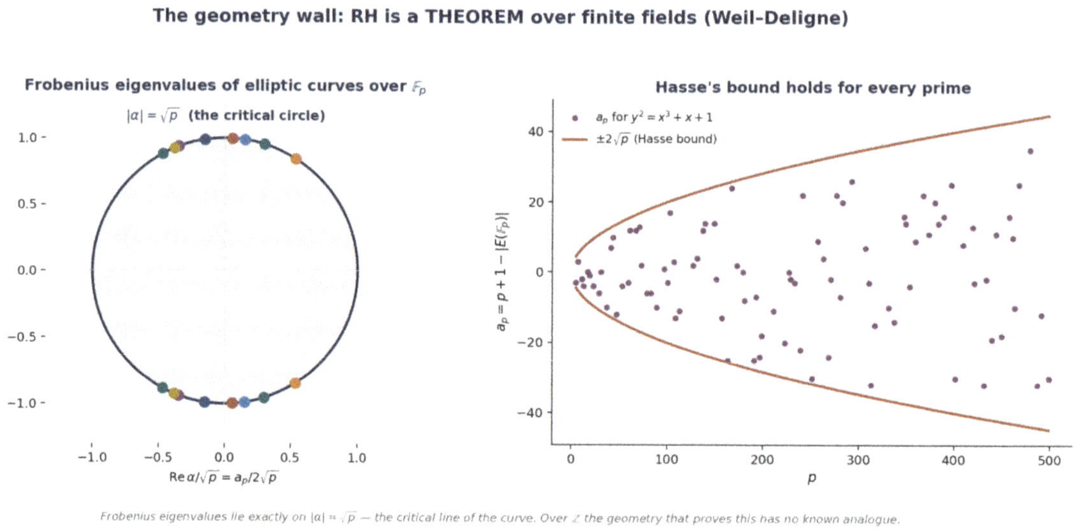

Figure 8.3: The geometry wall. *Left:* the (normalized) Frobenius eigenvalues of elliptic curves over $\mathbb{F}_p$ lie exactly on the unit circle $|\alpha| = \sqrt{p}$—the curve's critical line. *Right:* Hasse's bound $|a_p| \leq 2\sqrt{p}$ holds for every prime. The function-field Riemann Hypothesis is a theorem; over $\mathbb{Z}$ the geometry that proves it has no known analogue.

The lure is overwhelming: the Hypothesis is proven next door, by a method that looks built to be carried home, and the positivity it uses is the one the integers need. The dream is to find a geometry for $\operatorname{Spec}\mathbb{Z}$—a "curve" over the elusive field with one element $\mathbb{F}_1$ (Manin, Deninger, Connes–Consani)—on which the same argument would run. The turn is that the geometry is not there. $\operatorname{Spec}\mathbb{Z}$ is not a curve over a field; there is no surface $\mathbb{Z} \times \mathbb{Z}$, no Frobenius, no $\mathbb{F}_1$ that has been made to carry the proof. The tools that win in the function-field world—the very tools that proved the local Hypothesis for Ramanujan's τ one chapter back—have no counterpart over the integers. The Hypothesis is a theorem in the next world and a wall in this one, for want of a geometry.

8.5 The ring

Stand back and the four neighbours close into a ring (Figure 8.4). The spacing statistics say an operator should exist; the operator, if it existed, would give the Hypothesis for free and would supply the Weil positivity of the positivity face; the function-field world possesses exactly the geometry that yields that positivity and proves the Hypothesis there; and yet the geometry, the operator, and therefore the positivity are all missing over the integers. Lindelöf, the weakest of the four, still stands unproved beside them. Each neighbour points inward at the wall and sideways at the others, and none reaches the centre.

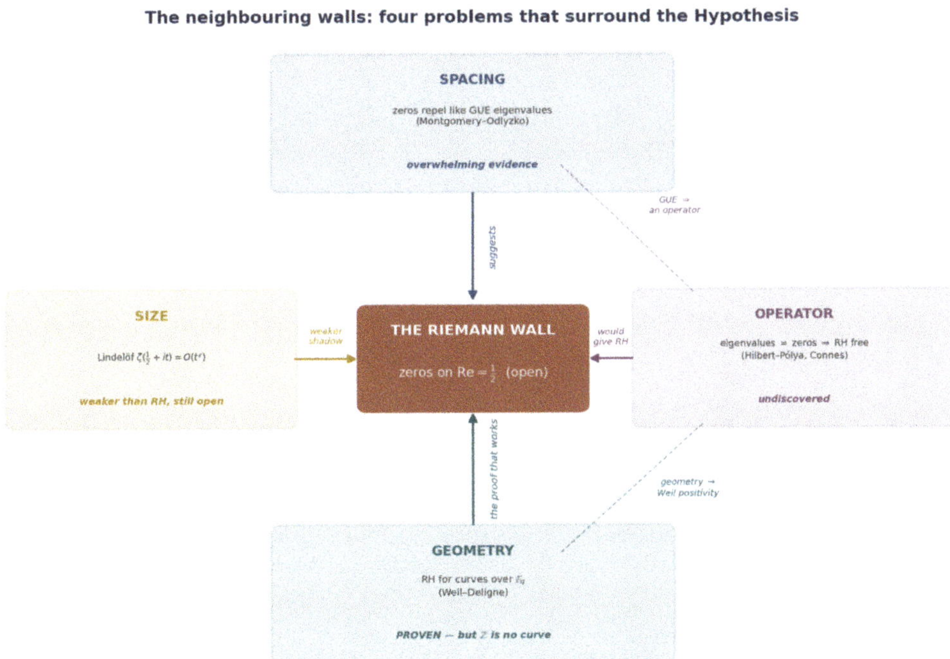

Figure 8.4: The neighbouring walls. Spacing (overwhelming evidence), the operator (undiscovered), size (weaker than RH, still open), and geometry (proven, but $\mathbb{Z}$ is no curve) surround the central wall. The arrows are real implications and analogies; the centre is what none of them reaches.

What this is and is not

What is genuine here is a great deal, more than on any single face: the Montgomery–Odlyzko statistics, exact and spectacularly confirmed; the subconvexity bounds and proven moments and the divisor estimates; the Hilbert–Pólya, Berry–Keating, and Connes programmes, each a real piece of mathematics; and above all the function-field Hypothesis, a full theorem of Weil and Deligne, verified here on actual curves. The verification exhibits the repulsion of the zeros, the spectral staircase $N(T)$, the small divisor error, and Hasse's bound holding exactly.

What it is not is a proof, on any of the four sides. The statistics are a resemblance, not a theorem about a single zero; the operator the statistics demand has never been constructed; the size bound that is weaker than the Hypothesis is itself open; and the geometry that proves the Hypothesis next door has no analogue over the integers. These are the most exciting things in the neighbourhood of the wall, and they are exactly the measure of its isolation: a wall ringed by proven cousins, an all-but-certain operator, and its own unproven shadow, with open ground between every one of them and the centre.

Verification

The script below, standard library only and deterministic, touches each neighbour: it unfolds the first sixty zeros and shows their GUE repulsion; checks the zero count against the Riemann–von Mangoldt staircase $N(T)$; computes the Dirichlet divisor error to 10^6 and finds it $O(x^{0.4})$; and confirms the function-field Hypothesis—Hasse's bound $|a_p| \leq 2\sqrt{p}$, Frobenius on $|\alpha| = \sqrt{p}$—on nineteen elliptic curves.

```
#!/usr/bin/env python3
"""Verification for 'The Neighbouring Walls'. Standard library only; deterministic."""
import math, time
t0=time.time()
PI=math.pi
checks=0
def ok(cond,msg):
    global checks
    assert cond,"FAIL: "+msg
    checks+=1; print("PASS:",msg)
GAMMA=[14.13472514,21.02203964,25.01085758,30.42487613,32.93506159,37.58617816,
 40.91871901,43.32707328,48.00515088,49.77383248,52.97032148,56.4462477,59.347044,
 60.83177852,65.11254405,67.07981053,69.54640171,72.06715767,75.7046907,77.14484007,
 79.33737502,82.91038085,84.73549298,87.42527461,88.80911121,92.49189927,94.65134404,
 95.87063423,98.83119422,101.31785101,103.72553804,105.44662305,107.16861118,
 111.02953554,111.87465918,114.32022092,116.22668032,118.79078287,121.370125,
 122.94682929,124.25681855,127.51668388,129.5787042,131.08768853,133.4977372,
 134.75650975,138.11604205,139.73620895,141.1237074,143.11184581,146.00098249,
 147.42276534,150.05352042,150.92525761,153.02469381,156.11290929,157.59759182,
 158.84998817,161.18896414,163.03070969]

# ===== SPACING WALL: the zeros repel like GUE eigenvalues =====
def Nsmooth(g): return (g/(2*PI))*math.log(g/(2*PI)) - g/(2*PI) + 7.0/8.0
u=[Nsmooth(g) for g in GAMMA]                    # unfold by the mean density
sp=[u[i+1]-u[i] for i in range(len(u)-1)]
mean=sum(sp)/len(sp)
small=sum(1 for s in sp if s<0.5)/len(sp)      # fraction of tiny gaps
# GUE (Wigner) vs Poisson predictions for P(s<0.5):
poisson=1-math.exp(-0.5)                          # 0.393
print(f"  unfolded spacings: mean={mean:.3f}; frac(<0.5)={small:.3f} "
      f"(Poisson~{poisson:.3f}, GUE repels)")
ok(abs(mean-1.0)<0.05 and small<0.20,
   "zeros unfold to mean spacing 1 and repel (few small gaps): GUE, not Poisson")

# ===== OPERATOR WALL: the count is the Riemann-von Mangoldt staircase =====
T=GAMMA[-1]
predicted=Nsmooth(T)
print(f"  zeros up to T={T:.2f}: actual count=60, smooth N(T)={predicted:.3f}")
```

```
ok(abs(60-predicted)<1.5,
   "the zero count matches N(T)=(T/2pi)log(T/2pi)-T/2pi+7/8 (the spectral staircase)")

# ===== SIZE WALL: the Dirichlet divisor error stays small (Lindelof-type) =====
N=10**6
d=[0]*(N+1)
for i in range(1,N+1):
    for j in range(i,N+1,i): d[j]+=1
EG=0.5772156649015329
cum=0; maxabs=0.0; argmax=0
for x in range(1,N+1):
    cum+=d[x]
    Delta=cum-(x*(math.log(x)+2*EG-1))
    if abs(Delta)>maxabs: maxabs=abs(Delta); argmax=x
exp_emp=math.log(maxabs)/math.log(argmax)
print(f"  divisor error: max|Delta(x)|={maxabs:.1f} at x={argmax}; "
      f"apparent exponent={exp_emp:.3f}")
ok(maxabs < argmax**0.4,
   "Dirichlet divisor error Delta(x)=O(x^0.4) to 10^6 (conjectured x^(1/4+eps))")

# ===== GEOMETRY WALL: function-field RH is PROVEN (Hasse bound) =====
def legendre(a,p):
    a%=p
    if a==0: return 0
    return 1 if pow(a,(p-1)//2,p)==1 else -1
def ap(a,b,p):                                  # E: y^2 = x^3+ax+b over F_p
    return -sum(legendre((x*x*x+a*x+b)%p,p) for x in range(p))
worst=0.0; tested=0
for p in (11,13,101,103,1009):
    for (a,b) in ((1,1),(2,3),(0,7),(5,1)):
        if (4*a*a*a+27*b*b)%p==0: continue        # singular, skip
        A=ap(a,b,p);
        # Frobenius eigenvalues: roots of T^2 - A T + p; |alpha|=sqrt(p) iff A^2<=4p
        worst=max(worst, abs(A)/(2*math.sqrt(p)))
        tested+=1
print(f"  Hasse |a_p|<=2 sqrt(p): {tested} curves, worst ratio={worst:.4f}")
ok(worst<=1.0,
   "function-field RH is a THEOREM: |a_p|<=2 sqrt(p), Frobenius on |alpha|=sqrt(p)")

print(f"\nALL {checks} CHECKS PASSED   ({time.time()-t0:.1f}s)")
```

Chapter 9

The Irresistible Twin Corridor

No structure in the elementary theory pulls harder than the twin corridor. Fold a period of the wheel and the twins fall into eight clean lanes, each mirrored across the fold, each recurring as far as one cares to count; the empty stretches between them keep a fixed rhythm; and the first composite to break a corridor does so just where one might have guessed. The arrangement looks so plainly engineered that it seems it must be a key to the oldest question about it—whether the twins go on forever. This chapter follows that pull to its end. The corridors are real, the symmetry exact, the recurrence genuine—and every part of it is reflection and equidistribution against a single constant fixed once and for all. The lure is honest. It simply is not a lever, and the work of the chapter is to feel it completely and then say so.

Scope

Of all the channels that ride the wheel, the twin corridor is the one that most invites belief. Fold a single period of 210 at its midpoint and the twin primes fall into eight clean lanes, each mirrored across the fold, each recurring period after period as far as one cares to count; the empty stretches between the lanes keep a fixed rhythm; and the smallest composite that breaks a corridor does so at a place one could almost have predicted. The arrangement is so plainly ordered that it seems it must be a key to the oldest open question about it—whether the twins go on forever.

This chapter follows that pull deliberately, and to its end. The first half draws the corridor out in full and lets the order accumulate: the lanes, the reflection that folds fifteen corridors to eight, the fixed rhythm of the deserts, the recurrence across periods. The second half turns. The groupings the eye seizes on are reflection symmetry and equidistribution and nothing besides; the whole recurrence is one constant, the twin-prime constant, computed once and repeating; and the wall—the parity barrier—stands exactly where it has always stood. The aim is to feel the lure completely and then to name, precisely, why it is a lure and not a lever.

In the arc of this book this is the second of the three great walls, reached from the same base as the first. The wheel hands over the corridor and its density just as, in the opening chapter, it handed ζ over to the explicit formula—one foundation, climbed by a different route. And the wall the corridor reaches is not the twins' alone: the parity barrier that stops it is the same obstruction the Liouville sum meets on the oscillation face, the point past which a sieve can fix a pattern's density exactly and still never certify that the pattern recurs. The twin wall is climbed from the foundation of the opening chapter and turned back for a reason that rhymes, in the end, with the Riemann wall's own.

9.1 The corridors, drawn

Begin with the picture (Figure 9.1). Five periods of 210 stand side by side, and each is folded at its midpoint 105: the left column climbs $-1, 0, 1, \dots$ up toward 105 while the right column descends $211, 210, 209, \dots$ down to meet it, so that every row carries a residue r beside its reflection $210 - r$, with the dashed axis between the columns marking the fold. Twin primes are filled green; a prime with no twin is left plain; the composite spokes—the subprimes—are orange; the rim, the positions divisible by $2, 3, 5$ or 7, is faint.

The twins live only in the corridors. A residue r can hold a twin only if both r and $r + 2$ are coprime to 210, and there are exactly fifteen such residues—the fifteen twin corridors. No twin ever touches the rim, the positions divisible by $2, 3, 5$ or 7; the corridors are the only doors.

The fold halves the count. The reflection $r \mapsto 210 - r$ carries the corridor at r to the corridor at $208 - r$—the pair $(r, r + 2)$ reflects to $(208 - r, 210 - r)$—so the fifteen corridors pair off into seven mirrors, $11 \leftrightarrow 197$, $17 \leftrightarrow 191$, $29 \leftrightarrow 179$, $41 \leftrightarrow 167$, $59 \leftrightarrow 149$, $71 \leftrightarrow 137$, $101 \leftrightarrow 107$, together with one self-paired seam: the corridor $(209, 211)$ that straddles the boundary between periods, where $211 \equiv 1$ opens the next one, marked $209 \leftrightarrow 1$. Eight folded corridors in all, shaded and named down the left margin of the figure, and a twin in the left column sits at the very same height as a twin-admissible row in the right.

And the lanes recur. Read across the five periods and the same eight lanes light up again and again, the twins thinning as the numbers grow—roughly as $1/(\log x)^2$, the density falling but the lanes never moving. It is hard, looking at this, not to feel that the recurrence is doing something: that the lanes are rails the twins are obliged to run on, period after period, and that the obligation might somehow be made to yield the infinitude of twins. Hold that feeling. It is exactly the right feeling to have, and we will account for every part of it.

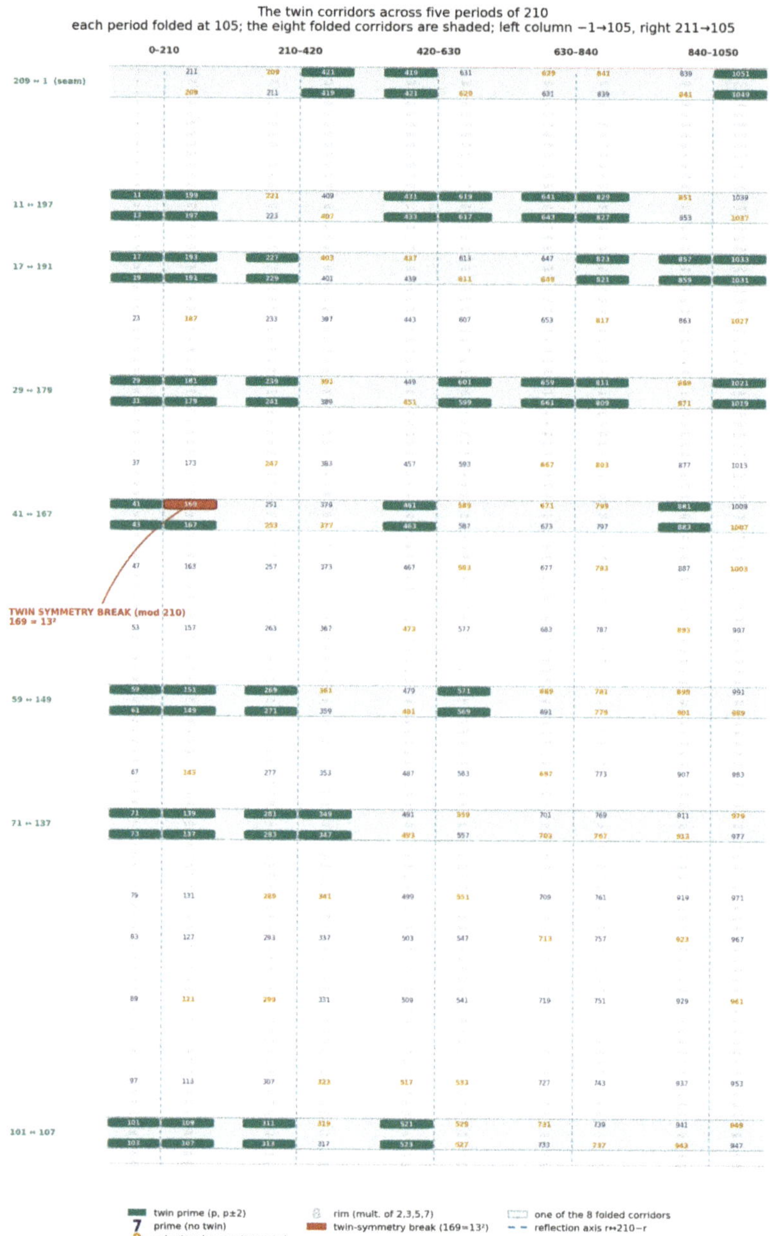

Figure 9.1: The twin corridors across five periods of 210. Each period is folded at 105: the left column runs $-1 \to 105$ and the right $211 \to 105$, so a residue r sits beside its reflection $210-r$ across the dashed axis. Twin primes are green, lone primes plain, the subprimes (composite spokes) orange, the rim faint. The fifteen twin corridors fold to eight shaded, named lanes—seven mirror pairs $r \leftrightarrow 208-r$ $(11 \leftrightarrow 197, \ldots, 101 \leftrightarrow 107)$ and the self-paired seam $(209, 211)$ shown as $209 \leftrightarrow 1$. In the $41 \leftrightarrow 167$ lane the genuine twin $(41, 43)$ stands on the left, while on the right the slot $(167, 169)$ carries the red twin-symmetry break $169 = 13^2$, so the twin the reflection would place there never forms. The lanes recur period to period and the twins thin; the alignments are reflection symmetry and equidistribution, not a correlation.

9.2 The rhythm of the deserts

Between the lanes lie the deserts—the runs of consecutive integers that hold no twin at all. In the lower half of the period they come in a fixed sequence of lengths: $3, 3, 9, 9, 15, 9, 27$. After the twin $(5, 7)$ three barren integers, then after $(11, 13)$ three more; then the deserts widen to nine, nine, fifteen, nine, and finally the great stretch of twenty-seven empty integers from 74 to 100 before the corridor at 101 opens again.

This rhythm is not a sample. It is fixed by where the admissible residues fall mod 210, and so it is identical in every period: the deserts of the thousandth period have the same lengths in the same order as the deserts of the first. What changes from period to period is only whether a given corridor, when it comes, holds two primes—the lanes and the deserts themselves are permanent furniture. The order is real, and it is the order of a fixed set of residues, drawn once.

The same corridors, read through the least prime factor, show the skeleton they ride on (Figure 9.2). Fold two periods and colour every cell by its smallest prime factor: the $2-3-2$ backbone of the rim, with $\div 5$ and $\div 7$ threaded through it, all of it reflection-symmetric across each fold because 210 is divisible by $2, 3, 5, 7$. The twins sit as green rungs on that backbone, the lone primes pale beside them, and the subprimes—least factor ≥ 11—stand out in purple, $169 = 13^2$ among them. The barren runs between the green rungs are the deserts, in their fixed lengths; and the last-digit columns flanking each period show the wheel's other mirror, $1 \leftrightarrow 9$ and $3 \leftrightarrow 7$, folding in exact step. Two reflections—the residue fold and the last-digit fold—run through one fixed skeleton.

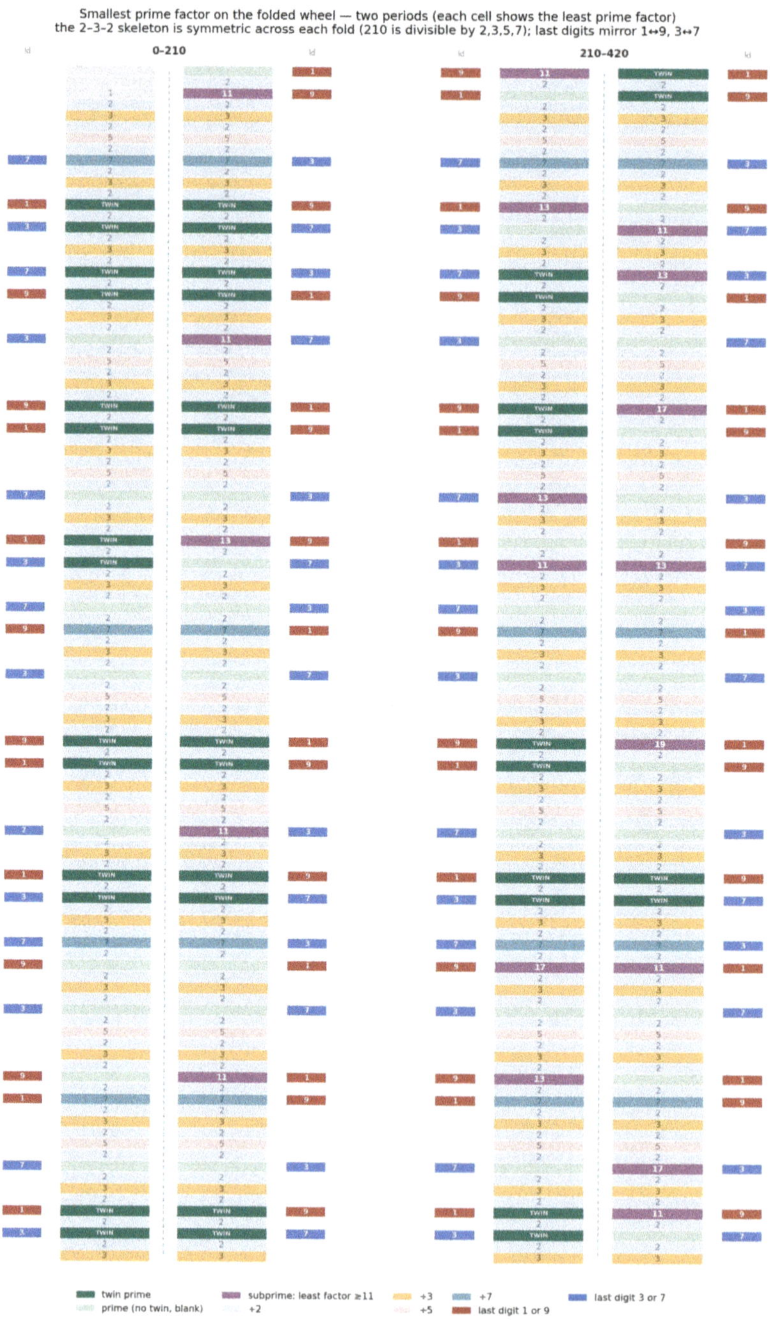

Figure 9.2: The same fold, read by least prime factor, over two periods. Each cell shows its smallest prime factor: ÷2 pale blue, ÷3 orange, ÷5 pink, ÷7 teal, and the subprimes (least factor ≥ 11) purple. The $2-3-2$ skeleton is reflection-symmetric across each fold and identical from one period to the next; twins are green, lone primes pale green, and the last-digit columns on either side mirror $1 \leftrightarrow 9$ and $3 \leftrightarrow 7$. The green twins ride the fixed backbone, the deserts are the barren runs between them, and the purple subprimes—$169 = 13^2$ the first to fall inside a corridor—multiply as the twins thin.

9.3 Where a corridor breaks from within

The reflection suggests a tidy expectation: if (41, 43) is a genuine twin, its mirror corridor at 167 ought to be one too. It is not. The corridor is there—both 167 and 169 are coprime to 210, a legitimate twin-admissible slot—but $169 = 13^2$. The slot holds a prime and a square, not two primes. In the figure this is the lane where the left column is green twice over while the right carries a lone prime, 167, alongside the red break $169 = 13^2$.

This is the first place a composite breaks a twin corridor from within, and it happens exactly when the smallest prime outside the wheel, 13, first lands its square inside a corridor. It is the twin entry on the wheel's first-break ladder—11^2 for the bare primes, $11 \cdot 13$ for the cousins, 13^2 for the twins—and it is the first honest crack in the symmetry: the reflection organizes the corridors perfectly and says nothing whatever about whether a given corridor is filled by primes.

9.4 The groupings are equidistribution, not linkage

Now the turn. Spend long enough with the figure and the eye begins to find groupings: a corridor and its mirror both lit in one period, then both lit again in the next; clusters of green that seem to travel together. It is natural—irresistible, even—to read those alignments as evidence of a binding, some correlation by which corridors hold one another's twins in place.

They are not. Tested against the only honest baseline, a corridor and its mirror are co-realised—both holding a twin in the same period—no more often than any two corridors chosen at random. Across several thousand periods the mean correlation between mirror corridors and the mean correlation across all corridor pairs come out statistically the same, both indistinguishable from zero (the verification below reports +0.002 against −0.000). The alignments are the visible sum of two things already named and nothing else: the reflection symmetry, which is fixed, and the rough equidistribution of twins among the corridors, which is simply what equidistribution looks like. There is no third thing. The groupings carry no signal.

This is the hardest sentence in the chapter to believe while looking at the picture, which is exactly why it has to be said while looking at the picture.

9.5 What the corridor is made of

So what is the recurrence, if not a mechanism? It is a constant. The Hardy–Littlewood conjecture gives the expected number of twins up to x as

$$\pi_2(x) \sim 2C_2 \int_2^x \frac{dt}{(\log t)^2}, \qquad C_2 = \prod_{p \geq 3} \left(1 - \frac{1}{(p-1)^2}\right) = 0.6601618\ldots,$$

the twin-prime constant. The density of twins in the corridors is this single number, C_2, times the same Mertens thinning the primes already feel. The lanes that recur period after period, the rhythm of the deserts, the whole sense of an engineered structure—all of it is the singular series C_2, an Euler product over the primes computed once and then repeated.

That is the deflation in one line. The recurrence is not structure held in reserve to be spent on the twin question; it *is* the answer to a question already answered—how dense would twins be, if they are infinite—wearing the look of a lever on the question still open, whether they are. The Euler product is cashed. There is nothing behind the lanes but the constant that defines them.

9.6 The wall

The question the corridor cannot answer is the oldest one: are there infinitely many twins? The corridor says where they must live and how thickly they would lie, and stops. Brun proved that the sum of the reciprocals of the twins converges, so they are at most sparse; Hardy and Littlewood conjectured the exact density above; and there it has stood.

The genuine progress of the last decade did not come from the corridor. Zhang, and then Maynard and Tao, proved that infinitely many pairs of primes sit within a bounded gap—below seventy million at first (Zhang, 2013), reduced to 600 by Maynard and Tao and to 246 by the Polymath project—which is real passage into the dimension-two channel. But it came from sieve theory clever enough to bend the parity barrier, not from the wheel's geometry; the corridors neither suggested it nor are touched by it. Between *bounded gap* and *gap two* stands the parity barrier of Selberg, the same one unit of sieve dimension that separates the solved prime channel from the open twin channel—and the same barrier that, on the oscillation face, holds the Liouville sum just beyond a sieve's reach.

The corridor, then, is the wall's own description. It is an exact, complete, and beautiful account of where twins can be and how dense they are—which is to say, it is the singular series and the parity barrier, drawn as a picture. It is not, and never was, a way across.

What this is and is not

What is genuine here is all of it. The fifteen corridors, the eight folded lanes, the seven mirror pairs and the seam, the desert rhythm $3, 3, 9, 9, 15, 9, 27$, the break at 13^2, the constant $C_2 = 0.6601618\ldots$—every one is real, exact, and elementary, and most of it is older than a century. None of it is a trick.

What it is not is a key. The twin corridor is the most irresistible object in the elementary theory precisely because every part of its order is real: the symmetry does not lie, the recurrence does not lie, the deserts keep their rhythm honestly. The lure is not deception. It is the genuine pull of genuine structure, and the only error on offer is the last one—to mistake the structure for a mechanism, the singular series for a lever, the picture of the wall for a door in it. We have let the corridor pull as hard as it pulls and named, at each step, exactly what the pull is made of. That naming is the whole of the defence, and it is the only one there is.

Verification

Every count and constant in this chapter is closed-form or a direct, seed-free computation. The script below, using only the Python standard library, recomputes the fifteen corridors, the eight folded lanes, the desert rhythm, the 13^2 break, the twin-prime constant, and the absence of any mirror-corridor linkage, by assertion.

```
#!/usr/bin/env python3
"""Verification for 'The Irresistible Twin Corridor'.
Standard library only; deterministic (no random seeds)."""
from math import gcd

def sieve(n):
    s=bytearray([1])*(n+1); s[0]=s[1]=0
    i=2
    while i*i<=n:
        if s[i]:
            s[i*i::i]=bytearray(len(s[i*i::i]))
        i+=1
    return s

M=210
checks=0
def ok(cond,msg):
    global checks
    assert cond, "FAIL: "+msg
    checks+=1
    print("PASS:",msg)

# --- the fifteen twin corridors of the wheel ---
corridors=[r for r in range(M) if gcd(r,M)==1 and gcd((r+2)%M,M)==1]
ok(len(corridors)==15, f"twin corridors mod 210 number 15 (got {len(corridors)})")

# --- the fold r <-> 208-r : seven mirror pairs + one self-paired seam = eight lanes ---
S=set(corridors); lanes=set(); selfpaired=[]
for r in corridors:
    m=(208-r)%M
    if m==r: selfpaired.append(r)
    lanes.add(frozenset({r,m}))
ok(len(lanes)==8, f"the fifteen fold to eight lanes (got {len(lanes)})")
pairs=[l for l in lanes if len(l)==2]
ok(len(pairs)==7 and len(selfpaired)==1,
   f"seven mirror pairs + one seam (got {len(pairs)} pairs, seam {selfpaired})")
ok(selfpaired==[209], f"the seam is the corridor 209 (got {selfpaired})")

# --- the rhythm of the deserts in the lower half ---
sm=sieve(220)
tp=[p for p in range(3,104) if sm[p] and sm[p+2]]      # twin lower-members in [1,105]
gaps=[]
for (a,b) in zip(tp,tp[1:]):
    g=b-a-3             # integers strictly between a+2 and b
    if g>0: gaps.append(g)
ok(gaps==[3,3,9,9,15,9,27], f"twin-desert rhythm 3,3,9,9,15,9,27 (got {gaps})")

# --- the first break: 169 = 13^2 breaks the mirror of the twin (41,43) ---
ok(sm[167] and not sm[169] and 169==13*13, "167 prime, 169 = 13^2 composite")
```

```
ok(gcd(167,M)==1 and gcd(169,M)==1, "(167,169) is a genuine corridor (both coprime to 210)")
ok((208-41)%M==167, "corridor 167 is the mirror of corridor 41")
ok(sm[41] and sm[43], "(41,43) is a true twin; its mirror is broken from within")

# --- the constant the corridor is made of: the twin-prime constant C2 ---
sp=sieve(200000); primes=[p for p in range(3,200001) if sp[p]]
C2=1.0
for p in primes: C2*=1-1.0/((p-1)**2)
ok(abs(C2-0.6601618)<1e-4, f"twin-prime constant C2 = {C2:.7f} (~0.6601618)")

# --- the groupings are equidistribution, not linkage ---
# co-realisation of mirror corridor pairs vs. arbitrary corridor pairs, over many periods
N=4000; big=sieve(M*N+10)
ind={c:[1 if (big[M*k+c] and big[M*k+c+2]) else 0 for k in range(1,N)] for c in corridors}
def corr(a,b):
    n=len(a); ma=sum(a)/n; mb=sum(b)/n
    cov=sum((x-ma)*(y-mb) for x,y in zip(a,b))/n
    va=sum((x-ma)**2 for x in a)/n; vb=sum((y-mb)**2 for y in b)/n
    return cov/((va*vb)**0.5) if va>0 and vb>0 else 0.0
mirror=[(r,(208-r)%M) for r in corridors if r<(208-r)%M]
mc=sum(corr(ind[a],ind[b]) for a,b in mirror)/len(mirror)
allp=[(corridors[i],corridors[j]) for i in range(15) for j in range(i+1,15)]
ac=sum(corr(ind[a],ind[b]) for a,b in allp)/len(allp)
print(f"  mean correlation, mirror pairs : {mc:+.4f}")
print(f"  mean correlation, all pairs    : {ac:+.4f}")
ok(abs(mc-ac)<0.03 and abs(mc)<0.05,
   "mirror corridors co-realised no more than arbitrary pairs (no linkage)")

print(f"\nALL {checks} CHECKS PASSED")
```

Chapter 10

The Walls

This book has climbed a dozen methods, and met, each time, the same kind of wall: the place where the wheel's local law stops deciding the global fact. This closing chapter sets the three great walls side by side—the Riemann Hypothesis, the twin primes, and factoring—and says, for each, exactly what is known and exactly where the barrier stands. They are three faces of one situation. The primes give up their local structure freely—their residues, their densities, the singular series, the characters—and their global structure not at all: where the zeros lie, whether two primes recur forever, how to undo a product. What is known is genuine and often deep, and it was won by the very methods that then disclosed their own limits. What is not known is not for want of trying. At each wall the lever that would cross it turns out to be the thing it was trying to find.

Scope

This is a survey, not a new climb. The chapters before it each took one method from the wheel to a wall; here we stand back and look at the walls themselves. Three of them carry the weight of the subject. The Riemann Hypothesis is the wall of *location*: where, exactly, the zeros sit, and so how regular the primes are. The twin primes are the wall of *recurrence*: whether a pattern whose density we know exactly occurs forever. Factoring is the wall of *decomposition*: whether the primes, so easy to multiply together, can be pulled back apart. The three are distinct theorems with distinct barriers, and yet one situation runs through all of them. We will name, for each, what is proven, and then the one thing that is not, and why the not is so hard.

10.1 What the wheel gave, and where it stops

Look back along the climb and a single shape repeats. Each chapter took the wheel's local data and cashed it for a genuine, finished fact. The opening chapter cashed the Euler product and the family of L-functions $L(s,\chi)$—with the non-vanishing $L(1,\chi)\neq 0$ that carries Dirichlet's theorem—in exact closed form. The Ramanujan chapter cashed the Sato–Tate law and Deligne's bound, the proven local Hypothesis one floor up. The twin corridor cashed the singular series, the twin constant $2C_2 = 1.3203\ldots$, the local product that predicts the twin density to within a percent. And the five faces each cashed a local equivalent of the Hypothesis itself—an error term, a positivity, a completeness, an inequality, a heat-flow constant—while the neighbouring walls cashed the random-matrix statistics of the zeros and the proven Hypothesis over finite fields. Every one of these is local in the deepest sense: a statement about residues modulo the primes, assembled into a product, a count, or a transform, and then it is done.

And every one of them stopped at the same kind of edge. The local law gives the main term and falls silent on the fluctuation; gives the density and falls silent on the occurrence; gives the sieve and falls silent on the factorization. The wheel is the primes' local law, and the three walls are the three places that law runs out. Figure 10.1 sets them side by side with what is known on each.

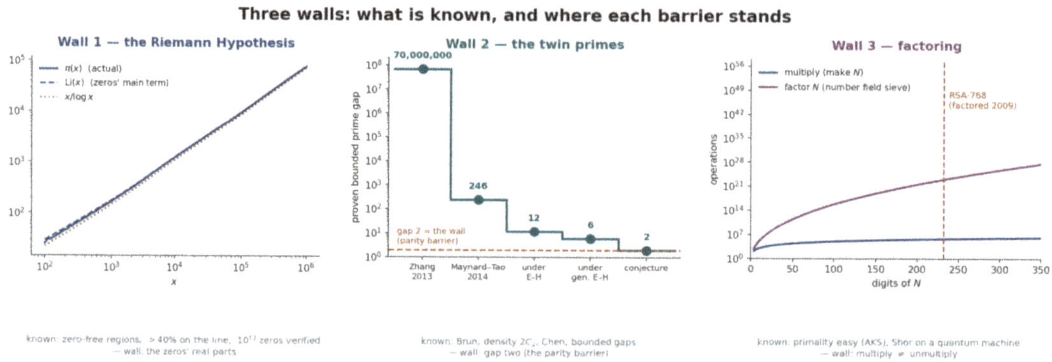

Figure 10.1: The three walls, with what is known on each. *Left:* for the Riemann Hypothesis, $\mathrm{Li}(x)$ tracks $\pi(x)$ far more closely than $x/\log x$—the better main term the zeros supply—and the residual error is what the Hypothesis is about. *Centre:* for the twins, the proven bound on prime gaps has fallen from Zhang's 70,000,000 to 246, and to 6 under a stronger hypothesis, but stops one step above gap two. *Right:* for factoring, making a number by multiplication is cheap while unmaking it by the number field sieve is not, the gap widening without bound in the number of digits.

10.2 The first wall: the Riemann Hypothesis

The Hypothesis is the wall of location: that every nontrivial zero has real part exactly $\frac{1}{2}$. By the explicit formula, this is the same as saying the primes are as regular as arithmetic permits—that the error in the prime number theorem is as small as it can possibly be. A great deal is known. Riemann's functional equation pins the zeros into a symmetric strip; de la Vallée Poussin's zero-free region, widened by Vinogradov and Korobov, is exactly what proves the prime number theorem with its error term; Selberg, then Levinson, then Conrey showed that a positive proportion—more than forty percent—of the zeros lie on the line; and the first many trillions of them have been checked, every one on the line, by direct computation. There are even fully elementary equivalents: Lagarias showed that the single inequality $\sigma(n) \leq H_n + e^{H_n} \log H_n$, for every n, is equivalent to the Hypothesis—and the verification confirms it holds throughout its range. The faces of the earlier chapters are all this one wall seen from different sides—the oscillation of the error term, the positivity of a functional, the completeness of a span, an extremal inequality, the constant of a heat flow—each provably equivalent to the Hypothesis and none of them a way past it; and the neighbouring walls ring it with a proven analogue over finite fields, an all-but-certain operator, and its own weaker shadow.

What is missing is any control of the real parts themselves. The zero-free regions creep toward the line and never reach it; the proportion on the line climbs and never reaches one; the computations verify and never conclude. The reason is the one this book has met at every turn: the dependence runs from the zeros to the primes, never back. The explicit formula expresses the primes through the zeros, so everything we know about the primes is downstream of the zeros, and nothing we know about the primes can reach up to constrain where the zeros are. The Hypothesis is the cleanest statement of the wall: we can describe the consequences of the zeros' location to any precision, and we cannot touch the location.

10.3 The second wall: the twins

The twin primes are the wall of recurrence, and here the book's own constant stands at the foot. We know the twins are sparse—Brun proved that the sum of their reciprocals converges—and we know exactly how sparse they would be: the density is the singular series $2C_2$, drawn from the wheel, matching the counts to a percent. We know more than density. Chen proved that infinitely often p and $p+2$ are a prime and a number with at most two prime factors—a near-twin, missing by the thinnest possible margin. And in the last decade the gap itself was bounded: Goldston, Pintz and Yıldırım opened the door, Zhang walked through it with a bound of seventy million, and Maynard and Tao, with the Polymath effort, brought it down to 246, to 12 under the Elliott–Halberstam conjecture, and to 6 under its generalized form. Infinitely many primes within a bounded gap is a theorem.

But the last step does not come. Between *bounded gap* and *gap exactly two* stands the parity barrier of Selberg—the same barrier the twin corridor met at its summit. A sieve cannot tell an even number of prime factors from an odd one; it cannot separate "two primes" from "a prime and a product of two primes," which is exactly Chen's almost-twin. The verification shows the blindness directly: the signed parity sum, the Liouville function over the first million integers, is tiny relative to its range, a quantity the sieve cannot see and cannot use. Chen's theorem is the sieve pressed to its very edge. Gap two is one parity beyond the edge.

10.4 The third wall: factoring

Factoring is the wall of decomposition, and it is a different kind of wall—not a barrier to proof but a barrier to computation, the one the modern world is built on. Its shape is an asymmetry. To *make* a composite is trivial: multiply two primes, an instant's work.

To recognise a prime is also easy now—the algorithm of Agrawal, Kayal and Saxena tests primality in polynomial time, and fast probabilistic tests predate it by decades. But to *unmake* a composite, to recover the primes from their product, the best classical method known, the number field sieve, takes time that grows sub-exponentially in the digits, fast enough that a number of a few hundred digits is beyond reach. Shor showed that a quantum computer would factor in polynomial time, so the wall is not absolute—but classically it stands, and public-key cryptography rests on it.

The wheel sees the asymmetry and cannot cross it. Its sieve strikes out every multiple of $2, 3, 5, 7$ at a glance, and identifies the subprimes—the composites coprime to 210, whose least factor is at least eleven—as the survivors. But it cannot factor a subprime; it knows only that the small primes are absent. The verification makes the point concrete: take two primes near ten million, multiply them, and the product has no factor below a hundred thousand, so the wheel's sieve learns nothing, while the product itself was formed in a single multiplication. Easy to make, hard to unmake. The primes are the multiplicative atoms, and the wall is that atoms combine freely and split only under great force.

10.5 One wall, three faces

Set the three together and the same situation shows three times (Figure 10.2). In each, the wheel— the primes' local law—is known and sits at the foot; in each, a global fact stands above it, walled; and in each, the barrier is that the lever which would cross is the very thing being sought. To cross the Riemann wall one would need to control the zeros' real parts, but the zeros are what govern the primes one is reasoning from. To cross the twin wall one would need to break parity, but breaking parity is precisely what isolating two primes means. To cross the factoring wall cheaply one would need to undo multiplication, but undoing multiplication is precisely factoring. The destination is the tool. That circularity is not a failure of ingenuity; it is the structure of the subject.

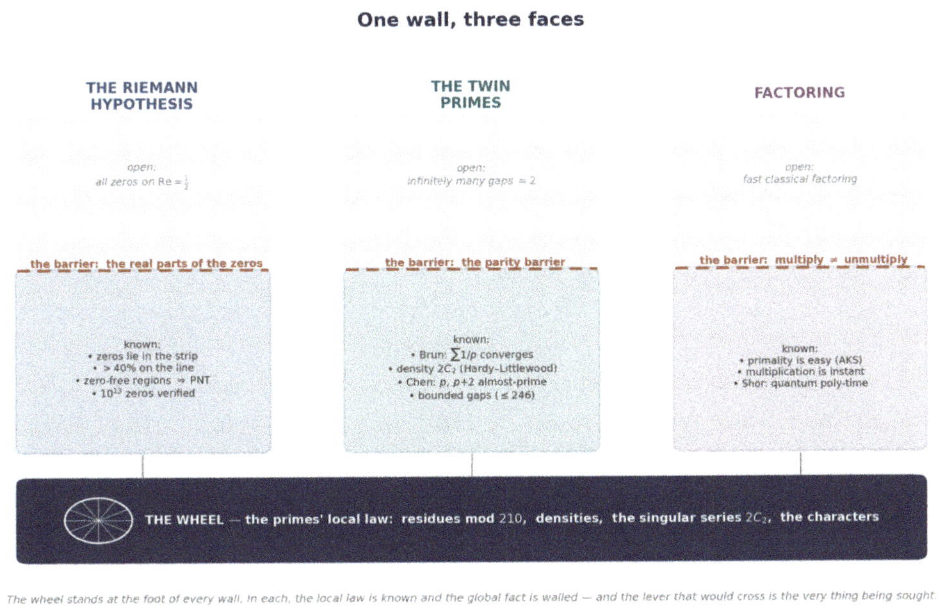

Figure 10.2: One situation, three faces. The wheel—residues, densities, the singular series, the characters—is the known foundation under every wall. On each, the local law is settled (below the barrier) and the global fact is open (above it): the real parts of the zeros, the parity barrier, the multiply–unmultiply asymmetry. In each the lever that would cross is the thing being sought.

10.6 What scaling the wall has meant

We have not crossed. That was never the promise. The promise was to climb each wall honestly, with the wheel at its foot, feeling the full pull of each method that seemed for a moment to lead over—and to name, at the top, exactly what the pull was made of. The gains along the way are real and were not small: the singular series and the corridor it draws; the non-vanishing of the L-functions in closed form; a positive proportion of the zeros on the line; primes recurring within a bounded gap; primality made easy; the factoring asymmetry on which a global infrastructure safely rests. Every one of these was won by a method that, pressed further, showed its own limit—the lure and the wall in the same hand.

That is the whole of the book's argument, and the whole of its honesty. The structures are genuine: the symmetry does not lie, the recurrence does not lie, the statistics of the zeros are real to every checked decimal. The only error ever on offer was the last one—to mistake a description of the wall for a door in it, the singular series for a lever, the spectral fingerprint for a hand, the local law for the global fact. We have let each wall pull as hard as it pulls, granted it every genuine thing it had to give, and located, each time, the one thing it lacked. The wheel turns at the foot of the wall. The wall still stands. And to know precisely where it stands, and precisely why, is what a century and more of the best mathematics has bought—which is not a crossing, and is not nothing.

What this is and is not

What is genuine here is the entire ledger of the known. The zero-free regions and the prime number theorem; the proportion of zeros on the line; Lagarias's elementary equivalent; Brun's convergence, the twin constant $2C_2$, Chen's near-twins, the bounded gaps of Zhang and of Maynard and Tao; the easy primality of Agrawal, Kayal and Saxena, and Shor's quantum factoring. Every item is proven, named, and real. None of it is a trick.

What it is not is a way over any of the three. The Riemann Hypothesis, the infinitude of twins, and fast classical factoring are open, and the verification marks the line precisely: an RH-equivalent inequality that holds as far as we look and is unproven beyond; a parity the sieve cannot see; a product cheap to form and costly to break. The three walls are three faces of the primes' refusal to hand over globally what they hand over locally for free. We have scaled each, with the wheel at its foot, and reported the view from the top without embellishment. That report is the book.

Verification

The known facts and the exact barriers are checked, where they can be made elementary, by the script below— standard library only, deterministic. It confirms an RH-equivalent inequality throughout its range, the better main term the zeros supply, the parity a sieve is blind to, and the make/unmake asymmetry of factoring.

```
#!/usr/bin/env python3
"""Verification for 'The Walls'. Standard library only; deterministic (no seeds)."""
from math import log, exp, isqrt

checks=0
def ok(cond,msg):
    global checks
    assert cond,"FAIL: "+msg
    checks+=1; print("PASS:",msg)

def sieve(n):
    s=bytearray([1])*(n+1); s[0]=s[1]=0
    i=2
    while i*i<=n:
        if s[i]: s[i*i::i]=bytearray(len(s[i*i::i]))
        i+=1
    return s

# ================= WALL 1: the Riemann Hypothesis =================
# (a) Lagarias's elementary inequality  sigma(n) <= H_n + exp(H_n) ln(H_n)
#     is EQUIVALENT to RH (Lagarias 2002). It holds throughout this range.
NL=100000
sigma=[0]*(NL+1)
for d in range(1,NL+1):
    for m in range(d,NL+1,d):
        sigma[m]+=d
H=0.0; lag=True; worst=0
```

```
for n in range(1,NL+1):
    H+=1.0/n
    rhs = 1.0 if n==1 else H+exp(H)*log(H)
    if sigma[n] > rhs+1e-9: lag=False; worst=n; break
ok(lag, f"Lagarias inequality holds for 1<=n<={NL} (RH-equivalent, unproven)")

# (b) the zeros' contribution: Li(x) tracks pi(x) far better than x/log x
X=1000000
pi=sum(sieve(X))
def Li(x):
    steps=200000; a=2.0; b=float(x); h=(b-a)/steps; tot=0.0
    for i in range(steps+1):
        w=0.5 if (i==0 or i==steps) else 1.0
        tot+=w/log(a+i*h)
    return tot*h
li=Li(X); rough=X/log(X)
print(f"  pi(10^6)={pi}; Li={li:.0f} (err {li-pi:+.0f})")
print(f"  x/logx={rough:.0f} (err {rough-pi:+.0f})")
ok(abs(li-pi) < abs(rough-pi)/10, "Li(x) beats x/log x (the zeros' main term)")

# ================= WALL 2: the twins (the parity barrier) =================
# A sieve cannot see the sign of Liouville lambda(n)=(-1)^Omega(n): cannot tell
# even from an odd number of prime factors -- exactly what separates a prime from a
# product of two primes. That blindness is the parity wall; the signed sum is tiny.
M=1000000
spf=list(range(M+1))
for i in range(2,isqrt(M)+1):
    if spf[i]==i:
        for j in range(i*i,M+1,i):
            if spf[j]==j: spf[j]=i
Om=[0]*(M+1)
for n in range(2,M+1):
    Om[n]=Om[n//spf[n]]+1
L=sum(-1 if Om[n]&1 else 1 for n in range(1,M+1))
print(f"  Liouville sum L(10^6) = {L}")
ok(abs(L) < 0.005*M, f"|L(10^6)|={abs(L)} small: the parity a sieve cannot see")

# ================= WALL 3: factoring (easy to make, hard to unmake) =================
def mr(n):
    if n<2: return False
    small=(2,3,5,7,11,13,17,19,23,29,31,37)
    for p in small:
        if n%p==0: return n==p
    d=n-1; s=0
    while d%2==0: d//=2; s+=1
    for a in small:
        x=pow(a,d,n)
        if x==1 or x==n-1: continue
        for _ in range(s-1):
            x=x*x%n
            if x==n-1: break
        else: return False
    return True
def nextprime(n):
    n+=1
    while not mr(n): n+=1
    return n
p=nextprime(10**7); q=nextprime(2*10**7)
N=p*q
ok(mr(p) and mr(q), f"primality is easy: {p} and {q} certified prime at once")
B=100000
ok(all(N%d!=0 for d in range(2,B+1)), f"sieve to {B} finds no factor of N (blind)")
ok(p*q==N and N>10**14, "N=pq is one multiply: easy to make, hard to unmake")
```

```
print(f"\nALL {checks} CHECKS PASSED")
```

Appendix A

Reformulations and Equivalences of the Riemann Hypothesis

The Riemann Hypothesis has been restated more ways than almost any open problem in mathematics, and the standing temptation is to file every restatement under one heading: equivalence. *They are not all equivalences. A clean catalogue has to keep three things apart—statements proven equivalent to the Hypothesis, approaches that would imply it or already assume it, and analogies that resemble it in another world—because conflating them is exactly the lever error the rest of this book has worked to name. A description is not a door; a hope is not a theorem; an analogy is not an identity. This appendix sorts the known reformulations by status, gives the sharp form of each genuine equivalence, flags two celebrated sharpenings that are simply false, and corrects two items that are not equivalences at all.*

Scope

This is a reference, sorted by what is actually known. The first part lists the genuine equivalences—the statements proven to hold if and only if every nontrivial zero has real part $\frac{1}{2}$—grouped by the kind of mathematics they come from. The second part records two famous strengthenings, of Mertens and of Pólya, that hold as far as anyone has computed and are nonetheless false: the sharpest reminders the subject offers that agreement in range is not truth. The third part places the spectral programme—Hilbert–Pólya, the random-matrix statistics, Berry–Keating, Connes—where it belongs, as approaches rather than equivalences; this is the spine of the companion volume, and is listed here only to be correctly labelled. The last part corrects two entries often miscatalogued: the function-field analogy, which is a different theorem in a different world, and the divisor bound, which is weaker than what is known unconditionally and so equivalent to nothing.

A.1 A map of the usual list

The fifteen statements most often gathered under "equivalences of RH" fall into the categories below.

#	Statement, as usually listed	Status
1	$\pi(x) = \mathrm{Li}(x) + O(x^{1/2+\varepsilon})$	equivalence
2	$\psi(x) = x + O(x^{1/2+\varepsilon})$	equivalence
3	Mertens $M(x) = O(x^{1/2+\varepsilon})$	equivalence
4	Liouville $L(x) = O(x^{1/2+\varepsilon})$	equivalence
5	Li's criterion, $\lambda_n \geq 0$ for all n	equivalence
6	Nyman–Beurling, 1 in an L^2 closure	equivalence
7	Hilbert–Pólya self-adjoint operator	would imply RH
8	Riemann ξ in the Laguerre–Pólya class	equivalence
9	Random-matrix (GUE) pair correlation	assumes RH; evidential
10	Berry–Keating $H = xp$	conjectural model
11	Divisor bound $O(x^{1/2+\varepsilon})$	*not* equivalent (trivial)
12	Möbius exponential sum	nuanced (see text)
13	Connes' noncommutative geometry	programme
14	Weil conjectures over finite fields	analogy
15	Quantum-chaos correspondence	analogy

A.2 The genuine equivalences

The error-term family (1–4). These four are one phenomenon—square-root cancellation—in four costumes. The prime counting function against the logarithmic integral, Chebyshev's ψ against x, the Mertens sum $M(x) = \sum_{n\leq x} \mu(n)$, and the Liouville sum $L(x) = \sum_{n\leq x} \lambda(n)$: each being $O(x^{1/2+\varepsilon})$ is equivalent to the Hypothesis. The sharp forms are due to Schoenfeld: on RH, $|\pi(x) - \mathrm{Li}(x)| < \frac{1}{8\pi}\sqrt{x}\log x$ for $x \geq 2657$, and $|\psi(x) - x| < \frac{1}{8\pi}\sqrt{x}(\log x)^2$. The verification confirms that $M(x)$ and $L(x)$ stay well inside an $x^{0.6}$ envelope to 10^6.

The function-theoretic family (5, 8). Li's criterion makes RH a positivity: the numbers $\lambda_n = \sum_\rho \left(1 - (1 - 1/\rho)^n\right)$ are all ≥ 0 exactly when the Hypothesis holds. Membership of the Riemann ξ in the Laguerre–Pólya class is RH almost by definition—that class is the closure of real polynomials with only real zeros, so ξ lies in it precisely when its zeros are all real. The modern sharpening is the de Bruijn–Newman constant Λ: RH is equivalent to $\Lambda \leq 0$, and the inequality $\Lambda \geq 0$ was proven by Rodgers and Tao in 2020, so RH is now equivalent to $\Lambda = 0$—the Hypothesis, if true, is true by the thinnest possible margin.

The Hilbert-space family (6). The Nyman–Beurling criterion makes RH an approximation problem: the constant function 1 lies in the $L^2(0,1)$-closure of the span of the dilated fractional parts $\{x/n\}$ exactly when the Hypothesis holds. Báez-Duarte's refinement (2003) sharpened this to a single, cleaner approximating sequence and is the form most used today.

The arithmetic family (additions). Two elementary inequalities deserve a place the usual list omits. Robin's inequality (1984): for every $n > 5040$, $\sigma(n) < e^{\gamma} n \log\log n$, and this holds for all such n if and only if RH holds; the verification checks it past its last exception, throughout $5040 < n \leq 10^6$. And Lagarias's inequality (2002), $\sigma(n) \leq H_n + e^{H_n} \log H_n$ for every n, with equality only at $n = 1$, likewise equivalent to RH—this is the one cashed in the Walls chapter. To these one may add the Riesz criterion, the Franel–Landau equidistribution of Farey fractions, and the Redheffer matrix, whose determinant equals $M(n)$.

Positivity (the genuine core of the spectral programme). Weil's explicit-formula criterion: RH holds if and only if a certain quadratic functional, built from the explicit formula, is positive. This is a true equivalence, and it is the part of the Connes programme that is settled mathematics rather than aspiration.

The fractal-geometric family (an addition the usual list omits). A genuine equivalence reaches the catalogue from a corner the standard fifteen rarely touch: the geometry of fractal strings, due to Lapidus and his collaborators. A *fractal string* is nothing more than a bounded open subset of the line, recorded by the decreasing sequence of its interval lengths $\ell_1 \geq \ell_2 \geq \cdots$; its *geometric zeta function* is $\zeta_{\mathcal{L}}(s) = \sum_j \ell_j^s$, and the poles of that function are its *complex dimensions*. The Cantor string—the gaps of the middle-thirds set, one length $1/3$, two of $1/9$, four of $1/27$—has $\zeta_{\mathcal{L}}(s) = 1/(3^s - 2)$, whose poles sit at $s = \frac{\log 2}{\log 3} + i\,n\,\frac{2\pi}{\log 3}$: a vertical lattice whose real part is the set's fractal dimension and whose imaginary parts encode the log-periodic oscillations of its geometry (Figure A.1). The complex dimensions are to a fractal's geometry what the zeros are to the primes—the frequencies of an explicit ("tube") formula for its volume. The Riemann zeta enters because a string can be *heard*: the frequencies of its intervals are k/ℓ_j, and their spectral zeta function factors as $\zeta(s)\,\zeta_{\mathcal{L}}(s)$, making ζ the exact bridge from geometry to spectrum. On that bridge Lapidus and Maier proved the equivalence: the inverse spectral problem—whether a string's spectrum fixes its Minkowski geometry—has an affirmative answer for every dimension $D \in (0,1)$ *except* $D = \frac{1}{2}$ if and only if the Riemann Hypothesis holds. The exceptional dimension is $\frac{1}{2}$ for the obvious reason, and RH becomes the statement that $\frac{1}{2}$ is the single dimension at which one cannot hear a fractal string's shape. Like every entry in this section it relocates the wall—here into fractal geometry—without lowering it.

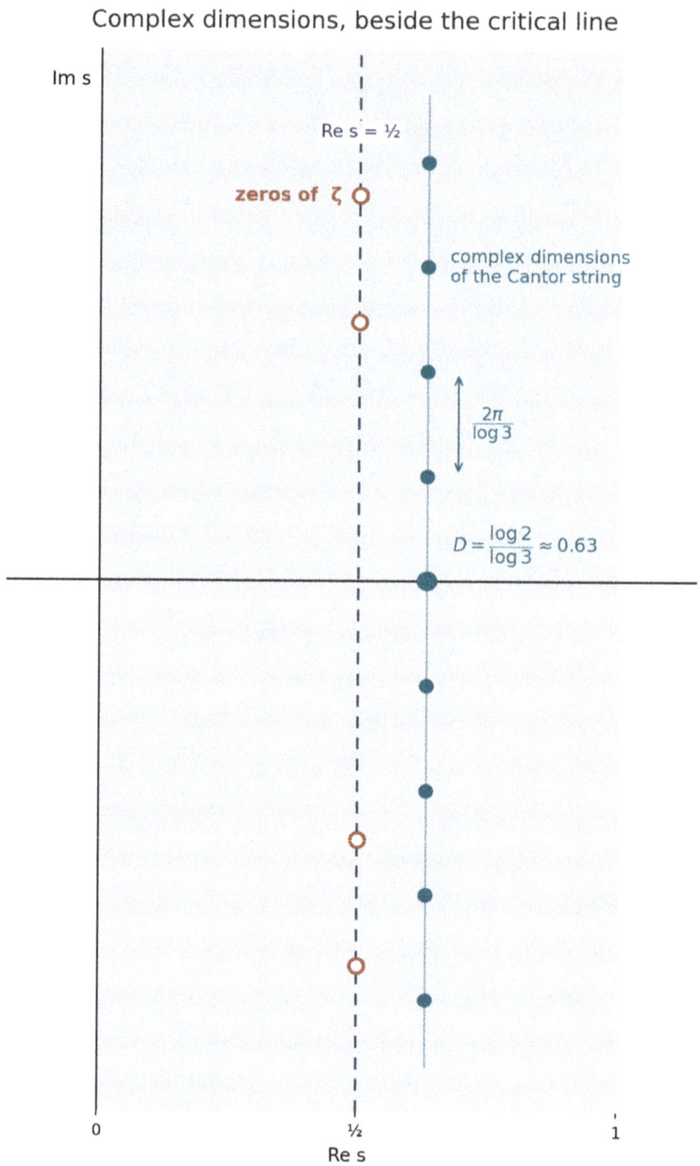

Figure A.1: Complex dimensions, beside the critical line. The Cantor string's complex dimensions—the poles of $1/(3^s - 2)$—form an exact vertical lattice on $\operatorname{Re} s = \log 2/\log 3$, its fractal dimension, spaced by $2\pi/\log 3$; they run an explicit formula for the string's geometry as the zeros of ζ run one for the primes. The lattice is perfectly periodic because the Cantor set is exactly self-similar—the zeros of ζ (on $\operatorname{Re} s = \frac{1}{2}$) are not a lattice, which is one way to say the primes are not. By Lapidus–Maier, $\frac{1}{2}$ is the lone dimension at which the inverse spectral problem can fail.

A.3 Two famous sharpenings that are false

The error-term family carries a warning the catalogue should never drop. Drop the ε from the Mertens bound and one gets the *Mertens conjecture*, $|M(x)| \leq \sqrt{x}$. It holds for every x anyone has ever computed, including the entire range of the verification here—and it is false, disproven by Odlyzko and te Riele in 1985 without exhibiting a single counterexample. Replace the Liouville bound by its tempting strong form and one gets *Pólya's conjecture*, $L(x) \leq 0$ for $x \geq 2$. It too holds throughout the verification's range, and it too is false: the first x with $L(x) > 0$ is 906,150,257. Both are the lure that dissolves, in its starkest form—a pattern exact across every accessible case and untrue past the horizon of computation. That the genuine equivalences keep the ε is not pedantry; it is the whole difference between a theorem and a coincidence.

A.4 Approaches, not equivalences: the spectral programme

Items 7, 9, 10, 13, 15 are the heart of the companion volume's spectral chapter, and not one of them is a proven equivalence.

The Hilbert–Pólya conjecture (7) would *imply* RH—a self-adjoint operator has real eigenvalues—but the implication runs one way, and the reverse is vacuous in the worst sense: if RH holds, the diagonal operator on the zero heights is already such an operator, the made-to-order construction whose emptiness the spectral chapter dwells on. The random-matrix correspondence (9) is the reverse case: Montgomery's pair-correlation result *assumes* RH and describes the finer statistics of zeros taken to be on the line; the agreement with the GUE eigenvalue statistics, confirmed numerically by Odlyzko, is the strongest evidence in the subject and is evidence, not equivalence. Berry and Keating's $H = xp$ (10) is a conjectural model; Connes' noncommutative geometry (13) is a programme, whose genuine equivalence is the Weil positivity already listed; and the quantum-chaos correspondence (15) is an analogy. Each is serious. None is an iff.

A.5 An analogy, and a non-equivalence

The function-field mirror (14). The Riemann Hypothesis for varieties over finite fields is a theorem—proven by Weil in the curve case and by Deligne in general—and it is the deepest source of confidence in the classical Hypothesis. But it is an analogy in another category, a statement about a different class of zeta functions, and it neither implies nor is implied by RH for $\zeta(s)$. It belongs on the list as inspiration, not as equivalence.

The divisor bound (11). This one is simply miscatalogued. The Dirichlet divisor error $\Delta(x) = \sum_{n \leq x} d(n) - x \log x - (2\gamma - 1)x$ is known *unconditionally* to be $O(x^{1/3})$ (Voronoï), with the open conjecture $O(x^{1/4+\varepsilon})$ and a proven lower bound of order

$x^{1/4}$. The bound $O(x^{1/2+\varepsilon})$ written in the list is therefore weaker than what is already proven without any hypothesis—trivially true, equivalent to nothing. The verification shows $\Delta(x)$ staying far below $\sqrt{x}$ across the whole range. The divisor problem is a hard and separate question, independent of RH.

The Möbius sum (12). This entry needs splitting. At $\theta = 0$ the sum $\sum_{n \le x} \mu(n)e(n\theta)$ is exactly the Mertens function, and its $O(x^{1/2+\varepsilon})$ bound is item 3, a genuine equivalence. Uniform over rational θ it becomes a statement about μ in arithmetic progressions, equivalent to the *generalized* Riemann Hypothesis, not the classical one. And under its usual name, the "Möbius randomness principle" is Sarnak's disjointness conjecture—$\sum_{n \le x} \mu(n)f(n) = o(x)$ for f from any zero-entropy system—a far broader statement that is not equivalent to RH at all.

What this is and is not

What is genuine here is a substantial body of exact equivalences: the four error-term forms, Li's positivity, Nyman–Beurling and Báez-Duarte, the Laguerre–Pólya/de Bruijn–Newman characterization, Robin's and Lagarias's inequalities, the Riesz and Franel–Landau and Redheffer criteria, and Weil's positivity, and the Lapidus–Maier inverse spectral problem for fractal strings. Each is proven equivalent to the Hypothesis, and each is named here in its sharp form.

What it is not is a list of fifteen equivalences. The spectral programme would imply RH or already assumes it; the function-field Hypothesis is an analogy in another world; the divisor bound is trivially true; the Möbius entry is three different statements wearing one name. And two of the most natural-looking strengthenings, the Mertens and Pólya conjectures, are outright false while holding across every computable case—which is the appendix's sharpest lesson and the book's: a reformulation is only as good as its proof of equivalence, and a pattern that holds as far as the eye can see is, until proven, only a pattern. Sorting these by status is not bookkeeping. It is the whole discipline of the subject in miniature.

Verification

The script below, using only the Python standard library and no random seeds, confirms the genuine error-term equivalences in range, exhibits the two false sharpenings holding throughout that same range, checks Robin's inequality past its last exception, and shows the divisor error sitting far below the threshold that would make its bound meaningful.

```
#!/usr/bin/env python3
"""Verification for 'Reformulations and Equivalences of RH'.
Standard library only; deterministic (no seeds)."""
from math import log, isqrt
import array, time
t0=time.time()
N=10**6
```

```
GAMMA=0.5772156649015329
EG=2.718281828459045**GAMMA   # e^gamma

checks=0
def ok(cond,msg):
    global checks
    assert cond,"FAIL: "+msg
    checks+=1; print("PASS:",msg)

# ---- smallest-prime-factor sieve -> mu, Omega ----
spf=list(range(N+1))
i=2
while i*i<=N:
    if spf[i]==i:
        for j in range(i*i,N+1,i):
            if spf[j]==j: spf[j]=i
    i+=1
mu=array.array('b',[0])*(N+1); mu[1]=1
Om=array.array('b',[0])*(N+1)
for n in range(2,N+1):
    x=n; om=0; sq=True
    while x>1:
        p=spf[x]; c=0
        while x%p==0: x//=p; c+=1; om+=1
        if c>1: sq=False
    Om[n]=om%2
    mu[n]=0 if not sq else (1 if om%2==0 else -1)

# ---- divisor sieve -> sigma(n) and divisor count ----
sigma=array.array('q',[0])*(N+1)
dct=array.array('i',[0])*(N+1)
for d in range(1,N+1):
    for m in range(d,N+1,d):
        sigma[m]+=d; dct[m]+=1

# ===== genuine error-term equivalences hold in range (RH-consistent) =====
M=0; L=0; maxM=0; maxL=0; polya_ok=True; mert_ok=True
for x in range(1,N+1):
    M+=mu[x]; L+= 1 if Om[x]==0 else -1
    if abs(M)>maxM: maxM=abs(M)
    if abs(L)>maxL: maxL=abs(L)
    if abs(M) > isqrt(x): mert_ok=False        # Mertens conjecture |M(x)|<=sqrt(x)
    if x>=2 and L>0: polya_ok=False            # Polya conjecture L(x)<=0
ok(maxM < N**0.6 and maxL < N**0.6,
   f"M(x), L(x) stay O(x^0.6) to 10^6 (max|M|={maxM}, max|L|={maxL})  [RH-equivalent]")

# ===== two FALSE sharpenings, holding throughout this range =====
ok(mert_ok,
   "Mertens conj. |M(x)|<=sqrt(x) holds to 10^6 -- yet FALSE (Odlyzko-te Riele)")
ok(polya_ok,
   "Polya conj. L(x)<=0 holds to 10^6 -- yet FALSE (first sign change 906150257)")

# ===== Robin's inequality (equivalent to RH) holds past its last exception 5040 =====
robin_ok=True
for n in range(5041,N+1):
    if sigma[n] >= EG*n*log(log(n)): robin_ok=False; break
ok(robin_ok,
   "Robin sigma(n)<e^g n loglog n for 5040<n<=10^6 (equiv to RH; unproven beyond)")

# ===== the Dirichlet divisor problem is NOT RH-equivalent =====
# D(x)=sum d(n)= x log x + (2g-1)x + Delta(x); |Delta(x)| is O(x^{1/3}) UNCONDITIONALLY
# (Voronoi) -- far below x^{1/2+eps}, so the stated bound is trivially true.
D=0; maxDelta=0.0
```

```
for x in range(1,N+1):
    D+=dct[x]
    delta=D-(x*log(x)+(2*GAMMA-1)*x)
    if abs(delta)>maxDelta: maxDelta=abs(delta)
print(f"  divisor: max|Delta|={maxDelta:.0f}, sqrt(1e6)=1000, ratio {maxDelta/N**0.5:.3f}")
ok(maxDelta < N**0.5,
   "divisor error << sqrt(x) unconditionally: that bound is trivial, not RH-equivalent")

print(f"\nALL {checks} CHECKS PASSED   ({time.time()-t0:.1f}s)")
```

Exercises

The exercises are collected here, by chapter, rather than interrupting the text. Difficulty is marked [routine], [moderate], [hard], with [computational] for those that extend a chapter's verification program, [discussion] for those that ask for an argument in prose, and [conceptual] for those that ask for a precise statement of a concept or thesis. The recurring *two-part turn* referred to throughout is the pattern of every chapter: a method grown from the wheel climbs to the wall, and the lever that would cross it proves to be the very thing being sought.

Chapter 1 The Wheel and the Wall

1. [routine] Show that the number of residues coprime to 210 is $\prod_{p|210}(p-1) = 1 \cdot 2 \cdot 4 \cdot 6 = 48$, and that the density 48/210 equals $\prod_{p|210}(1 - 1/p)$.

2. [routine] A *subprime* is a composite coprime to 210 whose least prime factor is at least 11. Find the smallest three.

3. [moderate] From the Euler product $\zeta(s) = \prod_p(1 - p^{-s})^{-1}$, derive $1/\zeta(s) = \sum_n \mu(n)n^{-s}$, identifying μ as the multiplicative function this forces.

4. [moderate] Using the functional equation $\xi(s) = \xi(1-s)$ and $\overline{\xi(s)} = \xi(\bar{s})$, show the nontrivial zeros are symmetric about the line $\operatorname{Re} s = \frac{1}{2}$ and about the real axis, so they come in quadruples unless they sit on the line.

5. [discussion] The explicit formula writes $\psi(x)$ as x minus a sum over the zeros. Explain why this makes "the location of the zeros" and "the size of the error $\psi(x) - x$" two names for one thing, and why that circularity is the shape of every later chapter.

Chapter 2 One Floor Up: Ramanujan's τ

1. [routine] Verify Ramanujan's congruence $\tau(p) \equiv p^{11} + 1 \pmod{691}$ for $p = 2$, given $\tau(2) = -24$.

2. [moderate] Deligne's bound gives $|\tau(p)| \leq 2p^{11/2}$. Writing $\tau(p)/p^{11/2} = 2\cos\theta_p$, compute θ_2 from $\tau(2) = -24$, and explain why $\theta_p \in [0, \pi]$ is the "angle on the circle" analogue of a zero on the critical line.

3. [moderate] Explain in what precise sense Deligne's theorem is "the Riemann Hypothesis proved one floor up": which object plays the rôle of ζ, which map plays the rôle of the zeros, and why $|\alpha| = p^{11/2}$ is the analogue of $\operatorname{Re} s = \frac{1}{2}$.

4. [discussion] The *local* Hypothesis for Δ is a theorem (Deligne); the *global* Hypothesis for $L(s, \Delta)$ is open. Distinguish the two statements and say why proving the first did not settle the second.

5. [hard] Sato–Tate (the θ_p are semicircle-distributed) is a theorem for Δ. Does it imply the global Hypothesis for $L(s, \Delta)$? Explain why a law for the *distribution* of the θ_p does not *locate* the zeros of $L(s, \Delta)$.

Chapter 3 The Oscillation Face

1. [computational] Extend the verification: with a Möbius sieve compute $M(x) = \sum_{n \le x} \mu(n)$ for $x = 10$ and $x = 100$, and check $|M(x)| \le \sqrt{x}$ in each case.

2. [moderate] Derive the Dirichlet series $\sum_n \lambda(n) n^{-s} = \zeta(2s)/\zeta(s)$ for the Liouville function $\lambda(n) = (-1)^{\Omega(n)}$.

3. [moderate] Each of $\pi - \mathrm{Li}$, $\psi - x$, M, and the Liouville sum is $O(x^{1/2+\varepsilon})$ if and only if RH. Using the explicit formula, explain why all four are "the same oscillation."

4. [discussion] The Mertens conjecture $|M(x)| \le \sqrt{x}$ is false, yet no counterexample is known below about 10^{16}. Reconcile "false" with "holds throughout all computation," and say what this teaches about numerical evidence for the Hypothesis.

5. [discussion] Pólya conjectured the Liouville sum stays ≤ 0 for $x \ge 2$; the first sign change is at 906,150,257. Why is such a distant counterexample consistent with the zeros picture, and why does "checked to 10^8" prove nothing?

Chapter 4 The Positivity Face

1. [routine] For a zero $\rho = \frac{1}{2} + it$ on the line, show $\left|1 - 1/\rho\right| = 1$ by computing $\left|1 - 1/(\frac{1}{2} + it)\right|^2$.

2. [moderate] Show the Möbius map $w = 1 - 1/s$ sends $\operatorname{Re} s = \frac{1}{2}$ to $|w| = 1$, $\operatorname{Re} s > \frac{1}{2}$ to $|w| < 1$, and $\operatorname{Re} s < \frac{1}{2}$ to $|w| > 1$.

3. [moderate] Li's coefficients are $\lambda_n = \sum_\rho \left[1 - (1 - 1/\rho)^n\right]$. Explain why on-line zeros ($|1 - 1/\rho| = 1$) keep the terms bounded and the $\lambda_n \ge 0$, while an off-line pair ($|1 - 1/\rho| > 1$) must eventually force some $\lambda_n < 0$.

4. [computational] Confirm $\lambda_1, \ldots, \lambda_{10}$ are positive and increasing for the on-line zeros; then insert the off-line quadruple $\{0.1 \pm i,\ 0.9 \pm i\}$ and find the first n with $\lambda_n < 0$.

5. [discussion] Weil positivity ($W(g,g) \geq 0$ for all $g \Leftrightarrow$ RH) is the target of the Connes programme. Why is exhibiting positivity for many test functions not a proof, and how does this point at the operator wall of Chapter 8?

Chapter 5 The Completeness Face

1. [routine] Verify the bridge identity $\zeta(s) = \dfrac{s}{s-1} - s\displaystyle\int_1^\infty \{x\}\, x^{-s-1}\, dx$ at $s = 2$, where $\{x\}$ is the fractional part.

2. [moderate] Show $\langle 1, \rho_{1/k}\rangle = \frac{1}{k}(1 - \gamma + \ln k)$ for the building block $\rho_\theta(t) = \{\theta/t\}$ on $(0,1)$.

3. [moderate] Explain the Mellin-transform bridge: why the dilations $\{\theta/t\}$ carry $\zeta(s)/s$, so their closed span reaches "down to the rightmost zero," making completeness (reaching the constant 1) equivalent to RH.

4. [computational] Extend the chapter's computation: form the Nyman–Beurling distance d_N for $N = 2, \ldots, 8$, confirm it decreases, and compare its decay to $C/\log N$.

5. [discussion] Báez-Duarte showed integer dilations suffice (one sequence with $\|b_N\| \to 0 \Leftrightarrow$ RH). Why does the distance shrinking, even at a rate set by the zeros, not amount to a proof?

Chapter 6 The Extremal Face

1. [routine] Compute $\sigma(n)/n$ for the primorials $n = 2, 6, 30, 210$ and compare each with $e^\gamma \log\log n$; observe that 6 and 30 are among the twenty-seven Robin exceptions (ratio > 1), while 210 already satisfies the bound (ratio $0.918 < 1$).

2. [routine] Verify Lagarias's inequality $\sigma(n) \leq H_n + e^{H_n}\log H_n$ at $n = 1$ (with equality) and at $n = 2, 3, 4$ (strictly).

3. [moderate] Show $\sigma(n)/n = \prod_{p^a \| n}(1 + p^{-1} + \cdots + p^{-a})$ and explain why the maximizers use many small primes—the primorial cascades.

4. [computational] Extend the divisor sieve to find the largest $n \leq 10^6$ that violates Robin's inequality, and confirm there are exactly twenty-seven exceptions.

5. [discussion] The margin beneath Robin's ceiling is $\psi(y) - y$ in divisor costume. Explain why testing Robin on the champions is the same square-root cancellation as the oscillation face, so it offers no new route to the wall.

Chapter 7 The Regularity Face

1. [routine] For the toy $f_t(z) = z^2 + (c - 2t)$, find the zeros and show they are real iff $t \geq c/2$; identify the threshold $\Lambda = c/2$.

2. [routine] For $c = 0$ (so $\Lambda = 0$), show the zero is real at $t = 0$ and complex for every $t < 0$—the "only barely so" edge.

3. [computational] Confirm that $H_0(z) = \int \Phi(u) \cos(zu)\, du$ vanishes at $z = 2\gamma_1$ for the first Riemann ordinate (recall Φ is even, so evaluate at $|u|$).

4. [moderate] Rodgers and Tao proved $\Lambda \geq 0$, and RH $\Leftrightarrow \Lambda \leq 0$. Deduce RH $\Leftrightarrow \Lambda = 0$ and explain Newman's remark that the Hypothesis, if true, is only barely so.

5. [discussion] The upper bounds on Λ descend toward 0 (Polymath15: 0.22). Explain why reaching 0 from above is not "almost proving RH" but *is* proving RH.

Chapter 8 The Neighbouring Walls

1. [computational] Unfold the first sixty zeros by $N(T)$, form the nearest-neighbour spacings, and show the mean is near 1 with few small gaps—the repulsion of GUE, not the clumping of Poisson.

2. [moderate] The semiclassical count for $H = xp$ reproduces $N(T) = \frac{T}{2\pi} \log \frac{T}{2\pi} - \frac{T}{2\pi} + \frac{7}{8}$. Evaluate $N(T)$ at $T = 163.03$ (the 60th ordinate) and compare with the count 60.

3. [routine] For $y^2 = x^3 + x + 1$ over $\mathbb{F}_{101}$, compute $a_p = p + 1 - \#E(\mathbb{F}_p)$ and check Hasse's bound $|a_p| \leq 2\sqrt{p}$; confirm $a_p^2 \leq 4p$, so the Frobenius eigenvalues lie on $|\alpha| = \sqrt{p}$.

4. [conceptual] The Lindelöf Hypothesis $\zeta(\frac{1}{2} + it) = O(t^{\varepsilon})$ is implied by RH and is weaker. Explain why subconvexity exponents creeping toward 0 do not reach it, and why even this weaker wall is open.

5. [discussion] The four neighbours form a ring: the statistics demand an operator, the operator would give RH and supply Weil positivity, and the function-field world owns the geometry that yields that positivity. Say what is missing, over $\mathbb{Z}$, on each side.

Chapter 9 The Twin Corridor

1. [computational] Compute the partial products $\prod_{3 \leq p \leq P}(1 - (p-1)^{-2})$ for growing P and watch them approach $C_2 = 0.6601618\ldots$; the twin constant is $2C_2$.

2. [moderate] Brun proved the sum of reciprocals of the twin primes converges, while $\sum_p 1/p$ diverges. What does the contrast say about how thinly the twins are spread?

3. [conceptual] State the parity barrier: a sieve cannot separate an even from an odd number of prime factors. Why does this stop the bounded-gap method one step short of gap two?

4. [routine] The bounded-gap record fell from 70,000,000 (Zhang) to 246 (Maynard–Tao), to 12 under Elliott–Halberstam, and to 6 under its generalized form. Explain why "bounded gaps" is weaker than "infinitely many twins."

5. [discussion] Chen's theorem—infinitely often p and $p+2$ are a prime and a number with at most two prime factors—sits at the edge of the sieve. Why is it the most the parity barrier allows, and why is gap two one parity step beyond?

Chapter 10 The Walls

1. [computational] Verify Lagarias's inequality for $1 \leq n \leq 100$ and identify the unique case of equality.

2. [computational] Build $N = pq$ from two large primes, then run trial division to 10^5 to confirm the sieve finds no factor though N was a single multiplication.

3. [conceptual] State the "one wall, three faces" thesis: for the Riemann Hypothesis, the twins, and factoring, name in each the local law that is known and the global fact that is walled.

4. [conceptual] Show that in each of the three walls the lever that would cross it equals the destination: controlling the zeros needs the zeros, breaking parity is isolating two primes, undoing multiplication is factoring.

5. [discussion] The recurring turn is "the structure you build is the thing you seek." Choose two faces from Part II and one neighbour from Part III and state the turn in each.

Solutions to the Exercises

These are answer sketches, not full write-ups: enough to grade against and to run a discussion from. Exercise numbers match the *Exercises* section at the back of the book. The recurring *two-part turn* is the pattern of every chapter—a method grown from the wheel climbs to the wall, and the lever that would cross it proves to be the very thing being sought.

Chapter 1 The Wheel and the Wall

1. By the Chinese remainder theorem the units mod 210 number $\prod_{p|210}(p-1) = 1 \cdot 2 \cdot 4 \cdot 6 = 48$. The density is $\frac{48}{210} = \frac{1}{2} \cdot \frac{2}{3} \cdot \frac{4}{5} \cdot \frac{6}{7} = \prod_{p|210}(1 - 1/p)$, each factor deleting the multiples of one prime.

2. $121 = 11^2$, $143 = 11 \cdot 13$, $169 = 13^2$. (Each is coprime to 210 with least prime factor ≥ 11; the next are $187 = 11 \cdot 17$ and $209 = 11 \cdot 19$.)

3. $1/\zeta(s) = \prod_p(1 - p^{-s})$. Expanding, n^{-s} appears with coefficient $(-1)^k$, where k is the number of distinct primes of n, if n is squarefree and 0 otherwise—that is $\mu(n)$. Hence $1/\zeta(s) = \sum_n \mu(n)n^{-s}$.

4. $\xi(s) = \xi(1-s)$ sends a zero ρ to $1-\rho$ (reflection across $\mathrm{Re} = \frac{1}{2}$); real coefficients give $\overline{\xi(s)} = \xi(\bar{s})$, so $\bar{\rho}$ is also a zero. Thus zeros occur as $\{\rho, 1-\rho, \bar{\rho}, 1-\bar{\rho}\}$—a genuine quadruple off the line, collapsing to the conjugate pair $\{\rho, \bar{\rho}\}$ when $\rho = 1 - \bar{\rho}$, i.e. on the line.

5. Each zero contributes a term of size $x^{\mathrm{Re}\,\rho}$ to $\psi(x) - x = -\sum_\rho x^\rho/\rho - \cdots$, so the *size* of the error and the *real part* of the zeros are one datum. To bound the error one must place the zeros, and the error is the only handle on them: the lever is the destination. Every face repeats this.

Chapter 2 One Floor Up: Ramanujan's τ

1. $\tau(2) = -24$; $2^{11} + 1 = 2049 \equiv 2049 - 2 \cdot 691 = 667 \pmod{691}$; and $-24 \equiv 691 - 24 = 667$. Equal.

2. $\tau(2)/2^{11/2} = -24/45.255 = -0.5304 = 2\cos\theta_2$, so $\cos\theta_2 = -0.2652$ and $\theta_2 \approx 1.84$ rad ($\approx 105°$). The conjugate eigenvalues are $p^{11/2}e^{\pm i\theta_p}$ on the circle of

radius $p^{11/2}$; θ_p is the angular position, the GL(2) analogue of a zero's height on the line.

3. The object is $L(s, \Delta)$, with local Euler factor $(1 - \tau(p)p^{-s} + p^{11-2s})^{-1}$ whose inverse roots $\alpha, \beta = p^{11/2}e^{\pm i\theta_p}$ are the Frobenius eigenvalues. The local Hypothesis is $|\alpha| = p^{11/2}$, the fixed-modulus condition that mirrors a fixed real part. Deligne proved it as eigenvalues of Frobenius on the cohomology of the associated (Kuga–Sato) variety—the Weil conjectures.

4. Local: each Euler factor's roots have $|\alpha| = p^{11/2}$ (a bound on the coefficients, proven). Global: all nontrivial zeros of the continued $L(s, \Delta)$ lie on its critical line (open). The first is a statement in the region of convergence; the second is about the analytic continuation. Bounding coefficients does not locate continued zeros.

5. No. Sato–Tate equidistributes the angles θ_p by the semicircle law—a statement about their *average distribution*. The zeros of $L(s, \Delta)$ are a continuation feature; equidistribution constrains average size, not the real part of any zero. Distribution is not location.

Pitfall: *the recurring error is to read Deligne's $|\tau(p)| \leq 2p^{11/2}$ (local, proved) as the global Hypothesis for $L(s, \Delta)$. Ex. 4 exists to separate them.*

Chapter 3 The Oscillation Face

1. μ on $1, \ldots, 10$ is $1, -1, -1, 0, -1, 1, -1, 0, 0, 1$, summing to $M(10) = -1$, and $|-1| \leq \sqrt{10} \approx 3.16$. Likewise $M(100) = 1 \leq 10$.

2. λ is completely multiplicative with $\lambda(p) = -1$, so the local factor is $\sum_{k\geq 0}(-1)^k p^{-ks} = (1 + p^{-s})^{-1}$. Hence $\sum_n \lambda(n)n^{-s} = \prod_p(1 + p^{-s})^{-1} = \prod_p \frac{1-p^{-s}}{1-p^{-2s}} = \zeta(2s)/\zeta(s)$.

3. The explicit formula writes each of $\pi - \mathrm{Li}$, $\psi - x$, M, and the Liouville sum as a sum over the zeros, term size $x^{\mathrm{Re}\,\rho}$. So each is $O(x^{1/2+\varepsilon})$ iff every $\mathrm{Re}\,\rho = \frac{1}{2}$ (RH), and "the oscillation" is literally the same zero-sum, dressed by μ, λ, or Λ.

4. The disproof (Odlyzko–te Riele) is non-constructive and the first counterexample is astronomically large, so $|M| \leq \sqrt{x}$ can survive every feasible computation while being false. Numerical agreement to any reachable height is not proof; the zeros' fine conspiracies surface only far out.

5. The Liouville sign is governed by the same zero-sum; a slow bias can persist to $\sim 9 \times 10^8$ before flipping. "Checked to 10^8" certifies only the range tested, and the relevant scales for such questions can be vastly larger (compare Skewes's number for $\pi - \mathrm{Li}$).

Chapter 4 The Positivity Face

1. $|1-1/\rho|^2 = |\rho-1|^2/|\rho|^2$. For $\rho = \frac{1}{2}+it$, $|\rho-1|^2 = \frac{1}{4}+t^2 = |\rho|^2$, so the ratio is 1.

2. $|w|^2 = |s-1|^2/|s|^2 = \dfrac{(\sigma-1)^2+t^2}{\sigma^2+t^2}$ for $s=\sigma+it$. Then $|w|<1 \Leftrightarrow (\sigma-1)^2 < \sigma^2 \Leftrightarrow \sigma > \frac{1}{2}$, with equality at $\sigma = \frac{1}{2}$ and $|w|>1$ for $\sigma < \frac{1}{2}$.

3. Write $w_\rho = 1-1/\rho$. On the line $|w_\rho| = 1$, so w_ρ^n stays on the unit circle and $1-w_\rho^n$ is bounded; summed over conjugate pairs the contributions are ≥ 0 and $\lambda_n \geq 0$. An off-line zero with Re $< \frac{1}{2}$ has $|w_\rho|>1$, so w_ρ^n grows geometrically and $-w_\rho^n$ eventually drives some $\lambda_n < 0$.

4. The on-line $\lambda_1, \ldots, \lambda_{10}$ come out positive and increasing $(0.019, 0.076, \ldots, 1.87)$. Inserting $\{0.1\pm i,\ 0.9\pm i\}$ (with $|1-1/(0.1+i)| = 1.339 > 1$) turns λ_n negative first at $n=7$.

5. $W(g,g) \geq 0$ for *all* g is RH; any finite or explicit family of g tests only those directions. The universal statement is the spectral side of a trace formula—the missing operator of Chapter 8. Positivity on samples is evidence, not the theorem.

Pitfall: *in Ex. 1 students expand numerically; the clean route is* $|1-1/\rho|^2 = |\rho-1|^2/|\rho|^2$, *after which the line* $\rho = \frac{1}{2}+it$ *gives* $|\rho-1| = |\rho|$ *in one step.*

Chapter 5 The Completeness Face

1. At $s=2$, $\dfrac{s}{s-1} = 2$, and the identity reads $\zeta(2) = 2 - 2\int_1^\infty \{x\}x^{-3}dx$. Computing the integral per unit interval gives ≈ 0.1776, so the right side is $2 - 2(0.1776) = 1.6449 = \pi^2/6$. Both sides agree.

2. With $u = \theta/t$ ($\theta = 1/k$), $\langle 1, \rho_{1/k}\rangle = \int_0^1 \{\theta/t\}dt = \theta\int_\theta^\infty \{u\}u^{-2}du$. Split at 1: on $[\theta,1]$, $\{u\} = u$ gives $\int_\theta^1 du/u = \ln k$; and $\int_1^\infty \{u\}u^{-2}du = 1-\gamma$. Hence $\langle 1, \rho_{1/k}\rangle = \frac{1}{k}(1-\gamma+\ln k)$.

3. The dilations have Mellin transform proportional to $\zeta(s)/s$, so a function lies in their closed span only where that transform allows—reaching every point to the right of the rightmost zero. Reaching the constant 1 therefore requires no zero with Re $> \frac{1}{2}$, i.e. RH.

4. The distances decrease $0.181 \to 0.083 \to 0.055 \to 0.049 \to 0.048 \to \cdots$; a fit against $C/\log N$ matches, with C governed by $\sum 1/|\rho|^2$.

5. Báez-Duarte's $\|b_N\| \to 0$ is equivalent to RH, and the very rate of approach is set by the zeros. Observing $d_N \to 0$ numerically is not proving the limit is 0; the limit *being* 0 is the Hypothesis.

Chapter 6 The Extremal Face

1. $\sigma/n = 1.5,\ 2,\ 2.4,\ 2.743$ for $n = 2, 6, 30, 210$. At $n = 210$, $e^{\gamma}\log\log n = 1.781 \cdot 1.677 = 2.987$, so the Robin ratio is $2.743/2.987 = 0.918 < 1$.

2. $n = 1$: $\sigma = 1$, $H_1 = 1$, RHS $= 1 + e \cdot \ln 1 = 1$—equality. $n = 2, 3, 4$: RHS $\approx 3.32, 5.62, 7.98$ exceed $\sigma = 3, 4, 7$.

3. $\sigma(n)/n = \prod_{p^a \| n}(1 + p^{-1} + \cdots + p^{-a}) < \prod_{p|n} \frac{p}{p-1}$. Each prime contributes near $p/(p-1)$, largest for the smallest p; so maximizers take the smallest primes with descending exponents—the primorial cascades (super/colossally abundant numbers).

4. Extending the divisor sieve to 10^6: the last violator is $n = 5040$ and the exceptions number exactly twenty-seven.

5. For a colossally abundant n, $\sigma(n)/n \approx \prod_{p\le y}(1 - 1/p)^{-1} \sim e^{\gamma}\log y$ (Mertens, exact), with $\log y \sim \log\log n$; the residual margin is $\sum_{p\le y}\log p - y = \psi(y) - y$, the Chebyshev error. So Robin's truth past 5040 is the same $\sqrt{y}$ cancellation as the oscillation face—one wall in divisor costume, no new route.

Pitfall: *in Ex. 1 the threshold $n > 5040$ is not a consequence of $\log\log n$ being small for small primorials. The primorials 6 and 30 genuinely violate Robin's inequality (ratio > 1), as do the rest of the twenty-seven exceptional values up to 5040; the bound holds for every $n > 5040$ because 5040 is the largest exception, not because of any $\log\log n$ size effect. Among the primorials, 210 is the first to satisfy the bound (ratio $0.918 < 1$).*

Chapter 7 The Regularity Face

1. $f_t(z) = z^2 + (c - 2t)$ has zeros $z = \pm\sqrt{2t - c}$, real iff $2t - c \ge 0$, i.e. $t \ge c/2$. So $\Lambda = c/2$.

2. $c = 0$: zeros $\pm\sqrt{2t}$. At $t = 0$ a real double zero; for $t < 0$, $\pm i\sqrt{2|t|}$, complex. Real exactly at and after $t = 0$, complex for every $t < 0$.

3. $H_0(z) = \int \Phi(u)\cos(zu)\,du$ with Φ even (evaluate at $|u|$, $n = 1\ldots6$); its first sign change falls at $z = 2\gamma_1 = 28.27$, matching the first ordinate.

4. $\Lambda \ge 0$ (Rodgers–Tao) and RH $\Leftrightarrow \Lambda \le 0$ force RH $\Leftrightarrow \Lambda = 0$. Since $\Lambda \ge 0$ always, RH puts Λ exactly at its floor: cool the flow by any amount and a zero leaves the line—"only barely so."

5. Reaching 0 from above means proving the flow is already real-rooted at $t = 0$, which *is* the Hypothesis. Each bound $\frac{1}{2} \to \cdots \to 0.22$ narrows the bracket, but the endpoint 0 is RH itself; there is no "almost."

Chapter 8 The Neighbouring Walls

1. Unfolding by $N(T)$ gives mean spacing ≈ 0.999 and only $\approx 3.4\%$ of gaps below half the mean, against 39% for Poisson: repulsion, the GUE signature.

2. $N(163.03) = \frac{163.03}{2\pi} \log \frac{163.03}{2\pi} - \frac{163.03}{2\pi} + \frac{7}{8} = 25.948 \cdot 3.256 - 25.948 + 0.875 \approx 59.4$, against the count 60—agreeing within the small $S(T)$ fluctuation.

3. For $y^2 = x^3 + x + 1$ over $\mathbb{F}_{101}$ the point count is 105, so $a_p = 101 + 1 - 105 = -3$. Then $|a_p| = 3 \leq 2\sqrt{101} \approx 20.1$ and $a_p^2 = 9 \leq 4 \cdot 101 = 404$, so the Frobenius eigenvalues lie on $|\alpha| = \sqrt{101}$.

4. Each subconvexity gain is hard-won and bounded away from 0; t^{ε} is the infimum, attained by no finite exponent. And Lindelöf, though implied by RH, is equivalent to an open statement about the horizontal spread of the zeros, so it stands unproved on its own.

5. Missing over $\mathbb{Z}$: the operator (none built; xp unregularized), the geometry ($\operatorname{Spec} \mathbb{Z}$ is no curve over a field—no surface, no Frobenius, no $\mathbb{F}_1$), and hence the positivity (the Weil functional is not shown ≥ 0). The statistics are evidence only, and Lindelöf is weaker yet open. Each side lacks the one object that would close it.

Chapter 9 The Twin Corridor

1. The partial products $\prod_{3 \leq p \leq P}(1 - (p-1)^{-2})$ decrease to $C_2 = 0.6601618\ldots$, agreeing to several digits well before P reaches the thousands; the twin constant is $2C_2 = 1.3203\ldots$.

2. $\sum 1/p$ diverges while $\sum 1/(\text{twin primes})$ converges (to Brun's constant ≈ 1.902). The twins are therefore far thinner than the primes—almost all primes are not twins.

3. A sieve's weights cannot read $\Omega(n) \bmod 2$, so it cannot certify "exactly one prime factor" against "two." The bounded-gap method places a prime in a short window but cannot force the second member to be prime rather than a product; gap two needs precisely that parity decision.

4. Bounded gaps give infinitely many pairs within a fixed distance H (now 246, 12 under Elliott–Halberstam, and 6 under its generalized form), not distance exactly 2, and not even a fixed recurring gap. Infinitely many twins is the pinned case $H = 2$, which the sieve cannot reach.

5. Chen's theorem extracts everything up to parity—p with $p+2$ prime or a product of two primes—but cannot delete the second alternative, since that deletion is a parity determination. Gap two is exactly that forbidden step.

Pitfall: *students sometimes claim Chen "almost proves" the twins; stress that the remaining step is not small but is the parity barrier itself.*

Chapter 10 The Walls

1. Lagarias's inequality holds for every $1 \le n \le 100$; equality occurs only at $n = 1$.
2. With $N = pq$ a product of two large primes, trial division to 10^5 returns no factor, though N was formed by one multiplication—easy to make, hard to unmake.
3. **RH:** local law = the wheel and Euler product (residues, densities), known; global fact = the real parts of the zeros, walled. **Twins:** local law = the singular series $2C_2$ (predicted density), known; global = whether the count is infinite, walled. **Factoring:** local law = multiplication and primality (AKS), easy; global = recovering the factors, walled.
4. Controlling the zeros' real parts needs the prime error, which *is* the zeros (explicit formula); isolating two primes is removing the semiprime alternative, which *is* a parity determination; factoring cheaply is undoing a multiplication, which *is* factoring. In each, lever = destination.
5. Open discussion. For instance: positivity (a positive λ_n sequence $\Leftrightarrow$ no off-line zero, but checking needs the zeros), completeness (the span reaches 1 $\Leftrightarrow$ no zero right of the line, at a rate the zeros set), and the operator neighbour (an operator with the zeros as spectrum would give RH, but building it independently is the wall). Each: the structure built is the thing sought.

Glossary

Admissible slot (twin corridor) A residue r modulo 210 for which both r and $r+2$ are coprime to 210 — a position a twin-prime pair can occupy. There are exactly fifteen.

AKS primality test A deterministic polynomial-time algorithm (Agrawal–Kayal–Saxena, 2004) deciding whether a number is prime; it shows that recognising primes is computationally easy.

a_p For an elliptic curve over $\mathbb{F}_p$, the quantity $a_p = p + 1 - \#E(\mathbb{F}_p)$. Hasse's bound $|a_p| \leq 2\sqrt{p}$ is the Riemann Hypothesis for the curve.

Báez-Duarte criterion A strengthening (2003) of Nyman–Beurling using only integer dilations: RH holds iff one explicit sequence of L^2 distances tends to zero.

Berry–Keating model The proposal that the quantisation of the classical Hamiltonian $H = xp$ underlies the Riemann zeros; it reproduces their average counting function $N(T)$ but yields no proven operator.

Brun's theorem The sum of the reciprocals of the twin primes converges (Brun, 1919), so the twins are sparse; Brun's sieve is the technique behind it.

Chebyshev bias The faint, persistent tilt by which some residue classes lead the prime race; governed by the zeros of Dirichlet L-functions.

Chebyshev functions $\theta(x) = \sum_{p \leq x} \log p$ and $\psi(x) = \sum_{p^k \leq x} \log p$. The error $\psi(x) - x$ is the prime-counting fluctuation governed by the zeros.

Chen's theorem Infinitely often p and $p+2$ are a prime and a number with at most two prime factors (Chen, 1973): the sieve pressed to its very edge.

Colossally abundant number An integer maximising $\sigma(n)/n^{1+\varepsilon}$; with the superabundant numbers, the extremal candidates for Robin's inequality, built from small primes.

Complex multiplication The extra symmetry an elliptic curve carries when its endomorphisms exceed the integers — for the lemniscate, multiplication by i from the Gaussian integers; the structure making the lemniscate the analogue of the circle.

Conductor The smallest modulus to which a Dirichlet character is genuinely attached; the wheel's seven real characters are organized by conductor into a Boolean lattice, the conductor a product of a subset of $\{3, 5, 7\}$.

Conductor lattice The wheel's characters arranged by conductor, forming the Boolean algebra of subsets of $\{3, 5, 7\}$.

Critical line / strip The line $\operatorname{Re} s = \frac{1}{2}$ and the strip $0 < \operatorname{Re} s < 1$ in which the nontrivial zeros of ζ lie; RH asserts they all lie on the line.

de Bruijn–Newman constant Λ The threshold time in the heat flow on Ξ: the flow is real-rooted iff $t \geq \Lambda$. RH $\Leftrightarrow \Lambda \leq 0$; Rodgers–Tao proved $\Lambda \geq 0$, so RH $\Leftrightarrow \Lambda = 0$. (Distinct from von Mangoldt's $\Lambda(n)$.)

Dedekind zeta function The zeta function of a number field, summed over its ideals; for $\mathbb{Q}(i)$ it factors as $\zeta_{\mathbb{Q}(i)} = \zeta \cdot L(\chi_4)$, encoding how the primes split in $\mathbb{Z}[i]$.

Deligne's bound The proof (1974) of the Ramanujan bound $|\tau(p)| \leq 2p^{11/2}$ — the Riemann Hypothesis for the associated variety, the local Hypothesis "one floor up."

Dirichlet character χ A periodic, completely multiplicative function modulo q; it builds the L-function $L(s, \chi)$.

Dirichlet divisor problem Bounding $\Delta(x) = \sum_{n \leq x} d(n) - x(\log x + 2\gamma - 1)$, conjectured $O(x^{1/4+\varepsilon})$; a size-and-exponent cousin of the Lindelöf Hypothesis.

Equidistribution The asymptotic fact that the primes fall in equal share $1/48$ across the 48 spokes (the prime number theorem for arithmetic progressions).

Euler–Mascheroni constant γ The limit of $H_n - \log n \approx 0.5772$, appearing in Robin's inequality and in Mertens' theorem.

Euler product $\zeta(s) = \prod_p (1 - p^{-s})^{-1}$: the wheel made into a function, binding the primes to ζ.

Explicit formula Von Mangoldt's identity writing the prime count $\psi(x)$ as x minus a sum over the nontrivial zeros ρ of ζ (and, twisted by a character, of L-functions); the exact duality, the hinge that carries information between the primes and the zeros.

Frobenius The Galois symmetry attached to a prime, recording how it splits in an extension; for $\mathbb{Q}(i)/\mathbb{Q}$ it is the character χ_4. For a curve over $\mathbb{F}_q$ its eigenvalues α determine the zeta function, and $|\alpha| = q^{1/2}$ is the function-field Riemann Hypothesis.

Functional equation The symmetry $\xi(s) = \xi(1 - s)$ of the completed zeta function, reflecting the critical strip about $\operatorname{Re} s = \frac{1}{2}$; the wheel's mirror $r \mapsto 210 - r$ is its elementary shadow.

Gaussian integers ($\mathbb{Z}[i]$) The numbers $a + bi$ with a, b integers; a rational prime splits among them when $p \equiv 1 \pmod 4$, stays prime when $p \equiv 3 \pmod 4$, and 2 ramifies.

Generalized Riemann Hypothesis (GRH) The Riemann Hypothesis stated for every Dirichlet L-function at once, not for ζ alone; strictly stronger, and equally open.

GUE The Gaussian Unitary Ensemble of random Hermitian matrices, whose eigenvalue spacings (pair correlation $1 - (\sin \pi x/\pi x)^2$) match those of the Riemann zeros.

Hardy–Littlewood (twin) constant $C_2 = \prod_{p \geq 3}(1 - 1/(p-1)^2) = 0.6601618\ldots$; the twin constant is $2C_2 = 1.3203\ldots$, the singular series giving the conjectural density of twin primes.

Harmonic number H_n $H_n = \sum_{k \leq n} 1/k = \log n + \gamma + o(1)$; the elementary ingredient of Lagarias's criterion.

Hasse bound $|a_p| \leq 2\sqrt{p}$ for elliptic curves over $\mathbb{F}_p$ (Hasse, 1930s): the Riemann Hypothesis for an elliptic curve, and a theorem.

Hecke L-function The L-function of a Hecke character; in the complex-multiplication setting Hecke proved its continuation and functional equation, yet its Riemann Hypothesis is open exactly as for ζ.

Hilbert–Pólya conjecture The idea that the ordinates of the nontrivial zeros are the eigenvalues of a self-adjoint operator, which would force them real and prove RH; no such operator is known.

Kronecker–Weber theorem Every abelian extension of $\mathbb{Q}$ lies in a cyclotomic field — the reason the circle's division points capture the simplest arithmetic over $\mathbb{Q}$.

Lagarias's criterion $\sigma(n) \leq H_n + e^{H_n} \log H_n$ for all $n \geq 1 \Leftrightarrow$ RH (Lagarias, 2002): the most elementary statement of the Hypothesis.

Laguerre–Pólya class The real entire functions that are limits of real-rooted real polynomials; RH $\Leftrightarrow \Xi$ belongs to this class.

Lemniscate The figure-eight curve $r^2 = \cos 2\theta$; its division points generate the abelian extensions of $\mathbb{Q}(i)$ by complex multiplication, as roots of unity do for $\mathbb{Q}$.

Lemniscate sine The function $\mathrm{sl}(u)$ inverting the lemniscatic arc-length integral; doubly periodic on a square lattice, obeying $\mathrm{sl}(iu) = i\,\mathrm{sl}(u)$, the complex multiplication by the Gaussian integers.

Level repulsion The vanishing probability of nearly-coincident zeros, the signature of GUE statistics and the opposite of independent (Poisson) points.

L-function $L(s,\chi)$ The series $\sum \chi(n)n^{-s} = \prod(1-\chi(p)p^{-s})^{-1}$; the family the wheel generates. Non-vanishing $L(1,\chi) \neq 0$ gives primes in arithmetic progressions.

Li coefficients λ_n The numbers appearing in Li's criterion (distinct from Liouville's $\lambda(n)$).

Li's criterion RH $\Leftrightarrow$ $\lambda_n = \sum_\rho[1-(1-1/\rho)^n] \geq 0$ for all $n \geq 1$ (Li, 1997); the Möbius map $w = 1-1/\rho$ sends the critical line exactly to the unit circle.

Lindelöf Hypothesis $\zeta(\frac{1}{2}+it) = O(t^\varepsilon)$; implied by RH, strictly weaker, and still open.

Liouville function $\lambda(n)$ $(-1)^{\Omega(n)}$, the parity of the number of prime factors; its summatory function underlies the (false) Pólya conjecture. A sieve cannot see its sign.

Mellin transform The transform carrying the dilations $\{\theta/t\}$ to $\zeta(s)/s$ in the Nyman–Beurling criterion.

Mertens function $M(x)$ $\sum_{n\leq x}\mu(n)$; the bound $M(x) = O(x^{1/2+\varepsilon})$ is equivalent to RH. The stronger Mertens conjecture $|M| \leq \sqrt{x}$ is false (Odlyzko–te Riele, 1985).

Mirror The reflection $r \mapsto 210-r$, an involution pairing the 48 spokes into 24 mirror pairs; the elementary shadow of the functional equation $s \mapsto 1-s$.

Möbius function $\mu(n)$ Zero if n has a squared factor, else $(-1)^{\#\text{primes}}$; it gives $1/\zeta(s) = \sum \mu(n)n^{-s}$.

Möbius transformation A conformal motion $z \mapsto (az+b)/(cz+d)$, a rotation of the sphere under stereographic projection; the integer ones form the modular group. Despite the name, unrelated to the Möbius function or strip.

$N(T)$ The Riemann–von Mangoldt count of zeros up to height T: $\frac{T}{2\pi}\log\frac{T}{2\pi} - \frac{T}{2\pi} + \frac{7}{8} + \cdots$; the spectral staircase.

Number field sieve The fastest known classical factoring algorithm, running in time sub-exponential in the number of digits.

Nyman–Beurling criterion RH $\Leftrightarrow$ the constant 1 lies in the closed span of the dilations $\{\theta/t\}$ in $L^2(0,1)$: a completeness statement.

One-level density The averaged behaviour of the lowest zeros across a family of L-functions, used to read off the family's symmetry type.

Parity barrier Selberg's obstruction: a sieve cannot distinguish an even from an odd number of prime factors, and so cannot, alone, deliver twin primes — it stops the bounded-gap method one step short of gap two.

Position and spectral planes The two planes the explicit formula links: the primes live in position, their L-function zeros in the spectral s-plane; the formula is the only exact passage between them.

Primon gas The model whose particles are the primes, with energy levels $\log n$, whose bosonic partition function is ζ; its fermionic superpartner yields the Möbius function. It reproduces ζ but has the wrong spectrum for the zeros.

Primorial A product of the first primes $(2, 6, 30, 210, \dots)$; $210 = 2{\cdot}3{\cdot}5{\cdot}7$ is the fourth, the wheel's modulus, and the extremal numbers of Robin's face are primorial cascades.

Principal character The trivial Dirichlet character, whose L-function is essentially ζ; the one character whose contribution the explicit formula predicts directly.

Riemann Hypothesis (RH) The conjecture that every nontrivial zero of ζ has real part exactly $\frac{1}{2}$.

Riemann zeta function $\zeta(s)$ $\sum n^{-s}$ for $\operatorname{Re} s > 1$, continued to the plane; its nontrivial zeros encode the distribution of the primes.

Robin's inequality $\sigma(n) < e^{\gamma} n \log\log n$ for all $n > 5040 \Leftrightarrow$ RH (Robin, 1984); below the threshold the inequality fails at exactly twenty-seven integers, the largest being 5040.

Sato–Tate The semicircle law for the Frobenius angles (and the angles of τ); proven for elliptic curves and for Δ.

Selberg trace formula An exact identity on a hyperbolic surface equating the Laplacian's spectrum to the lengths of its closed geodesics — the explicit formula made geometric, and the one setting where a zeta's zeros are realized as eigenvalues of a self-adjoint operator.

Shor's algorithm A quantum algorithm (1994) that factors in polynomial time, so the factoring wall is not absolute — though it stands classically.

Singular series The arithmetic factor in the Hardy–Littlewood conjectures predicting the density of a prime pattern; drawn directly from the wheel. For twin primes it is the $p = 2, 3, 5, 7$ product made concrete as fifteen admissible slots out of 210.

Subconvexity Bounds on $|\zeta(\frac{1}{2}+it)|$ below the convexity exponent $\frac{1}{4}$ (Weyl $\frac{1}{6}$, toward $\frac{13}{84}$), creeping toward the Lindelöf value.

Subprime A composite coprime to 210 — hence with least prime factor at least 11 — a wheel survivor that is not prime; the shadow the primes cast through multiplication. The smallest is $121 = 11^2$.

Sum of divisors $\sigma(n)$ $\sum_{d|n} d$; the function of the extremal face (Robin and Lagarias).

Symmetry type The classification of a family of L-functions by which random-matrix ensemble its low zeros mimic: unitary, orthogonal, or symplectic.

Trace formula An identity equating a sum over a system's spectrum to a sum over its closed orbits; for the zeros, the primes play the role of the orbits.

Turning Advancing one full period: the spoke r on turn k carries the integer $210k+r$.

Two-part test The recurring criterion for a candidate Riemann operator: it must have the zeros as its spectrum, and must not be built from them in the first place.

von Mangoldt function $\Lambda(n)$ $\log p$ if $n = p^k$, else 0; it gives $\psi(x) = \sum_{n\leq x} \Lambda(n)$. (Distinct from the de Bruijn–Newman constant Λ.)

Weil conjectures The Riemann Hypothesis for varieties over finite fields; proven by Weil (curves, 1948) and Deligne (the general case, 1974).

Weil positivity The positivity of an explicit functional, equivalent to RH and the target of the Connes programme.

Wheel The residues modulo $210 = 2{\cdot}3{\cdot}5{\cdot}7$ arranged around a circle; the $\varphi(210) = 48$ residues coprime to 210 are its live *spokes*, the survivors that encode the primes' local law.

Windowed difference The contrast between an observable averaged at the zero ordinates and the same observable averaged at the midpoints between them — a basic detection statistic.

$\xi(s)$ and $\Xi(z)$ The completed, entire zeta function with $\xi(s) = \xi(1-s)$, and its symmetric form $\Xi(z) = \xi(\frac{1}{2} + iz)$, the Fourier transform of the heat kernel Φ.

Bibliography

This is a guide to the sources behind the three books, gathered into one reading list and grouped by theme rather than by book. It is a reading list, not a complete record; the aim is to point the reader to the original results named in the text.

Foundations and classical analytic number theory

- G. H. Hardy and E. M. Wright, *An Introduction to the Theory of Numbers*, 6th ed., Oxford University Press, 2008.
- H. Davenport, *Multiplicative Number Theory*, 3rd ed., Springer, 2000.
- H. L. Montgomery and R. C. Vaughan, *Multiplicative Number Theory I: Classical Theory*, Cambridge University Press, 2007.
- G. Tenenbaum, *Introduction to Analytic and Probabilistic Number Theory*, 3rd ed., AMS, 2015.
- C. F. Gauss, *Disquisitiones Arithmeticae* (1801), Section VI — the periods of decimal fractions and the order of 10 modulo the denominator. Trans. A. A. Clarke, Yale University Press, 1966.
- B. Riemann (1859), the founding memoir: $\zeta(s)$, the functional equation, the explicit formula, and the Hypothesis.
- P. G. L. Dirichlet (1837), characters and L-functions, primes in arithmetic progressions, and $L(1,\chi) \neq 0$.
- J. Hadamard (1896) and C.-J. de la Vallée Poussin (1896), independent proofs of the prime number theorem from the absence of zeros on $\operatorname{Re} s = 1$.
- H. von Mangoldt (1895), the explicit formula for $\psi(x)$.
- F. Mertens (1874), the theorems on sums and products over primes — the setting for the twin constant.
- L. Schoenfeld (1976), explicit bounds for $\pi(x)$ and $\psi(x)$ under the Hypothesis.

The prime race and its biases

- H. Cramér (1936), the probabilistic model of the gaps between consecutive primes. *Acta Arith.* **2**, 23–46.
- M. Rubinstein and P. Sarnak (1994), Chebyshev's bias. doi:10.1080/10586458.1994.10504289
- A. Granville and G. Martin (2006), prime number races. doi:10.2307/27641834
- R. J. Lemke Oliver and K. Soundararajan (2016), unexpected biases in the distribution of consecutive primes. doi:10.1073/pnas.1605366113

The explicit formula and the pair correlation of the zeros

- A. P. Guinand (1948), a summation formula in the theory of prime numbers.
- A. Weil (1952), the explicit formula and the positivity statement equivalent to the Hypothesis (distinct from the Weil *conjectures* below).
- H. L. Montgomery (1973), the pair correlation of the zeros of ζ, and its agreement with the eigenvalue statistics of large random unitary (GUE) matrices.
- A. M. Odlyzko (1987), the numerical distribution of spacings between zeros and their GUE statistics. doi:10.1090/S0025-5718-1987-0866115-0

The zeros as a spectrum: random matrices, quantum chaos, and the operator dream

- O. Bohigas, M. J. Giannoni, and C. Schmit (1984), chaotic quantum spectra and the universality of level-fluctuation laws. doi:10.1103/PhysRevLett.52.1
- D. Hilbert and G. Pólya (folklore, c. 1910s), the conjecture that the zeros are the eigenvalues of a self-adjoint operator (no single published source).
- M. C. Gutzwiller (1971), the trace formula relating classical periodic orbits to quantum energy levels — the template for reading the primes as orbits. doi:10.1063/1.1665596
- A. Selberg (1956), the trace formula on hyperbolic surfaces — the rigorous analogue behind the orbit picture.
- M. V. Berry and J. P. Keating (1999), the Riemann zeros and eigenvalue asymptotics, and the xp model. doi:10.1137/S0036144598347497
- M. V. Berry (2012), hearing the music of the primes: auditory complementarity and the siren song of zeta.
- A. Connes (1999), the trace-formula approach to the Hypothesis and the positivity criterion equivalent to it. doi:10.1007/s000290050042

- J. Keating and N. Snaith (2000), random-matrix moments of ζ.
- L. de Branges, the approach via Hilbert spaces of entire functions; J. B. Conrey and X.-J. Li (2000), the located obstruction to it. doi:10.1155/S1073792800000489
- Z. Rudnick and P. Sarnak (1996), zeros of principal L-functions and random matrix theory — the n-level correlations and the independence of distinct L-functions. doi:10.1215/S0012-7094-96-08115-6
- N. M. Katz and P. Sarnak (1999), symmetry types of families of L-functions, and the function-field model. doi:10.1090/S0273-0979-99-00766-1
- H. Iwaniec, W. Luo and P. Sarnak (2000), one-level densities and the behaviour of low-lying zeros. numdam.org
- A. E. Özlük and C. Snyder (1999), the symplectic one-level density of the quadratic family of Dirichlet L-functions. doi:10.4064/aa-91-3-209-228
- M. O. Rubinstein (2001), the n-level density of the same quadratic family. doi:10.1215/S0012-7094-01-10916-2
- Y.-H. He, K.-H. Lee, T. Oliver and A. Pozdnyakov (2022), "murmurations" in families of elliptic-curve L-functions. arXiv:2204.10140
- G. Tang, B. N. Bhaskar, P. Shah and B. Recht (2013), atomic-norm super-resolution of line spectra off the grid. arXiv:1207.6053

The primon gas and the physics of the zeros

- B. L. Julia (1990), statistical theory of numbers — the free Riemann ("primon") gas.
- D. Spector (1990), supersymmetry and the Möbius inversion function.
- I. Bakas and M. J. Bowick (1991), curiosities of arithmetic gases. doi:10.1063/1.529511
- S. A. Hartnoll and M. Yang (2025), the conformal primon gas at the end of time. doi:10.1007/JHEP07(2025)281
- D. Schumayer and D. A. W. Hutchinson (2011), physics of the Riemann hypothesis. doi:10.1103/RevModPhys.83.307
- F. Dyson (2009), birds and frogs — the quasicrystal viewpoint on the zeros.

Equivalences and criteria for the Hypothesis (the five faces)

- X.-J. Li (1997), the positivity criterion $\lambda_n \geq 0$ (doi:10.1006/jnth.1997.2137); E. Bombieri and J. Lagarias (1999), the arithmetic form of the λ_n (doi:10.1006/jnth.1999.2392).

- B. Nyman (1950) and A. Beurling (1955), the completeness criterion in $L^2(0,1)$; L. Báez-Duarte (2003), the integer-dilation strengthening.
- G. Robin (1984), the inequality $\sigma(n) < e^{\gamma} n \log\log n$; T. H. Gronwall (1913), the maximal order of $\sigma(n)$; J. Lagarias (2002), the elementary harmonic-number equivalent.
- N. G. de Bruijn (1950) and C. M. Newman (1976), the heat flow and the de Bruijn–Newman constant Λ; G. Csordas, W. Norfolk and R. Varga, the kernel and its flow; B. Rodgers and T. Tao (2020), $\Lambda \geq 0$; the Polymath15 project (2019), $\Lambda \leq 0.22$.

One floor up: Ramanujan's τ and the finite-field Hypothesis

- S. Ramanujan (1916), the function $\tau(n)$ and its conjectured properties.
- L. J. Mordell (1917), the multiplicativity of τ.
- P. Deligne (1974), the proof of the Weil *conjectures* and the Ramanujan bound $|\tau(p)| \leq 2p^{11/2}$ — the Riemann Hypothesis as a theorem over finite fields. numdam. org
- L. Clozel, M. Harris, N. Shepherd-Barron, R. Taylor and collaborators (2008–2011), the Sato–Tate theorem.
- A. Weil (1948), the Riemann Hypothesis for curves over finite fields; H. Hasse (1936), the elliptic-curve bound.

Zeros on the line, and the size of zeta

- G. H. Hardy (1914), infinitely many zeros on the line; A. Selberg (1942), N. Levinson (1974), and J. B. Conrey (1989), positive proportions of zeros on the line.
- E. Lindelöf, the size hypothesis; J. Bourgain and others, subconvexity bounds.

The twin corridor and the sieve

- G. H. Hardy and J. E. Littlewood (1923), the circle method, the prime k-tuple conjecture, and the singular series. doi:10.1007/BF02403921
- V. Brun (1919), the convergence of the twin reciprocals and the sieve that proves it.
- J.-R. Chen (1973), the almost-twin theorem; A. Selberg, the parity problem.
- D. Goldston, J. Pintz and C. Yıldırım (2009), small gaps between primes; Y. Zhang (2013), bounded gaps (doi:10.4007/annals.2014.179.3.7); J. Maynard (2015) and the Polymath8 project, the gap 246 (doi:10.4007/annals.2015.181.1.7).

Factoring and computation

- P. W. Shor (1994/1997), the polynomial-time quantum algorithm for factoring. doi:10.1137/S0097539795293172
- M. Agrawal, N. Kayal and N. Saxena (2004), *PRIMES is in P.* doi:10.4007/annals.2004.160.781
- A. K. Lenstra and H. W. Lenstra, Jr. (eds., 1993), the number field sieve.

Cautions: conjectures that failed

- A. M. Odlyzko and H. J. J. te Riele (1985), the disproof of the Mertens conjecture. doi:10.1515/crll.1985.357.138
- C. B. Haselgrove (1958), the disproof of the Pólya conjecture.

Expository and further reading

- B. Mazur and W. Stein, *Prime Numbers and the Riemann Hypothesis*, Cambridge University Press, 2016.
- M. R. Watkins, *Fourier analysis and number theory* (expository web resource).
- L. Csóka (2017), the Fourier transform of the non-trivial zeros of the zeta function. arXiv:1712.08434

www.ingramcontent.com/pod-product-compliance
Lightning Source LLC
LaVergne TN
LVHW071340210926
848486LV00003B/198
* 9 7 9 8 9 9 6 7 3 2 5 9 3 *